IMAGINING OTHER WORLDS

IMAGINING OTHER WORLDS
Explorations in Astronomy and Culture

Edited by
Nicholas Campion & Chris Impey

SOPHIA CENTRE PRESS

Sophia Centre Press
University of Wales, Trinity Saint David
Ceredigion, Wales SA48 7ED, United Kingdom.
www.sophiacentrepress.com

Cover Image: Giacomo Balla, *Mercurio che passa davanti al sole (Mercury Passing before the Sun)*,
1914, opaque watercolor over graphite on textured wove paper adhered to canvas. Philadelphia
Museum of Art. Gift of Sylvia and Joseph Slifka, 2004. © DACS 2017

Publisher's Cataloging-in-Publication
(Provided by Cassidy Cataloguing Services, Inc.).

Names: Campion, Nicholas, editor. | Impey, Chris, editor.

Title: Imagining other worlds : explorations in astronomy and culture / edited by
              Nicholas Campion & Chris Impey.

Description: Ceredigion, Wales, United Kingdom : Sophia Centre Press, 2018. | Series:
              Studies in cultural astronomy and astrology ; vol. 9

Identifiers: ISBN: 978-1-907767-11-1 (paperback) | 978-1-907767-61-6 (ebook)

Subjects: LCSH: Astronomy--History. | Astronomy--Social aspects. | Science and
              civilization.

Classification: LCC: QB15 .I434 2018 | DDC: 520.9--dc23

Printed by Lightning Source.

CONTENTS

For Ron Olowin
(1945–2017)

FOREWORD

Nicholas Campion and Chris Impey

HUMANS HAVE PROBABLY IMAGINED OTHER WORLDS for as long as they have been conscious. It is in the nature of the imagination that it takes us to different kinds of existence where the rules of material, waking, Earth-bound life no longer exist. The earliest evidence suggests that a fascination with other worlds preoccupied our ancestors tens of thousands of years ago. For example, some readings of Palaeolithic cave art see the cave wall itself as the boundary between this world and another, and archaic shamanic traditions sometimes have the shaman travelling to the stars. From there we can move to the Egyptian pharaoh's post-mortem ascent to the stars, or the soul's celestial journey in Plato's *Republic*. Via Dante and Jules Verne we then arrive at the present and the fabulous success of the *Star Trek* and *Star Wars* franchises, entertaining adults and children alike with tales of civilizations on far flung planets. Still, we imagine other worlds. We have therefore named this volume, selected from lectures presented at the ninth Conference on the Inspiration of Astronomical Phenomena, *Imagining Other Worlds*. It seems a fitting way to sum up one of the themes which ran through the conference papers as well as though popular culture.

The conferences on the Inspiration of Astronomical Phenomena (http://www.in-sap.org/) have been organised periodically since 1994 to bring together academics from science and the humanities, along with artists and independent scholars, in order to exchange ideas and information on the ways in which astronomy inspires people in the past and present, and in all cultures. The ninth conference (http://sophia-project.net/conferences/insapix/) was organised at the suggestion of Valerie Shrimplin and ran from Monday 24th to Thursday 27th August, 2015, in the grand surroundings of the Old Hall in London's Gresham College. Valerie was then working at Gresham College, home to the distinguished line of Gresham Professors of Astronomy, dating back to 1597. A high point of the conference was the Gresham College public lecture given by Lord Rees of Ludlow, Astronomer Royal, and Emeritus Professor of Astronomy at the College. Lord Rees' lecture is published as the Introduction to this volume, followed by a discussion panel consisting of former Gresham Professors of Astronomy, including Ian Morison, Michael Rowan-Robinson, Carolin Crawford and Andrew Fabian.

The rest of the conference presentations consisted of the stimulating mix of papers which we have come to expect from INSAP conferences. The presentations crossed disciplines, cultures, and epochs, ranging from the history of astronomy to the technological state-of-the-art with a demonstration of viritual reality from the Royal Observatory Edinburgh.

We hope that the chapters in this volume will appeal to everyone who is fascinated by the planets, stars and universe, and by human attempts to portray, represent and describe them. This book is for everyone who imagines other worlds.

Lastly, we would like to recognise the ongoing work of Rolf Sinclair, one of the founders and guiding lights of INSAP, and we are deeply saddened by the loss of our good friend Ron Olowin, INSAP Chair during the conference, who died shortly before the tenth INSAP meeting at Santiago de Compostella in September 2017.

Nicholas Campion,
Associate Professor of Cosmology in Culture,
Principal Lecturer, Faculty of Humanities and the Performing Arts,
University of Wales Trinity Saint David.

Chris Impey,
University Distinguished Professor and Associate Dean
of the College of Science at the University of Arizona.

A COSMIC PERSPECTIVE:
FOUR CENTURIES OF EXPANDING HORIZONS

Professor Lord Rees of Ludlow FRS

Introduction

Astronomy is a fundamental science. It is also the grandest of the environmental sciences, and the most universal – indeed the starry sky is the one feature of our environment that has been shared, and wondered at, by all cultures throughout human history. Today, it is an enterprise that involves a huge range of disciplines: mathematics, physics and engineering, of course; but others too. The INSAP conference series celebrates this diversity, bringing together astronomers, anthropologists, historians, and artists. It also celebrates our shared experience of the starry sky; Van Gogh's 'Starry Night' has been the emblem of INSAP since its inception.

We want to understand the exotic objects that our telescopes have revealed. But also to understand how the cosmic panorama, of which we are a part, emerged from our universe's hot dense beginning. This is a brilliant time for young researchers in astronomy. The pace of advance has crescendoed rather than slackened; instrumentation and computer power have improved hugely.

Our Solar System and Space Exploration

I will start with a flashback to Isaac Newton. He must have thought about space travel. Indeed, there is a famous picture, in the English edition of his *Principia*, which depicts the trajectory of cannon balls being fired from a mountaintop. If they are fired fast enough, their paths curves downward no more sharply than the Earth's surface curves away underneath them: the cannon-balls go into orbit. This is still the neatest way to teach the concept of orbital flight.

Newton knew that, for a cannon-ball to achieve an orbital trajectory, its speed must be 25000 km /hour. That speed was not achieved until 1957 with the launch of Sputnik 1. Four years later, Yuri Gagarin went into orbit. Eight years after that we had the Moon landings. The Apollo programme was a heroic episode in the history of humanity. But it was all over more than forty years ago – you have got to be middle-aged to remember when men walked on the Moon; it is ancient history to the younger generation. If the momentum had been maintained there would be footprints on Mars by now. Actually, people have done no more than circle the Earth in low orbit – more recently, in the International Space Station.

Space technology has burgeoned – for communication, environmental monitoring, navigation by GPS, and so forth. We depend on it every day. For astronomers, it is revealed the far infrared, the UV, the X-ray, and the gamma ray sky. Unmanned probes to other planets have beamed back pictures of varied and distinctive worlds. The most recent has been ESA's Rosetta comet mission, which landed a small probe on the comet itself, to check, for instance, if isotopic ratios in the cometary ice are the same as in the Earth's water – crucial for deciding where that water came from. NASA's 'New Horizons' probe has passed Pluto, and is now heading into the Kuiper Belt. Rosetta was launched ten years ago; its design was frozen five years before that. It is robotic technology dating from the 1990s – that is the greatest frustration for the team that has been dedicated to it for so long because present-day designs would have far greater capabilities.

During this century, it is likely that the entire Solar System will be explored and mapped by flotillas of tiny robotic craft. On a larger scale, robotic fabricators may build vast lightweight structures floating in space (solar energy collectors, for instance), perhaps mining raw materials from asteroids or the Moon.

Will people follow them? Robotic advances will erode the practical case for human spaceflight. Nonetheless, I hope people will follow the robots, though it will be as risk-seeking adventurers rather than for practical goals. The most promising developments are spearheaded by private companies. For instance SpaceX, led by Elon Musk, who also makes Tesla electric cars, has launched unmanned payloads and docked with the Space Station. He hopes soon to offer orbital flights to paying customers. Wealthy adventurers are already signing up for a week-long trip round the far side of the Moon – voyaging further from Earth than anyone has been before (but avoiding the greater challenge of a Moon landing and blast-off). I am told they have sold a ticket for the second flight but not for the first flight. We should surely cheer on these private enterprise efforts in space – they can tolerate higher risks than a western government could impose on publicly-funded civilians, and thereby cut costs.

Some people now living will walk on Mars – as an adventure, and as a step towards the stars. They may be Chinese. Indeed, if China wishes to assert its super-power status by a 'space spectacular' it would need to aim for Mars. Just going to the Moon, in a re-run of what the US achieved fifty years earlier, would not proclaim parity.

Perhaps the future of manned spaceflight, even to Mars, lies with privately-funded adventurers, prepared to participate in a cut-price programme far riskier than any government would countenance when civilians were involved – perhaps even one-way trips. (The phrase 'space tourism' should however, be avoided. It lulls people into believing that such ventures are routine and low-risk. And if that is the perception, the inevitable accidents will be as traumatic as those of the US Space Shuttle were. Instead, these cut-price ventures must be 'sold' as dangerous sports, or intrepid exploration).

By 2100, groups of pioneers may have established bases independent from the Earth – on Mars, or maybe on asteroids. But do not ever expect mass emigration from Earth. Nowhere in our Solar System offers an environment even as clement as the Antarctic or the top of Everest. Space does not offer an escape from Earth's problems.

What are the long-term hopes for space travel? The most crucial impediment today stems from the intrinsic inefficiency of chemical fuel, and the consequent requirement to carry a weight of fuel far exceeding that of the payload. Launchers will get cheaper when they can be designed to be more fully reusable. However, so long as we are dependent on chemical fuels, interplanetary travel will remain a challenge. A space elevator would help. Nuclear power could be transformative. By allowing much higher in-course speeds, it would drastically cut the transit times to Mars or the asteroids (reducing not only astronauts' boredom, but their exposure to damaging radiation).

Another compelling question – is there life out there already? Prospects look bleak in our Solar System, though the discovery of even the most vestigial life-forms – on Mars, or in oceans under the ice of Europa or Enceladus – would be of crucial importance, especially if we could show they had an independent origin. The prospects brighten if we widen our horizons to other stars – far beyond the scale of any probe we can now envisage.

Exoplanets and Stars

Perhaps the hottest current topic in astronomy stems from the realization that many other stars – perhaps even most of them – are orbited by retinues of planets, like the Sun is. The planets are not detected directly but inferred by precise measurement of their parent star. There are two methods.

(A) First, if a star is orbited by a planet, then both planet and star move around their centre of mass, called the barycentre. The star, being more massive, moves slower. The tiny periodic changes in the star's Doppler Effect can be detected by very precise spectroscopy. By now, more than 5000 extra-solar planets, or exoplanets, have been inferred in this way. We can infer their mass, the length of their 'year', and the shape of their orbit. This evidence pertains mainly to 'giant' planets, objects the size of Saturn or Jupiter. Detecting Earth-like planets – hundreds of times less massive – is a real challenge. They induce motions of merely centimeters per second in their parent star.

(B) A second technique works better for smaller planets. A star dims slightly when a planet is 'in transit' in front of it. An earth-like planet transiting a sun-like star causes a fractional dimming, recurring once per orbit, of about one part in 10,000. The Kepler spacecraft pointed steadily at a 7-degree-across area of sky for more than three years – monitoring the brightness of over 150000 stars, at least twice every hour, with precision of one part in 100,000. It has already found more than 2000 planets, many no bigger than the Earth. And of course it only detects transits of those whose orbital plane is nearly aligned with our line of sight. We are especially interested in possible 'twins' of our Earth – planets the same size as ours, on orbits with temperatures such that water neither boils nor stays frozen. Some of these have already been identified in the sample, suggesting that there are billions of earth-like planets in the Galaxy.

The real goal, of course, is to see these planets directly – not just their shadows. That is hard. To realise just how hard, suppose an alien astronomer with a powerful telescope was viewing the Earth from 30 light years away – the distance of a nearby star. Our planet would seem, in Carl Sagan's phrase, a 'pale blue dot', very close to a star (our Sun) that outshines it by many billions: a firefly next to a searchlight. But if it could be detected, even just as a 'dot', several features could be inferred. The shade of blue would be slightly different, depending on whether the Pacific Ocean or the Eurasian land mass was facing them. The alien astronomers could infer the length of our 'day', the seasons, the gross topography, and the climate. By analysing the faint light, they could infer that it had a biosphere.

Within ten years, the huge E-ELT telescope planned to be built by the European Southern Observatory on a mountain in Chile (where the site has already been leveled) – with a mosaic mirror 39 metres across – will be drawing inferences like this about planets the size of our Earth, orbiting other Sun-like stars. Nearby, the 24-meter GMT telescope will also be inspecting Earth-like planets by direct imaging. What most people want to know is: Could there be life on them – even intelligent life? Here we are still in the realm of science fiction.

We know too little about how life began on Earth to lay confident odds. What triggered the transition from complex molecules to entities that can metabolise and reproduce? It might have involved a fluke so rare that it happened only once in the entire Galaxy. On the other hand, this crucial transition might have been almost inevitable given the 'right' environment. We just do not know – nor do we know if the DNA/RNA chemistry of terrestrial life is the only possibility, or just one chemical basis among many options that could be realized elsewhere. Moreover, even if simple life is widespread, we cannot assess the odds that it evolves into a complex biosphere. And, even it did, it might anyway be unrecognizably different. I would argue that the SETI programme is a worthwhile gamble – because success in the search would carry the momentous message that concepts of logic and physics are not limited to the hardware in human skulls.

It is too anthropocentric to limit attention to Earth-like planets even though it is prudent strategy to start with them. Science fiction writers have other ideas – balloon-like creatures floating in the dense atmospheres of Jupiter-like planets, swarms of intelligent insects, etc. Perhaps life can flourish even on a planet flung into the frozen darkness of interstellar space, whose main warmth comes from internal radioactivity (the process that heats the Earth's core).

We should also be mindful that seemingly artificial signals could come from super-intelligent (though not necessarily conscious) computers, created by a race of alien beings that had already died out. Indeed I think this is the most likely possibility, we may learn this century whether biological evolution is unique to our Earth, or whether the entire cosmos that teems with life – even with intelligence. Even if simple life is common, it is a separate question whether it is likely to evolve into anything we might recognize as intelligent or complex. Perhaps the cosmos teems even with complex life; on the other hand, our Earth could be unique among the billions of planets that surely exist. That would be depressing for the searchers. But it would allow us to be less cosmically modest: Earth, though tiny, could be the most complex and interesting entity in the entire Galaxy.

Back now to the physics, far simpler than biology. What has surprised people about the newly-discovered planetary systems is their great variety. However, the ubiquity of such systems was not surprising. We have learnt that stars form, via the contraction of clouds of dusty gas; and if the cloud has any angular momentum, it will rotate faster as it contracts, and spin off a dusty disc around the protostar. In such a disc, gas condenses in the cooler outer parts; closer in less volatile dust agglomerates into rocks and planets – this should be a generic process in all protostars.

Next, I will outline how the cosmogonic causal chain has been pushed back further – to the formation of galaxies, stars, atoms, and right back to the first nanosecond of the big bang.

First, what about stars and atoms? We see stars forming, in places like the Eagle Nebula, 7000 light-years away. And we see many star dying – as the Sun will in around 6 billion years, when it exhausts its hydrogen fuel, blows off its outer layers, and settles down to a quiet demise as a white dwarf.

More massive stars die explosively as supernovae, generally leaving behind a neutron star or black hole. The most famous is the Crab Nebula, the expanding debris from a supernova recorded by oriental astronomers in 1054 CE, with, at its centre, a neutron star spinning at 30 revs/second. Supernovae are important for us: if it was not for them we would not be here. By the end of a massive star's life, nuclear fusion has led to an onion skin structure – with hotter inner shells processed further up the periodic table. This material is then flung out in the supernova explosion. The debris then mixes into the interstellar medium and re-condenses into new stars, orbited by planets.

The concept of cosmic chemistry was developed primarily by Fred Hoyle and his associates. They analysed the specific nuclear reactions involved, and were able to understand how most atoms of the periodic table came to exist and why oxygen and carbon (for instance) are common, whereas gold and uranium are rare. Our Galaxy is a huge ecological system where gas is being recycled through successive generations of stars. Each of us contains atoms forged in dozens of different stars spread across the Milky Way, which lived and died more than 4.5 billion years ago, polluting the interstellar cloud in which the Solar System condensed.

Beyond our Galaxy: Cosmic Horizons

Let us now enlarge our spatial horizons to the extragalactic realm. We know that galaxies – some disc-like, resembling our Milky Way or Andromeda; others amorphous 'ellipticals' – are the basic constituents of our expanding universe. But how much can we actually understand about galaxies? Physicists who study particles can probe them, and crash them together in accelerators at CERN. Astronomers cannot crash real galaxies together. Galaxies change so slowly that in a human lifetime we only see a snapshot of each. However, we can do experiments in a 'virtual universe': computer simulations, incorporating gravity and gas dynamics.

We can redo such simulations making different assumptions about the mass of stars and gas in each galaxy, and so forth, and see which matches the data best. Importantly, we find, by this method and others, that all galaxies are held together by the gravity not just of what we see. They are embedded in a swarm of particles that are invisible, but which collectively contribute about five times as much mass as the ordinary atom – the dark matter.

We also can test ideas on how galaxies evolve by observing eras when they were young. The Hubble Telescope has been used to study 'deep fields', each encompassing a tiny patch of sky – just a few arc minutes across. You can see hundreds of smudges. These are all galaxies, some fully the equal of our own, but they are so far away that their light set out more than 10 billion years ago – they are being viewed when they have recently formed. What happened before there were galaxies? The key evidence here, dating back to Arno Penzias and Robert Wilson 50 years ago, is that intergalactic space is not completely cold. It is warmed to three degrees above absolute zero by weak microwaves, known to have an almost exact black body spectrum. This is the 'afterglow of creation' – the adiabatically cooled and diluted relic of an era when everything was squeezed hot and dense. It is one of several lines of evidence that have allowed us to firm up the 'hot big bang' model. The background radiation was last scattered when the temperature was 3000 degrees and the free electrons combined with nuclei to mainly hydrogen and helium atoms. This was after about 300,000 years of expansion. The helium and deuterium abundance was determined by nuclear reactions in the first few minutes, at temperatures of a few billion degrees.

Let us address an issue that might seem puzzling. Our present complex cosmos manifests a huge range of temperature and density, from blazingly hot stars, to the dark night sky. People sometimes worry about how this intricate complexity emerged from an amorphous fireball. It might seem to violate the second law of thermodynamics, which describes an inexorable tendency for patterns and structure to decay or disperse. The answer to this seeming paradox lies in the force of gravity. Gravity enhances density contrasts rather than wiping them out. Any patch that starts off slightly denser than average would decelerate more, because it feels extra gravity; its expansion lags further and further behind, until it eventually stops expanding and separates out. Many simulations have been made of parts of a 'virtual universe' – modelling a domain large enough to make thousands of galaxies. The calculations, when displayed as a movie, clearly display how incipient structures unfold and evolve. Within each galaxy-scale clump, gravity enhances the contrasts still further; gas is pulled in, and compressed into stars.

There is one very important point. The initial fluctuations fed into the computer models are not arbitrary – they are derived from the observed fluctuations in the temperature of the microwave background, which have been beautifully and precisely delineated over the whole sky by ESA's Planck Spacecraft. The amplitude of the temperature fluctuations is only one part in 100,000, but computing forward, they are amplified by gravity into the conspicuous structures in the present universe.

What about the far future of our universe? In 1998 cosmologists had a big surprise. It was by then well known that the gravity of dark matter dominated that of ordinary stuff – but also that dark matter plus baryons contributed only about thirty percent of the critical density. This was thought to imply that we were in a universe whose expansion was slowing down, but not enough to eventually be halted. But, rather than slowly decelerating, the Hubble diagram of Type 1a supernovae famously revealed that the expansion was speeding up. Gravitational attraction was seemingly overwhelmed by a mysterious new force latent in empty space which pushes galaxies away from each other.

Moreover there was independent evidence supporting this. According to Einstein's theory, a straightforward low-density universe would have negative curvature – the three angles of a big triangle would add up to less than 180 degrees. This can be tested from microwave background measurements. That is because there is a straightforward effect that makes the temperature ripples more conspicuous for a particular wavelength – about 300,000 light years. This so-called 'Doppler peak' was first revealed by a balloon-borne experiment called Boomerang, and has been confirmed by the Planck data. It is on an angular scale that is consistent with a flat universe. If we had just had the supernova Hubble diagram, some of us would not have been convinced. But these two interlinked and almost simultaneous discoveries together clinched the case. The issue now is the nature of the dark energy – is it time-independent, like Einstein's cosmological constant, or was it different in the past?

Long-range forecasts are seldom reliable, but the best and most 'conservative' bet is that we have almost an eternity ahead – an ever colder and ever emptier cosmos. Galaxies accelerate away and disappear over an 'event horizon' – rather like an inside out version of what happens when things fall into a black hole. All that is left will be the remnants of our Galaxy, Andromeda, and smaller neighbours. Protons may decay, dark matter particles annihilate, occasional flashes when black holes evaporate – and then silence.

Speculative Thoughts on the Very Early Universe

We can trace the cosmos back to 1 second after the initial instant. Indeed we can probably be confident back to a nanosecond: that is when each particle had about 50 Gev of energy – as much as can be achieved in CERN's Large Hadron Collider – and the entire visible universe was squeezed to the size of our Solar System. But questions like 'where did the fluctuations come from?' and 'why did the early universe contain the actual mix we observe of protons, photons and dark matter?' take us back to the even briefer instants when our universe was hugely more compressed still – when energies were 10^{16} Gev, where experiments offer no direct guide to the relevant physics.

The discourse hereafter becomes much more speculative. According to a popular theory, the entire volume we can see with our telescopes was at 10^{16} Gev, a hyperdense blob no bigger than an apple. And it had inflated from something at least a trillion times smaller than an atomic nucleus. The so-called 'inflationary universe' model is supported already by much evidence. But it may be useful to summarise the essential requirements for the emergence of our complex and structured cosmos from simple amorphous beginnings.

(i) The first prerequisite is of course the existence of the force of gravity – which (as explained earlier) enhances density contrasts as the universe expands, allowing bound structures to condense out from initially small-amplitude irregularities. It is a very weak force. On the atomic scale, it is about forty powers of ten weaker than the electric force between electron and proton. But in any large object, positive and negative charges almost exactly cancel, in contrast, everything has the same 'sign' of gravitational charge so when sufficiently many atoms are packed together, gravity wins. But stars and planets are so big because gravity is weak. Were gravity stronger, objects as large as asteroids (or even sugar-lumps) would be crushed. So, though gravity is crucial, it is also crucial that it should be very weak.

(ii) The second requirement is that there must be an excess of matter over anti-matter.

(iii) Another requirement for stars, planets and biospheres is that chemistry should be non-trivial. If hydrogen were the only element, chemistry would be dull. A periodic table of stable elements requires a balance between the two most important forces in the micro-world: the nuclear binding force (the 'strong interactions') and the electric repulsive force that drives protons apart.

(iv) Also, there must be stars – enough ordinary atoms relative to dark matter. (Indeed there must be at least two generations of stars: one to generate the chemical elements, and a second able to be surrounded by planets).

(v) The universe must expand at the 'right' rate – not collapse too soon, nor expand so fast that gravity cannot pull together the structures.

(vi) Moreover, there must be some fluctuations for gravity to feed on – sufficient in amplitude to permit the emergence of structures. Otherwise the universe would now be cold ultra-diffuse hydrogen: no stars, no heavy elements, no planets and no people. In our actual universe, the initial fluctuations in the cosmic curvature have an amplitude of 0.00001. According to inflationary models, this amplitude is determined by quantum fluctuations. Its actual value depends on the details of the model.

Here is another fundamental question: How large is physical reality? We can only see a finite volume and a finite number of galaxies. That is essentially because there is a horizon: a shell around us, delineating the distance light can have travelled since the big bang. However, that shell has no more physical significance than the circle that delineates your horizon if you are in the middle of the ocean. We would expect far more galaxies beyond the horizon.

There is no perceptible gradient in temperature or density across the visible universe: this suggests that, even if it is of finite extent, it stretches thousands of times further. But that is just a minimum. If space stretched far enough, then all combinatorial possibilities would be repeated. Far beyond the horizon, we could all have avatars. Be that as it may, even conservative astronomers are confident that the volume of space-time within range of our telescopes – what astronomers have traditionally called 'the universe' – is only a tiny fraction of the aftermath of our big bang.

Something else emerges from these theoretical speculations. Plausible models for the physics at the ultra-high energies where inflation could have occurred lead to so-called 'eternal inflation'. 'Our' big bang could be just one island of space-time in a vast cosmic archipelago – a multiverse. The requirements to evolve the complex and structured cosmos that we observe lead to some key questions.

(A) Is there one big bang, or many?

(B) If there are many, are they all replicas of each other, or do they 'ring the changes' on the laws and constants of physics, so that most are 'stillborn' and we find ourselves in one of the subset that allow complexity to emerge (so called 'anthropic selection')?

This is speculative physics – but it is physics, not metaphysics. There is hope of firming it up. Further study of the fluctuations in the background radiation will reveal clues. But, more important, if physicists developed a unified theory of strong and electromagnetic forces – and that theory is tested or corroborated in our low-energy world – we would then take seriously what it predicts about an inflationary phase and what the answers to the two questions above actually are. If the answer to the second question is 'yes', then what we call 'laws of nature' may in the grandest perspective be mere local bylaws governing our cosmic patch. Many patches could be still-born or sterile; the laws prevailing in them might not allow any kind of complexity. We therefore would not expect to find ourselves in a typical universe. Rather, we would be in a typical member of the subset where an observer could evolve. This is anthropic selection.

I described earlier newly discovered planets orbiting other stars. I would like to give a flashback to planetary science 400 years ago – even before Newton. At that time, Kepler thought that the Solar System was unique, and Earth's orbit was related to the other planets by beautiful mathematical ratios involving the Platonic regular solids. We now realise that there are billions of stars, each with planetary systems. Earth's orbit is special only insofar as it is in the range of radii and eccentricities compatible with life (e.g., not too cold and not too hot to allow liquid water to exist).

Maybe we are due for an analogous conceptual shift, on a far grander scale. Our big bang may not be unique, any more than planetary systems are. Its parameters may be 'environmental accidents', like the details of the Earth's orbit. The hope for neat explanations in cosmology may be as vain as Kepler's numerological quest.

If there is a multiverse, it will take our Copernican demotion one stage further – our solar system is one of billions of planetary systems in our Galaxy, which is one of billions of galaxies accessible to our telescopes. However, this entire panorama may be a tiny part of the aftermath of 'our' big bang – which itself may be one among billions. It may disappoint some physicists if some of the key numbers they are trying to explain turn out to be mere environmental contingencies, no more 'fundamental' than the parameters of the Earth's orbit round the Sun. In compensation, we would realize space and time were richly textured. But on scales so vast that astronomers are they not directly aware of it, any more than plankton whose 'universe' was a spoonful of water, would be aware of the world's topography and biosphere.

At a conference in Stanford there was a panel discussion where the panelists were asked how strongly they would bet on the multiverse concept. I said that, on the scale would you bet your goldfish, your dog or yourself, I was almost at the dog level. Andrei Linde said he was far more confident – after all he had devoted 25 years of his life to the eternal inflation idea. And the great theorist Steven Weinberg later said that he would happily bet my dog and Andre Linde's life.

We have made astonishing progress. Fifty years ago, cosmologists did not know if there was a big bang. Now, we can draw quite precise inferences back to a nanosecond. So in fifty years, debates that now seem flaky speculation may have been firmed up. But it is important to emphasise that progress will continue to depend, as it has up till now, 95 percent on advancing instruments and technology – less than five percent on armchair theory.

Concluding Perspectives

Finally, I want to draw back from the cosmos – even from what may be a vast array of cosmoses, governed by quite different laws – and focus back closer to the here and now. I am often asked – is there a special perspective that astronomers can offer to science and philosophy? We view our home planet in a vast cosmic context. And in coming decades we will know whether there is life out there. But, more significantly, astronomers can offer an awareness of an immense future. This awareness echoes the ethos of this conference – we find inspiration in astronomical phenomena on a vast range of scales of space and time.

The stupendous time-spans of the evolutionary past are now part of common culture (maybe not in Kentucky, or in parts of the Muslim world). Darwinism tells us how our present biosphere is the outcome of more than four billion years of evolution. But most people still somehow think we humans are necessarily the culmination of the evolutionary tree. That hardly seems credible to an astronomer – indeed, we are probably still nearer the beginning than the end. Our Sun formed 4.5 billion years ago, but it has got 6 billion more before the fuel runs out. It then flares up, engulfing the inner planets. And the expanding universe will continue – perhaps forever – destined to become ever colder, ever emptier. To quote Woody Allen, eternity is very long, especially towards the end.

Any creatures witnessing the Sun's demise 6 billion years hence will not be human – they will be as different from us as we are from a bug. Post-human evolution – here on Earth and far beyond – could be as prolonged as the Darwinian evolution that has led to us – and even more wonderful. Of course, this evolution is even faster now – it happens on a technological timescale, operating far faster than natural selection and driven by advances in genetics and in artificial intelligence (AI). We do not know whether the long term future lies with organic or silicon-based life.

My final thought is this. Even in this 'concertinaed' timeline – extending billions of years into the future, as well as into the past – this century may be a defining moment. Over most of history, threats to humanity have come from nature – disease, earthquakes, floods, and so forth. But this century is special. It is the first where one species has Earth's future in its hands, and could jeopardise life's immense potential. We have entered a geological era called the Anthropocene.

Our Earth, this 'pale blue dot' in the cosmos, is a special place. It may be a unique place. And we are its stewards at an especially crucial era. That is an important message for us all, whether we are interested in astronomy or not.

A Cosmic Perspective
A Panel Discussion with the Gresham
Professors of Astronomy

Professor Lord Rees of Ludlow FRS,
Professor Carolin Crawford,
Professor Ian Morison,
Professor Andrew Fabian,
Professor Michael Rowan-Robinson,

Chaired by Professor Christopher Impey

EDITORS' NOTE: A highlight of the INSAPIX conference was a public lecture and panel discussion organised by Gresham College in the grand surroundings of the Royal College of Surgeons. The lecture by Professor Lord Rees of Ludlow is the foreword to this volume. We are pleased to present the following panel discussion here.

CHRIS IMPEY: My name is Chris Impey. I am a Professor of Astronomy at the University of Arizona. I think Gresham College had been around for about 300 years when my university was founded, but I do have the pleasure in this coming April of giving a talk in the Gresham Visiting Professor series. You are in for a treat tonight because, in addition to what you have just heard from Lord Rees, and the follow-up questions from it, we now have a great assemblage of talent from previous Gresham professors. I am going to ask them to come up on stage now.

As they take their places, I will give them in their order of being Gresham professors. Martin Rees was Gresham professor in 1975 to 1976. Then we have got Michael Rowan-Robinson (1981–82), Andrew Fabian (1982–84), Ian Morison (2007–11), and Carolin Crawford (2011–15), and between them, everything is up for discussion. I am going to ask each of them to introduce themselves for a few minutes of comments, so you will get their perspective and a sense of their expertise, and then we will take questions which can be directed to either any one of them or the entire panel We will just start with Michael.

MICHAEL ROWAN-ROBINSON: I am Michael Rowan-Robinson. My main interests are infrared astronomy and cosmology, the subject of the last part of Martin's talk. It is interesting that, in different wavebands, we see completely different pictures of the universe. In the infrared, what we especially see is little particles of dust which are spread through interstellar space and through the solar system. The stuff of our bodies has probably spent quite a lot of time sitting in small grains of interstellar dust, made of carbon and nitrogen, carbon and oxygen and other key elements. So, with infrared astronomy, for example, in the solar system, what we are especially interested in is the solar system's debris disk. Martin talked about the planets, but the planets are only a tiny part of the solar system. There are about ten million objects, a kilometre in size or bigger, and the zodiacal dust is the debris from these that flows in towards the Sun. In Martin's talk, we saw the Milky Way with these dark patches which are the clouds when new stars are forming, and in the infrared we see these clouds shining brightly and we also see very distant galaxies undergoing huge bursts of star formation. So, those are the kind of things that we study in infrared astronomy.

On cosmology, I will just say one thing: I think that these new missions have given us a very detailed picture of the geometry of the universe and quite a lot of information about what it is made of. So, ordinary matter, the stuff we are made of, is not very important; dark matter is much more important, and there is also dark energy, whatever that may be, which governs the current motion of the universe, and this points back towards the inflationary picture at the beginning of the universe that Martin talked about. Personally, however, I am very sceptical about the multiverse and all that stuff – I think it is totally speculative, there is no evidence for it, so…

MARTIN REES: I agree! [Laughter]

MICHAEL ROWAN-ROBINSON: Okay. Well, that's all I wanted to say – thank you!

IAN MORISON: I am Ian Morison. I have been a radio astronomer for virtually fifty years now at the Jodrell Bank Observatory. I have many interests, one of which is of course is the search for life out there, and, for six years, I was involved in Project Phoenix, which to date has been the most sophisticated search for extra terrestrial intelligence yet.[1] Obviously, we never found it, otherwise you would know about it, and I would be quite famous, and neither of those is true! But in fact, radio astronomers have actually made quite a contribution to astronomy in the last twenty years. You have heard some of them, such as the discovery of the

[1] Project Phoenix, https://www.seti.org/seti-institute/project/details/project-phoenix [accessed 8 August 2017].

cosmic micro background. I am very pleased that our observatory helped build some of the receiver systems for the Planck spacecraft, and I and my colleagues have been working on the data for quite a long time. I do love cosmology as well. Funnily enough, even though the things I like most include like Einstein and cosmology, I am actually an experimental physicist and astronomer, I am not a theoretician, but I do enjoy theory. In the blurb about the very last lecture I gave as Gresham professor, I said perhaps we live in a multiverse that stretches beyond our imagination, and I was trying to imply that maybe the totality of the cosmos was beyond the ability of our human minds to comprehend.[2] I was actually quite pleased that, the following summer, I think in an interview in the *Sunday Times*, Lord Rees said much the same thing.[3] I sincerely hope I am wrong. Maybe we will find out! Will we find other life? I worry that in order for life to evolve, first there is a problem in getting from single-cell to multi-celled organisms, and then a planet has got to be relatively stable over a long period of time, in order to allow life to evolve further. Sadly, I suspect that that will not happen very often, but I sincerely hope it does.

CAROLIN CRAWFORD: I am the most recent Gresham Professor of Astronomy. My research career has been spent studying some of the largest structures in the Universe. I worked on enormous clusters of galaxies, gravitationally bound systems where hundreds or thousands of galaxies are all contained in a cluster, which is only a few tens of millions of light years across. In particular I look at the central cluster galaxies, the most massive and extreme galaxies anywhere in the Universe – and the most massive galaxies contain the biggest black holes. I used optical, infrared, and X-ray observations to study the symbiosis of that giant galaxy with its surrounding cluster, but I must confess that over the last ten years or so, I have moved away from the research and I have discovered an enthusism and a passion for communicating the science and the research. Now, along with my College lecturing, I have a role as a public astronomer in the Institute of Astronomy at Cambridge, where I can indulge my passion for explaining science. The best thing for me about what I do now is that it allows me to engage with a much wider

---

[2] Ian Morison, 'To Infinity and Beyond' Gresham College Lecture, Museum of London, 13 April 2011, https://www.gresham.ac.uk/lectures-and-events/to-infinity-and-beyond [accessed 8 August 2017].

[3] Editors note. Also see Ian Sample, 'Lord Rees: I've got no religious beliefs at all', *The Guardian*, 6 April 2011, https://www.theguardian.com/science/2011/apr/06/astronomer-royal-martin-rees-interview [accessed 8 August 2017].

breadth of astronomy than just the narrow topics that I was researching. It means I can now indulge my passion for the subject, whether it is just observing the night sky, talking about exoplanets or the latest missions around the solar system, or indeed reverting to talking about clusters of galaxies and some of that large-scale structure. I love all of astronomy and I love talking about all of astronomy.

ANDREW FABIAN: I am Andy Fabian. I was a Gresham professor back in the early-'80s when we started having our meetings in the Barbican, and I would give lectures in a cinema, which was a unique experience for me. Back at that time as well, Martin Rees was the Director of the Institute of Astronomy where Carolin Crawford and I work, and so he was my boss. I am now the Director of the Institute of Astronomy, so, in some sense, I am Martin's boss, although I would not dare tell him what to do! It certainly is a great privilege having Martin down the corridor to go and gossip to and find out what is really going on. My work is to look at X-rays from the cosmos and I study things like accreting black holes, trying to look at and understand how material acts under these very extreme conditions. I am interested in the whole breadth of astronomy, and I like looking at the night sky myself, and I think that there are many, many questions yet to be sorted out. In a way it sounds a little from Martin's talk, as though we understand everything, and everything will be fixed once we can put quantum mechanics and gravity together. I suspect that in a hundred years' time there will be many new issues which we have not dreamt of today. I am very optimistic about the future.

AUDIENCE QUESTION: It is really a double question, if I am allowed that. The first is, I have never really understood why the further away objects are, the faster they are receding. If the universe is expanding at a constant rate, should they not all be moving away at the same rate, whether they are nearby or further away? The second question is that my understanding was that the expansion of the universe started accelerating at some point in the universe's history, and I do not understand why that should be: what changed that caused the universe to start expanding at a higher rate?

MARTIN REES: I think the answer to the first question, if you imagine an infinite lattice filled with rods, in all three dimensions, the rods are like markers of space. Now imagine all the rods get longer, then the more distant ones are separated from us by more rods, and so the speed of their apparent recession is proportionate to the number of intervening rods. Space-time is invisible and this analogy allows us to visualize the expansion. The entire network is growing and the further away we look into the network, the faster it is expanding.

To your second question, it is true that the expansion rate is now speeding up: and this was discovered in 1998 and is indeed one of the big surprises of recent times. It seems that the universe, in the early stages, was expanding but the rate of expansion was slowing down, as you would expect, because everything exerts a gravitational pull on everything else, so you would expect that there would be a gradual decelerating pull. But it seems – and this was a big surprise – that there is some extra force latent in empty space which, when the density of matter becomes low enough, starts to dominate. The push due to empty space then itself overwhelms the gravitational pull which the galaxies exert on each other and instead we get an acceleration. There are theories which model this, but the fact that a repulsive force existed was something which was a big surprise. I think it is something which we will not understand until we understand the nature of empty space in a fundamental way. Everyone suspects that if we are to unify the very large and the very small, we will have to consider phenomena on a scale which is a billion, billion times smaller than an atomic nucleus. That is the scale where quantum effects and gravity open up. Most people suspect that if you were to chop up space, then you could not do it indefinitely. We know I can chop up this lectern, until we get down to atoms, and then we have to stop. But most people suspect that if you try to chop up space, when you get down to that very tiny scale, you will find that what looks like a point in our space is actually a tightly-wound origami in extra dimensions: this is what string theorists are talking about. Until we understand a theory like that, we will not understand why this mysterious repulsive force exists. It is one of the deepest mysteries and it may be something we will never understand, but we will try.

CHRIS IMPEY: Michael, if you think that the multiverse is speculative, you would probably also raise your eyebrows there too…?

MICHAEL ROWAN-ROBINSON: Well, I do not disagree with Martin that we are not likely to find an explanation for the acceleration of the universe unless we understand better the connection between quantum theory and general relativity, but I think one of the questions that you asked was why did the Big Bang get started, and I think that is an interesting question, but I think it is an example of a question that we just cannot answer because we have no evidence from the moment of the Big Bang. With our telescopes we see very distant galaxies, maybe a billion years after the Big Bang, and with the microwave background we go back to a few hundred thousand years after the Big Bang. From other evidence, like the manufacture of helium during the Big Bang, we get back to one second. At the moment we just cannot get back nearer to this moment of the Big Bang, so we do not know what its structure is or why it got started. People speculate about it. One idea that floats around is that the universe existed for an infinite time in a kind of shadowy quantum vacuum and then there was some unusual fluctuation which led to this expanding universe growing out of the fluctuation. So, you can think that it might be that time did not start at the Big Bang, that there was a time before with this shadowy quantum world, but that is just speculation. We do not have any evidence back to that point and so we just cannot really begin to answer that question. Theorists like to have a go at it, but without evidence, I do not see how we can make much progress. So, there are certain questions we have to set aside because we just do not have any evidence that will help us answer them.

AUDIENCE QUESTION: In the increasing complexity of the subject, I can see no beginning or end to this increasing complexity, even long after the Earth has been gobbled up by the Sun running out of fuel. The question that I would ask is: what alternative would you have in a religion to call a focal point, a God, Allah or whatever, because of this complexity, this ever-increasing unknown.

MICHAEL ROWAN-ROBINSON: I disagree that the unknown is increasing. On the contrary, what Martin portrayed in the history of science, and especially of astronomy, is that, all the time, things that appeared very mysterious, like the motion of the planets to start with, and then, in our own century, the universe and its history, gradually become clearer. When Martin and I started research, we knew nothing really about the universe and what its structure was. We knew about the expansion of the universe and that was about all. So, basically, as time goes on, as we study things about which we have some evidence – and that is the problem with the Big Bang itself, we do not have evidence about that – but where we have evidence, we can make progress and reduce this area of uncertainty. So, I just do not agree that anything about the universe we do not understand tells us that we need God to explain to it. There are some things we have not explained, there are some things we do not understand, and there is a lot of detail, for example, in how stars form, how galaxies form, how planets form. We have not got it all sorted by any means, but I think we will understand by studying these problems, by building new telescopes and space missions and so on. I do not see the need for God.

Chris Impey: If I might be allowed to re-cast the question, because Martin has written and spoken about the problem of a very simple unified beginning of the universe leading eventually to a large old universe that has a lot of complexity, including the complexity of all of us in this room talking about the subject. You have addressed some of these issues, such as the nature of the universe in terms of complexity or entropy. Is there any comment you would make on that?

Martin Rees: Well, I would like to say just two things: first, that the most complicated things are living things – they are more complicated than anything inanimate in the universe, in my opinion. To follow up on what Michael said, many of the issues which were controversial fifty years ago when we were students have now been settled, and the questions that we are tackling now could not even have been posed back then. That is the nature of science s a progressive enterprise: we are now trying to address questions which have been posed by the fact that our observations are now much more precise. It is 95% due to new equipment and better computers that we have made advances in knowledge. I think it is important for us to emphasise that armchair theory does not get you very far in this subject. In this sense we are no wiser than Aristotle was.

Ian Morison: You talked about a god. In his talk, Sir Martin mentioned the anthropic principle, why are we here. There is a wonderful book that he wrote some years ago called *Just Six Numbers*, which makes the point that if some of the parameters that define our universe were not as they were, we could not be here.[4] So, then there is a question of: why is this correct? Well, one possibility, and I do not think you can discount it, is that a single god created a universe fit for life. I quite like that possibility in a way because, if you were a wonderful musician, composer, artist, you might ask what is the point of making a wonderful piece of art if no one can see it? So, perhaps, if you were going to make a universe, you would definitely make it so that life like ours could arise to appreciate the wonders of that universe. Now, the alternative, of course, is a multiverse where there are myriads of universes all of which have perhaps different properties, and so some of them, by chance, will be right for life. You can perhaps ask: well, where are they? I do not think we will ever know. But there is a bit of analogy, based on the idea of string theory. Basically, one version of string held that there were ten dimensions, six of them were curled up to form the little strings that formed the particles you have

---

[4] Martin Rees, *Just Six Numbers: The Deep Forces that Shape the Universe* (London: Weidenfeld and Nicholson, 1999).

heard a bit about this evening. But Ed Witten, a wonderful physicist, combined them all into what he called M Theory, where there were eleven dimensions, so with an extra hidden dimension, and perhaps these universes all live in that extra dimension without knowing about each other. And finally, another analogy: if you had a knitting needle and three slices of bread, with some ants on the right side of each of those slices, to some sense they are defined to be in a two-dimensional universe – they can only move up, down and sideways – and they certainly would not know about the other ants on the other piece of bread. So, in perhaps the same way, we, in our third dimension, know about our own universe, but there is no way we conceive about the presence of others that might exist in a further dimension.

AUDIENCE QUESTION: In support of what Ian's just said, John Barrow, his predecessor was a great advocate of multiverses, and his lectures used to contain an awful lot of information and suggestions about them. But my question is: the Large Hadron Collider has not been mentioned this evening, and it is in its second phase as I understand it. They have discovered some negative subatomic particles, but how much evidence is there that this will compute to a quantity of dark matter sufficient to justify current percentages of the universe?

MARTIN REES: Of course, the LHC (Large Hadron Collider) discovered the Higgs particle, and that was the last piece of the jigsaw which gave people confidence that their picture of existing particles was on the right lines; but of course the hope is that it may discover some other particles. In particular, there are speculations about a class of so-called super-symmetric particles which do not exist in the everyday world but could have been made in the early universe and could make dark matter. There is a hope that the LHC may be able to create some of these particles by reproducing conditions that have not existed since the first nanosecond of the Big Bang. That would be great if that happens, but do not hold your breath. It could be that the LHC will not detect these particles and that the dark matter particles may be much, much heavier than anything that the LHC could produce. So, we hope the LHC will produce something but we cannot be sure.

CAROLIN CRAWFORD: I will just add to that, even if we are successful, and we should know perhaps within the next couple of years about whether we can identify a potential dark matter candidate from the LHC, that is only part of the battle. It might illuminate some of the physics, and it might give us a candidate for dark matter; but you then still have to prove that it exists in enough quantities, in the right places in the Universe, to account for the astronomical observations, and that it will be stable and long-lived enough to have survived since the Big Bang, to hang around and litter up our current universe. You have got to have the work from the LHC, but the evidence also needs to be backed up with the astronomical observations, and there is quite a lot of development yet to go in this field.

Chris Impey: I cannot resist asking, since Martin gets the UFO questions to which he alluded, Andy, do you get the questions about LHC making blackholes?

Andrew Fabian: Well, yes. However, fortunately, it is very unlikely that the LHC will make black holes, unless there are other dimensions, and we have already heard about other dimensions. There is also the idea of trying to explain why gravity is so weak – Martin told us that it is forty powers of ten weaker than say electric fields and so forth. The reason that gravity is so weak might be because it is leaked away into another dimension, and if you can penetrate that other dimension by slinging protons fast enough at each other, suddenly, their gravity becomes extremely strong and they turn into black-holes and so forth. So, that would be really exciting! There has already been a book written about it but I think it is highly unlikely that is going to happen. Other dimensions are something that we all think about occasionally, but for which there is currently zero observational evidence, I am an astronomer who likes to see something and then I will start believing it. It is great to speculate, but it is really good to see the observations.

Ian Morison: If you are worried about whether the LHC is going to destroy mankind and the world, there is a website, www.hasthelhcdestroyedtheearth.com – I went on it the other day to check and it said no! [Laughter]

Audience Question: Lord Rees mentioned that the future evolution of man might make us something quite different than what we are right now, and the question has been raised what caused the Big Bang, or what we refer to as creation. I would be curious to know if anyone on the panel has considered any of the thinking of Teilhard de Chardin who wrote about creation and the future of homo-sapiens, and if so, do they have any comments?

Martin Rees: I must say, I am not a great fan of Teilhard de Chardin because I remember the wonderful review of his book by Peter Medawar, who is one of my scientific heroes, who was rather sceptical about this. He said that '…before deceiving others, he has been to great pains to deceive himself'.[5] But seriously, he did have the idea that we were advancing towards some sort of common mind, the Noosphere, and maybe there is something in that because, if we think of what is going to happen in the future, then even within a century or two, it may well be that human brains will be surpassed by AI, artificial intelligence, and then of course, if the machines can themselves make something even better, then we would be getting into a situation where the dominant cogitations would no longer come from the wet hardware in human skulls, but will come from machines, and then you can imagine that this would be a sort of network. So, if we think of the far future, then indeed it could be that it would be an inorganic future where there will be machines of vastly superhuman powers. Incidentally, going back to SETI searches, I think if they are going to discover anything, it is far more likely to be something like that than an organic civilisation like we have on Earth.

MICHAEL ROWAN-ROBINSON: That is Fred Hoyle's *Black Cloud*.[6]

MARTIN REES: Yes.

CHRIS IMPEY: Not so far off. Probably the next but one Gresham Professor of Astronomy will be a bot!
[Laughter]
So, enjoy the flesh and blood versions while you can!

AUDIENCE QUESTION: Thank you, Lord Rees, for a lovely talk. You mentioned briefly in it dark matter and dark energy. We have just touched on dark matter, but I wonder if the panel could speak a little on the dark energy side. It is something that does not get a lot of public understanding and I wonder if that is because no one knows anything about it. Perhaps if you could give us a state of play on it and perhaps any strategies as to how it might be brought forward in knowledge.

MARTIN REES: Well, we know that this force exists, but we do not know much about it, except it seems to be fairly uniform. It seems to, as far as we can tell, be at equal strength in the past as it is now, and that is all we know. I think we will not understand it better for a very long time, but fortunately, it is not as important as dark

---

[5] In Peter Medawar, 'Review of Teilhard de Chardin's *The Phenomenon of Man*', *Mind* 70 (1961): pp. 99–105.
[6] Fred Hoyle, *The Black Cloud* (London: Heinemann, 1957).

matter in helping us to interpret stars and galaxies. Dark energy is important in intergalactic space, but whereas dark matter is crucial in determining the way galaxies form and their morphology, etc., dark energy is irrelevant to that and only affects the overall dynamics of the universe. So, although you sometimes see a pie diagram where dark energy is presented as the dominant contribution, it is not as important as dark matter in terms of helping us to understand what is going on in galaxies and stars.

MICHAEL ROWAN-ROBINSON: There are two ways of looking at dark energy. One is that it is the energy density of the vacuum, so the phrase 'dark energy' refers to the idea that it is the energy of the vacuum which has a repulsive effect and, on a large scale in the universe, causes the acceleration of the expansion of the universe. So, one idea, is that dark energy is connected to the vacuum, and therefore that there will probably be some particle connected to it and so on; a quantum explanation, perhaps. But there is another way of thinking about it in Einstein's General Relativity, instead of just having one constant of gravity (the constant of gravity that is to do with pulling things together), there is another constant of gravity, the cosmological constant, which is about pushing. So, on small scale, gravity pulls, on the large scale, gravity pushes, and in that way of looking at it, both effects are a microscopic features of gravity. In this sense accelerating expansion is not to do with the vacuum and it is not a dark energy; it is just the way gravity works on the large scale. But again, to understand dark energy fully, we would have to get into these bigger theories that combined quantum theory and gravity and then maybe we would start to understand why there are these two constants.

IAN MORISON: Einstein put in that constant as a fudge factor. He believed at the time, as many astronomers did, that the universe was static, that things just stayed where they were. But of course gravity is attractive, and eventually all galaxies would just collapse down into a heap. Einstein added the cosmological constant to basically balance it out and make a static universe, but later, he realised this was a totally unstable situation – if the galaxies got a bit further apart, then anti-gravity would win and they would fly apart, but if they got closer together, everything would collapse. So, Einstein said this was the worst blunder of his life. But of course, when it was discovered that the universe was expanding, he realised that he could actually have predicted that there would be an expanding or a contracting universe, and now of course, perhaps we know he was not quite so wrong after all.

CAROLIN CRAWFORD: To clarify, if there is anyone in the audience who is still confused about dark matter and dark energy (because they sound like they ought to be similar things) you should realise that dark matter is associated with gravity and thus it is associated with 'stuff'. It is required, as Martin said, to allow galaxies to congregate, and masses to grow in time, in order that we can observe the Universe that is around us. Dark energy, whether it is the repulsive gravitational force on large scales or some other property of the vacuum of space, is associated with the voids of space. This is why, in the early Universe, dark matter is so important: because stuff is close together, the gravity of the dark matter wins. But as the expansion of the Universe increases, the density of matter becomes more diluted as voids become bigger, and the strength of the repulsive force of dark energy grows bigger. Dark energy may not have been important for our past, but this is why it is important for our future, and determining the nature of dark energy, its relation to that vacuum and to the void of space, is really what will determine whether we are going to end up in a 'big rip' (in which the universe is torn apart by exponential expansion) or a 'big chill' (in which all matter and energy in the Universe disintegrates) twenty or however many billion years in the future.

ANDREW FABIAN: Just a simple thing to understand is the word 'dark' to a physicist or an astronomer means we do not understand it! [Laughter]

CHRIS IMPEY: Simple honesty there at the end of the line!

AUDIENCE QUESTION: From time to time in the history of physics, there have been statements to the effect that the big questions have been solved and essentially a closed system is close to being formed. Andrew Fabian, in his introduction, gave a sense of the excitement of knowing your ignorance. How much longer is there going to be a supply of inflationary fields of ignorance within physics and astronomy which will give your great-great-grandchildren the potential for careers in astrophysics?

ANDREW FABIAN: If I can start that one off: basically, there is dark energy and dark matter – they are things we do not understand. There are dark photons about which we know even less, and we could carry on. But I have found, throughout my career, that, we are continually finding new fundamental things to puzzle about. We can ask questions and start to give answers to things that we could not think of decades ago. I am very optimistic that we have not come to the end, that the physics books that we have got today will be looked on in 100 years' time or

200 years' time as being, 'gosh, did they really believe that kind of thing?' ! I remind you that in late Victorian times, people really believed in the ether, and then it was discovered that it does not exist. We are likely to have other theories keep rolling on. So, I think it is mainly that I see many things around that are really intriguing. We have got more and more telescopes, more things we can observe, and there is just so much more to look at.

Chris Impey: I would like to hear from all the panellists on this. Let me add a thought that some of the questions are too hard to answer. Is there some anthropocentric hubris that we enjoy at this time that causes us to imagine that we can answer all these big questions still on the table?

Carolin Crawford: Well, I would just like to remind everyone of that quote for which Donald Rumsfeld was ridiculed on 'unknown unknowns'. Actually I would say that I and so many other scientists rely and positively look forward to discovering those unknown unknowns. If they were not there, we would give up now. It is really exciting to think there is so much left to be discovered, so much left to be understood, and that, it is that that drives so many scientists, whether they are going to be my great-great-however-many grandchildren, I really think it is exciting, and we absolutely need that sense of unknown.

Ian Morison: Absolutely. The big new thing in radio astronomy is going to be the Square Kilometre Array – the headquarters are in fact at Jodrell Bank. There are a number of things that it is aimed to do – we can look at pulsars and all these various things – but one of them is to discover things that we have no idea exist. Almost every major new instrument that is ever been built in the world that has perhaps had a factor of ten advancement in its capabilities, has actually discovered things that were not known about before it was built. Let us just hope that this continues to happen in the future. There will be lots for our research students and their research students and so on in the future.

Michael Rowan-Robinson: I am going to express a concern, and that is that we have reached a point in astronomy where the telescopes we are building are at the limit of what can be built. With every advanced nation contributing together in projects such as the Square Kilometre Array, the ALMA (Atacama Large Millimetre Array) project in the submillimetre, the European Extremely Large Telescope with a 40-metre diameter. Similarly, with the space missions, basically we are making things as big as we can at the moment. To make anything 10 times bigger is going to be extraordinarily difficult. So we may be coming towards the end of a period, which we have been very lucky to live through, in which every waveband was opened up with new space missions or instruments, and wonderful discoveries were made, quite cheaply really at the beginning. Gradually we got to the point in each field where the missions and the telescopes are gigantic, they have to be international and on an immense scale, and so I cannot see that that can continue at the same rate of growth for the next century. I think we will have to be very selective about the questions we try to address. The Large Hadron Collider is another example. It is a vast international machine. How are they going to make a bigger machine than that? It is practically at the limit of what we can put together, unless we are willing to spend 30% of our income on science. Currently, I think we spend about 0.5%: this is what governments spend on your behalf on science. But if you were willing to say we will give up football and all sorts of expensive pleasures and funnel it all into the next big telescope, well, then maybe we can have some bigger telescopes, but it is not going to be quite the same as it has been for the past fifty years I think.

CHRIS IMPEY: That is an excellent question. I am going to let Martin have the last word on this and for the panel discussion. We will find out whether he is an optimist or a pessimist.

MARTIN REES: Well, first, I wish Rumsfeld had stopped at philosophy! But I am not quite as pessimistic as Michael. It is true that any particular sub-field does eventually get to this gigantism, but new fields start up – for instance, is there life on planets around other stars? This field did not exist twenty years ago. Also a lot could be done with small telescopes. Ideas of the origin of life could be done by individuals with clever ideas. So, I think there will be new subjects starting up where we can work without great resources, and I certainly hope so. As science advances, the periphery gets longer and the frontier with the unknown therefore gets more extensive, and I think planets around other stars and life on those planets is going to be a very important subject in this century, as well as the more fundamental physics areas.

There is also the question which was alluded to about whether there will be some things which we will never understand because our brains are not up to it, and I think we must be open minded about that because, just as we know a monkey cannot understand quantum theory, there may be some fundamental features of reality which are just beyond the powers of human brains to grasp. But I think there are many that we can grasp, and just as there is been huge progress in the last fifty years, there will be more in the next fifty years.

Finally, since the theory of the multiverse came up, I would like to end with an anecdote. At a conference where I was on a panel we were asked about how much would we bet on the multiverse, and I said, 'Well, on a scale, would you bet your goldfish or your dog or your life?' I was nearly at the dog level. The other person on the panel was Andrei Linde, the inventor of the eternal inflation idea – he has spent twenty-five years working on it. He said he would bet his life on the idea. Then, at a later meeting, the great theorist Stephen Weinberg said, when he was asked the question, he would happily bet Martin Rees' dog and Andrei Linde's life!

Chris Impey: That is great! Well, thanks for your excellent questions. You are in a minute going to be invited to Edward Lumley Hall, which is outside and on the left, for a reception, but first, please thank Lord Rees and the excellent panellists for a wonderful evening.

DREAMS OF DISTANT WORLDS

Chris Impey

ABSTRACT: Humans have dreamed of other worlds for thousands of years. At the dawn of science, over two thousand years ago, some Greek philosophers thought that the Earth might not be unique, but a geocentric cosmology squashed this speculation until the time of Galileo. Gradually, telescopic observations revealed planets and moons to be worlds in space, and robotic space probes in the past fifty years sharpened this awareness. In 1995, the first exoplanet was discovered. In twenty years, the inventory has grown to over 3500, of which hundreds are Earth-like and dozens are habitable. Astronomers project roughly a hundred million habitable 'Earths' in the Milky Way and the search for life on those worlds is one of the most compelling projects in science. Meanwhile, cosmologists paint a picture of an expanding universe where space-time we can see holds 10^{23} stars and their likely attendant planets. If inflation occurred, then other space-times are likely to exist, with properties that may or may not be hospitable to life. Artists and scientists have seized on these discoveries to visualize other worlds. The potential of a boundless, biological universe recasts what it means to be human.

Ancient Dreams

Anyone who missed the past thirty years by being in a coma or marooned on a desert island would have to make adjustments as they re-joined civilization. The largest would be due to dramatic progress in computers and information technology and the creation of the Internet. If their attention turned to space science and astronomy, they would be amazed by the progress in our knowledge.[1] Mars has turned from an arid desert into a planet of ancient lake beds and subterranean glaciers. The outer region of the Solar System has gone from being frigid and uninteresting to a place with a dozen potentially habitable worlds. They would learn about a horde of exoplanets, projecting to billions across the Milky Way galaxy. Their familiar view of the visible sky would be extended by images spanning the electromagnetic spectrum and revealing brown dwarfs, black holes, and other exotic worlds. In cosmology, they would learn that the limit to our vision has been greatly extended by new telescopes in space and large telescopes on the ground. Light takes a long time to travel through the vast spaces of the cosmos, so looking out in space is looking back int time. Astronomers have information reaching back across 99.99% of the 13.8-billion-year age of the universe. They would even encounter the idea that the visible universe might be one among many universes.

It took centuries for ancient ideas to expand into a universe filled with galaxies

---

[1] Neil deGrasse Tyson, Michael Strauss, and J. Richard Gott, *Welcome to the Universe: An Astrophysical Tour* (Princeton: Princeton University Press, 2016).

and stars and their attendant planets. The first philosopher known to have dreamed of worlds beyond ours was Anaxagoras. He is credited with the quote: 'The opportunity to understand the universe is the reason it is better to be born than not to exist.'[2] Almost nothing written by Anaxagoras has survived but we do know he was born around 500 BCE. in Clazomenae, a bustling port on the coast of present-day Turkey.[3] Before he moved to Athens and helped to make it the intellectual centre of the ancient world, and long before he was sentenced to death for his heretical ideas, we imagine him as an intense, austere young man far removed from the concerns of everyday life.

Radical ideas often come with a price. For daring to suggest that the Sun was larger than the Peloponnese peninsula, Anaxagoras was charged with impiety. He avoided the death penalty by going into exile in Asia Minor, where he spent the remainder of his life. Pluralism – the idea of multiplicity of worlds, including the possibility that some might harbour life – had its antecedents in work by Anaximander and Anaximenes, and in speculation passed down by the Pythagorean School.[4] But Anaxagoras was the first to embed the idea in a sophisticated and fully-fledged cosmology. By the time of the early atomists Leucippus and Democritus, plurality of worlds was a natural and inevitable consequence of their physics.[5] There were not just other worlds in space, but infinite worlds, some like this world and some completely unlike it.[6] In an era with no telescopes, it was a startling conjecture.

Copernican Revolution

For the next two thousand years the idea of the plurality of worlds ebbed and flowed as different philosophical arguments were presented and moulded to accommodate Christian theology.[7] The pluralist position was countered by the arguments of Plato, and particularly Aristotle, who held that the Earth was unique and so there

---

[2] Diogenes Laertius, *Lives of the Philosophers*, trans. Robert Drew Hick, (New York: G.P. Putnam's Sins, Loeb Classical Library, 1925), pp. 31–32.

[3] Patricia Curd, *Anaxagoras of Clazomenae: Fragments and Testimonia* (Toronto: University of Toronto Press, 2007), Preface..

[4] Geoffrey Kirk, John Raven, and Malcolm Schofield, *The Presocratic Philosophers: A Critical History with a Selection of Texts*, 2nd ed., (Cambridge: Cambridge University Press, 1983), pp. 100–62, 214–38.

[5] Michael Crowe, *The Extraterrestrial Life Debate, 1750–1900* (New York: Courier Dover, 1999), p. 3.

[6] James Warren, 'Ancient Atomists on the Plurality of Worlds', *Classical Quarterly* 54 (2004): pp. 354–65.

[7] Stephen Dick, *Plurality of Worlds: The Origins of the Extraterrestrial Life Debate from Democritus to Kant* (Cambridge: Cambridge University Press, 1982), pp. 142–75.

could be no other system of worlds.[8] The European cultures were not alone in conceiving of the plurality of worlds. Hindu and Buddhist traditions assume a multiplicity of worlds inhabited by intelligent entities. In one myth the god Indra says 'I have spoken only of those worlds within this universe. But consider the myriad of universes that exist side by side, each with its own Indra and Brahma, and each with its evolving and dissolving worlds.'[9] The Roman writers Cicero and Plutarch wrote of creatures living on the Moon,[10] and in the second century BCE, Lucian of Samosata wrote an extraordinary fantasy about an interplanetary romance. Lucian's *A True Story* satirized Homer and began with the advisory that readers shouldn't believe a single word of it.[11] Lucian and his fellow travellers are deposited by a water spout on the Moon, where they encounter a bizarre race of humans who ride on the backs of three-headed birds. This singular work is the precursor of modern science fiction.[12]

For over a millennium it was dangerous in Europe to espouse the idea of worlds in space with life on them. Throughout medieval times, the Catholic Church considered it heresy. In the late sixteenth century, Giordano Bruno ran headlong into this conflict. The lapsed Dominican monk deviated from the Catholic orthodoxy in a number of ways, but his espousal of the Copernican system, which displaced the Earth from the center of the universe, brought extra scrutiny. He believed that the stars were infinite in number, and that each hosted planets and living creatures.[13] Bruno was incarcerated for seven years before his trial and he was eventually tried and convicted of heresy. A statue in the Campo dei Fiori in Rome marks the place where he was burned at the stake in 1600 as an 'impenitent and pertinacious heretic'.[14] Religion had cast an ominous shadow over the plurality of worlds (FIGURE 3.1).

---

[8] Dick, *Plurality of Worlds*, pp. 23–43.

[9] Cited in Ronald Huntingdon, 'Mything the Point: ETIs in a Hindu/Buddhist Context', in *Extraterrestrial Intelligence: The First Encounter*, ed. J.L. Christian, (Buffalo, NY: Prometheus Books, 1976).

[10] Francis Godwin, *The Man in the Moone* (Calgary: Broadview Press, 2009), p. 42.

[11] H.W. Fowler and F.G. Fowler, *The Works of Lucian of Samasota* (Oxford: The Clarendon Press, 1905), Introduction.

[12] S.C. Fredericks, 'Lucian's True History as Science Fiction', *Science Fiction Studies* 3, no. 1 (1976): pp. 49–60.

[13] I. Rowland, *Giordano Bruno: Philosopher/Heretic* (New York: Farrar, Straus, and Giroux, 2008).

[14] Ken Croswell, *Planet Quest: The Epic Discovery of Alien Solar Systems* (Oxford: Oxford University Press, 1999), p. ix.

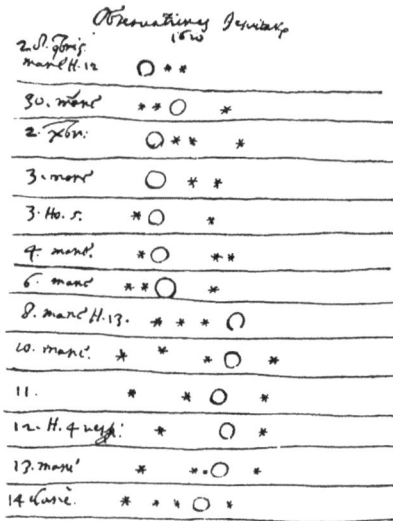

FIGURE 3.1. Galileo took a crucial step towards the idea of "other worlds" by showing that Jupiter had orbiting satellites just as the Moon orbits the Earth. This drawing announced his discovery in *Sidereus Nuncius* in 1610.

Nine years after Bruno's death, Johannes Kepler was working with data that would cement the Copernican model of the Solar System. He dusted off his student dissertation, where he defended the Copernican idea by imagining how Earth might look when viewed from the Moon. Kepler elaborated on this idea his youthful paper and added a dream narrative to turn it into a sophisticated scientific fantasy: *Somnium*.[15] Kepler was a scientist who wanted to realistically envisage space travel and aliens. His narrative is rich with comments on the problems created by acceleration and varying gravity. He speculates on the effect of environment on lunar creatures, and so foreshadows Darwin and Lyell.[16] Kepler had every reason to take refuge in a dream. He was frail, bow-legged, covered with boils, and cursed with myopia so severe that he never witnessed any of the celestial phenomena he enunciated so elegantly. *Somnium* was a crucial step in the progression towards rational speculation about other worlds.

The Copernican revolution displaced the Earth into motion around the Sun; another step was recognizing that the Earth is one of many worlds in space (FIGURE 3.2). The principle of mediocrity says that there's nothing unusual about the Earth's

---

[15] Johannes Kepler, *Somnium*, trans. E. Rosen, (New York: Dover Publications, 2003).

[16] Gayle Christianson, 'Kepler's *Somnium*: Science Fiction and the Renaissance Scientist', *Science Fiction Studies* 3, no. 1 (1976): pp. 76–90.

situation, or by extension, about the existence of life on this planet.[17] Within months of Kepler's dream piece, Galileo affirmed the Moon as a geological world, with topography similar in scale to the Earth. He also showed that Jupiter had orbiting moons and that the Milky Way resolved into points of light that seemed like distant versions of bright stars.[18] The word *world* was no longer confused with *kosmos*; it meant a potentially life-bearing planet orbiting the Sun or a distant star.[19] Speculation about life on the Moon was routine. However, theology and philosophy coloured the debate in two distinct ways. One was the principle of plenitude – everything in God's power must be realized, so inhabited worlds should be abundant. Another was teleology – the purpose and direction in nature that implies a Creator, who surely wouldn't have gone to the trouble of creating uninhabited worlds.[20]

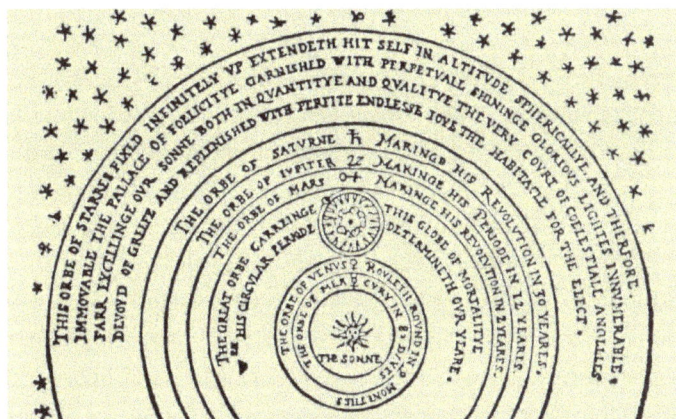

FIGURE 3.2. English astronomer Thomas Digges was the first to imagine a Copernican system where the stars might be distributed through infinite space in his 1576 book *A Perfect Description of the Celestial Orbs*.

Worlds Next Door

The planets have been objects of our imagination for millennia. We have anthropomorphized near neighbours Venus and Mars as the goddess of love and the god of war. Mars had pride of place as the place to project our expectations of life on other worlds. In 1854, the polymath William Whewell speculated that Mars had

---

[17] Andre Kukla, *Extraterrestrials: A Philosophical Perspective* (Lanham, MD: Rowman and Littlefield, 2010), p. 19.

[18] Galileo Galilei, *Discoveries and Opinions of Galileo*, trans. S. Drake, (New York: Anchor, 1957), p. 146.

[19] Crowe, *The Extraterrestrial Life Debate: 1750-1900*, p. 150.

[20] Arthur Lovejoy, *The Great Chain of Being* (Cambridge, MA: Harvard University Press, 1936).

continents, oceans, and life forms.[21] By the end of the nineteenth century, telescopes showed strange, nearly linear markings that were called canals, but spectroscopy had been used to show that Mars had no significant atmosphere. This didn't stop Percival Lowell from claiming the canals to be conduits carrying water to a dying Martian civilization in 1895,[22] and two years later H.G. Wells cemented the idea of life on Mars with his popular science fiction book *War of the Worlds*.[23] Although better known for Tarzan, Edgar Rice Burroughs got his start writing pulp stories about life on Mars.[24] On the eve of the Second World War, Orson Welles caused a local panic with his radio play based on H.G. Wells' novel, and Ray Bradbury credited the influence of Rice Burroughs on his epic 1950 work *The Martian Chronicles*.[25]

The first space probes to Mars brought all this speculation down to Earth with a bump. In the late 1970s, orbiters and landers found no trace of artificial features, and confirmed that the atmosphere was so thin and the surface was so cold that life couldn't exist in the soil. Over the past twenty years, a series of rovers have swung the pendulum back in favor of biology, sequestered in sub-surface aquifers where water is kept as a liquid by a combination of pressure from surface rock and radioactive heating from interior rock.[26] The inventory of water ice at the two poles and in buried glaciers nearer the equator is sufficient, if melted, to coat the planet with a layer of water a few inches deep. Mars is a marginally habitable world.[27]

The most important insight from the Solar System in the past thirty years is the fact that the moons of giant planets can be complex, interesting worlds.[28] A large moon like the Earth's is too small to be geologically active or to hold a significant atmosphere. In the outer Solar System, temperatures at the surface of a large satellite or moon are low enough that an atmosphere can be retained. Tidal heating from the planet adds a significant internal heat source to any moon that's close or on an elliptical orbit. As on Mars, water can be kept liquid under a hard cap of rock and ice

[21] William Whewell, *On the Plurality of Worlds* (London: J.W. Parker and Son, 1853), p. 187.
[22] W.G. Hoyt and W.G. Wesley, 'Lowell and Mars', *American Journal of Physics* 45, no. 3 (1977): pp. 316–17.
[23] D.Y. Hughes and H.M. Gelduld, *A Critical Edition of War of the Worlds: H. G. Wells's Scientific Romance* (Bloomington and Indianapolis: Indiana University Press, 1993).
[24] Edgar Rice Burroughs, *A Princess of Mars* (New York: A.C. McLurg, 1917).
[25] Ray Bradbury, *The Martian Chronicles* (New York: Doubleday, 1950).
[26] D.M. Harland, *Water and the Search for Life on Mars* (Berlin: Springer-Praxis, 2005).
27 J.M. Trigo-Rodriguez, F. Raulin, C. Muller, and C. Nixon, eds., *The Early Evolution of the Atmospheres of Terrestrial Planets* (Heidelberg: Dordrecht, 2013).
[28] D.A. Rothery, *Satellites of the Outer Planets: Worlds in Their Own Right* (Oxford: Oxford University Press, 1999).

even when an atmosphere is absent. There is a bewildering array of distant worlds in the Solar System. The most compelling moons are Jupiter's Europa, with an icepack overlying a water ocean kilometers deep (FIGURE 3.3), and Saturn's Titan, with its thick atmosphere and ethane-methane lakes. About one dozen moons are likely to have all of the ingredients required for life: liquid water, organic material, and a local energy source.[29]

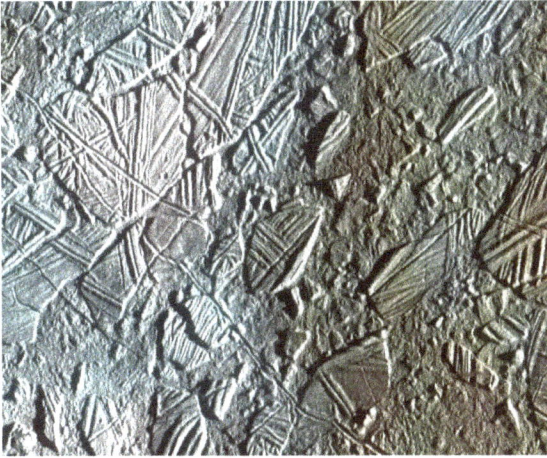

FIGURE 3.3. The fractured ice pack of Europa, one of the Galilean moons of Jupiter. Europa is a "water world" with geological heating from the interior and the possibility of life in the ocean under the ice pack (NASA).

The Exoplanet Era

After decades of false starts and frustration, the age of exoplanet discovery was ushered in on October 6, 1995, when Michel Mayor and Didier Queloz from the Geneva Observatory announced the discovery of an exoplanet half Jupiter's mass orbiting 51 Pegasi.[30] The exoplanet was unseen and was dtected by its gravitational tug on the parent star. The detection of exoplanets had been long anticipated, but the excitement of the discovery was tinged with confusion. 51 Pegasi b was a Jupiter-mass planet far closer its star than Mercury is to the Sun, whipping around a complete orbit in four days, at a scorching temperature of 1000 °C (1800 °F). Discoveries soon accelerated from one a month for the first few years to one a week at the turn of the century, reaching a current average of one new exoplanet every day. The

[29] M. Carroll, *Living Among Giants: Exploring and Settling the Outer Solar System* (Berlin: Springer, 2014).

[30] Michel Mayor and Didier Queloz, 'A Jupiter-Mass Companion to a Solar-Type Star', *Nature* 378 (1995): pp. 355–59.

first planets detected by the Doppler method were very surprising because they were massive and close to their stars, and most had orbital eccentricities larger than any of the planets in the Solar System. Such 'hot Jupiters' were unusual and completely unexpected (FIGURE 3.4).

FIGURE 3.4. Many of the first exoplanets to be discovered were in situations quite different from any planet in the Solar System. CoRoT-7b is not much larger than the Earth but is sixty times closer to its Sun-like star, with the surface facing the star molten rock (ESO/L. Calcada).

Kepler has blown the lid off the search for low-mass or Earth-like planets. This one-meter NASA telescope looks for the signature of a planet by its periodic partial eclipse of the parent star, a signature that only appears for systems where our line of sight is oriented in the equatorial plane. The team announced over 1,200 candidates in early 2011. By 2012, the number of candidates had grown to over 2,300, nearly 250 of which were less than 1.3 times the Earth's size.[31] In 2014, the Kepler team announced 1,000 confirmed exoplanets, with 2,900 additional candidates. It's just a matter of time before we detect clones of home: Earth-like planets in Earth-like orbits (FIGURE 3.5). Analogues of Earth orbiting Sun-like stars have a transit depth of 0.01% and a 1-in-200 probability of a suitable alignment. The transit geometries are much more favourable for planets around abundant small red dwarfs than around solar-type stars.

[31] Natalie Batalha et al., 'Planetary Candidates Observed by Kepler. III. Analysis of the First 16 Months of Data', The Astrophysical Journal Supplements 204 (2013): pp. 24–45.

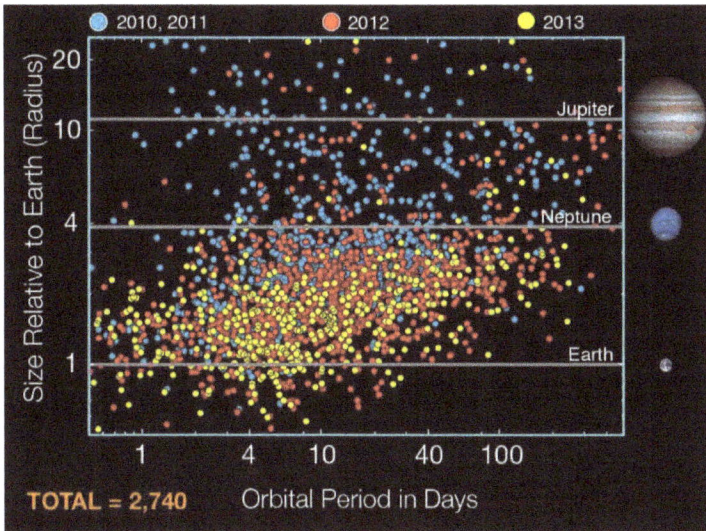

FIGURE 3.5. The Kepler satellite detects exoplanets by their eclipses of the parent star, and it has found most of the new exoplanets since 2010. When all the mission data is analyzed, Earth-sized planets on Earth-like orbit will be recovered (Kepler Science Team/NASA).

A major goal of this research is to determine the fraction of all stars that have planets. Doppler surveys in the pre-Kepler era showed that 10% of Sun-like stars have planets, with indications that the true fraction might be much higher and that rocky terrestrial planets might outnumber gas giants.[32] Recent research moves this estimate upwards, as both Doppler and transit methods reach sensitivities where Earth proxies are detectable. The biggest leverage on the total number of exoplanets comes from small, dim stars. Dwarf stars outnumber stars like the Sun by a factor of a hundred. Surveys indicate an incidence rate of 1.4 planets per star, of which 1.1 are between three and ten Earth masses, and 0.2 are in their stars' habitable zones.[33] Transit surveys like Kepler give similar results, with 0.9 planets per star in the range one half and four times the Earth's radius, of which 1 in 5 are within their dwarf star's habitable zones. Both techniques are pointing to a rapid rise in the incidence rate of terrestrial planets towards the low mass stars.[34] Microlensing, the only method that's

---

[32] Geoffrey Marcy et al., 'Observed Properties of Exoplanets: Masses, Orbits, and Metallicities', *Progress of Theoretical Physics Supplement* 158 (2005); pp. 24–42.

33 Mikko Tuomi et al, 'Bayesian Search for Low-mass Planets Around Nearby M Dwarfs: Estimates for Occurrence Rate Based on Global Detectability Statistics', *Monthly Notices of the Royal Astronomical Society*, in press, 2016.

[34] Andrew Howard, 'Observed Properties of Extrasolar Planets', *Science* 340 (2013): pp. 572–76.

not very strongly skewed towards massive planets close to their stars, indicates an average of 1.6 planets per star. The golden age of exoplanets is just beginning.

Living Worlds

Astronomers use conventional definition of habitability: the zone around a star within which water can be in stable liquid form on the surface of a rocky planet. This calculation is affected by the atmosphere; Venus and the Earth are similar in mass and size and would be equally detectable by the Doppler or transits methods, yet Venus is almost certainly uninhabitable due to a strong greenhouse effect. Also, habitable zones evolve as stars age and the amount of radiation they deliver changes, and planetary atmospheres also evolve due to geological activity (and of course, due to life). The definition is conservative because it supposes stellar radiation is the only energy source that can power biology. On the Earth, life is found above the boiling point of water and below its freezing point, in total darkness on the sea floor, and deep inside rock. Within the Solar System, there may be a dozen habitable "spots," many of which are in a cryogenic habitable zone where the icy and rocky surfaces conceal water kept liquid by pressure, geological heating and, in the case of moons around giant planets, tidal heating.

Exoplanet research is a young field, with many observational and theoretical puzzles to solve before we can confidently project the number of habitable worlds. Rough estimates suggest at least one planet per star across the Milky Way, or a total of 100 billion.[35] That projects to about 100 million terrestrial planets around Sun-like stars in the Milky Way, several million of which are both Earth-like and habitable. Biomarkers are required to take the huge step forward from demonstrating habitability to the first detection of life beyond the Earth. That detection might come in the form of a shadow biosphere on our planet, from trace fossils in a Mars rock, from exploration of targets in the Solar System, from a spectral signature in the atmosphere of an extrasolar planet (FIGURE 3.6), or from success in the long-standing campaign to detect signals from remote civilizations. Each possibility implies a different type of evidence, which must be matched against very uncertain criteria for the definition of success.

Worlds Without End

Most known exoplanets are in the proximate universe, within a few hundred light years of the Sun. This census can be reliably projected to population of potentially

---

[35] Arnaud Cassan et al., 'One or More Bound Planets per Milky Way Star from Microlensing Observations', *Nature* 481 (2012): pp. 167–69.

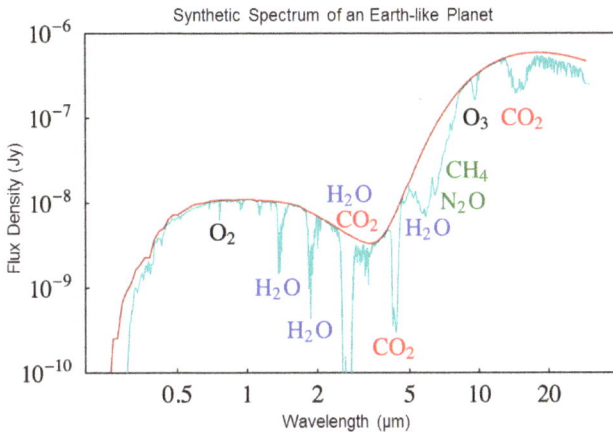

FIGURE 3.6. Synthetic spectrum of an Earth-like planet shows biomarkers or spectral signatures of water and photosynthetic organisms (Heike Rauer).

habitable locations in the entire Milky Way: roughly 100 million Earth-like planets, and several billion 'exomoons', free-floating planets, and locally habitable 'spots' in otherwise inhospitable worlds. Most of the 400 billion stars in the Milky Way are dim, low mass dwarfs, so a majority of the habitable worlds are in situations quite different from the Earth's. However, dreams of other worlds should not be confined to a corner of one galaxy among many. To the limit of vision of the Hubble Space Telescope in the Extreme Deep Field (XDF), the census in this small field projects to 120 billion galaxies across the sky, but there are so many dwarf galaxies just below the limit of detection and at high redshift that the total census might be close to 220 billion.[36] Extrapolating, the number of stars in the observable universe is 10^{23} to within a factor of two.[37] With at least one planet per star, this truly staggering number, a hundred thousand billion billion, is also a conservative estimate of the number of worlds in the cosmos.

In the big bang model, the expanding universe is governed by dark matter, which decelerates the expansion rate, and dark energy, which accelerates the expansion rate and which has been dominant for the past few billion years.[38] Many thousands of galaxies have been observed at a lookback time of nine billion years or more, where

---

[36] Harry Teplitz et al., 'Ultraviolet Imaging of the Hubble Ultra Deep Field with Wide-Field Camera 3', *The Astronomical Journal* 146 (2013): pp. 159–78.

[37] Pieter Van Dokkum and Charlie Conroy, 'A Substantial Population of Low-Mass Stars in Luminous Elliptical Galaxies', *Nature* 468 (2010): pp. 940–42.

[38] Jeremiah Ostriker and Simon Mitton, *Heart of Darkness: Unraveling the Mysteries of the Invisible Universe* (Princeton: Princeton University Press, 2013).

galaxies were moving away from us faster than light when the light we see now was emitted. The standard big bang model therefore implies that the physical universe, the totality of space-time, is larger than the space-time we can observe. It's possible to get constraints on space that lies beyond our horizon. The curvature of the observable universe is very small, based on observations of the microwave background radiation.[39] So space within the particle horizon is close to Euclidean. This allows for a lower bound on curvature beyond the particle horizon, including situations where global space-time is bounded. The result is a lower bound of a factor of 400.[40] Assuming a cosmological principle, that our accessible space-time is representative of the whole universe, the lower bound is about 10^{26} stars and a similar or larger number of planets.

The only modification to the big bang model in the past 35 years is inflation. Inflation is the exponential expansion of the universe a fraction of a second after the big bang, amplifying quantum fluctuations to become seeds for subsequent galaxy formation. The physical cause of inflation is not understood, but the mechanism only works as an explanation if there are 60 e-foldings of the infant universe, implying a totality of space-time 10^{24} times larger than the observable universe.[41] In chaotic inflation, there could be disconnected space-times that sample all possible initial conditions of the laws of physics.[42] This is the multiverse. As we reach the shimmering shores of conjecture, with so many potential worlds that every possible outcome is realized somewhere in the multiverse, let's return to what we know. After centuries of anticipation, astronomers have shown that the Earth is not unique.[43] The worlds resulting from star formation are likely to be more physically, chemically, and biologically diverse than the familiar real estate of the Solar System. So far we've only been discovering and counting these worlds; characterizing them and detecting life on them in the future. One of the oldest questions about our place in the universe may be on the brink of being answered.

---

[39] Charles Bennett et al., 'Nine Year Wilkinson Microwave Anisotropy Probe (WMAP) Observations: Final Maps and Results', *The Astrophysical Journal Supplement Series* 208 (2013): pp. 20–75
[40] Mihran Vardanyan, Roberto Trotta, and Joseph Silk, 'Applications of the Bayesian Model Averaging to the Size and Curvature of the Universe', *Monthly Notices of the Royal Astronomical Society* 413 (2011): pp. L91–L95.
[41] Alan Guth, 'Eternal Inflation and Its Implications', *Journal of Physics A* 40 (2007): pp. 6811–26.
[42] Bernard Carr, *Universe or Multiverse?* (Cambridge: Cambridge University Press, 2007).
[43] Chris Impey, *The Living Cosmos: Our Search for Life in the Universe* (Cambridge: Cambridge University Press, 2011).

ACKNOWLEDGEMENTS: Kudos go to Valerie Shrimplin and Nick Campion for organizing and hosting a cracking good conference. I'm grateful to Holly Smith, an INSAP stalwart, for collaborating with me on the 2013 book for Princeton University Press that was an inspiration for this contribution: *Dreams of Other Worlds*. I also recognize the organisers of the San Marino Conference in 2014 where I first aired some of the ideas in this paper. I acknowledge many conversations with colleagues over the years, at Steward Observatory and beyond, that have informed me about progress in planetary science and astrobiology.

MEMORIES UNLOCKED AND PLACES EXPLORED: STELLARIUM, TEMPORALITY AND SKYSCAPES

Daniel Brown

ABSTRACT: Skyscapes are a combination of landscape, sky and people. As a viewer explores their environment they start to realise how they dwell in a location filled with meanings, emotions and memories. This connection to skyscapes occurs most powerfully when temporality is encountered, not only through task and landscape exploration but also by watching cosmic rhythms unfolding in the skies above. This article will illustrate how the planetarium software Stellarium allows both the creative practitioner and viewer to simulate long term cycles of celestial motions as well as the powerful potential to include panoramic landscapes. Rather than constituting photo realistic imagery they should be seen as a canvas onto which one can express the place experience encountered. This allows for a creative and imaginary approach to express temporal aspects of the landscape, shaped by a dialogue between people, culture, seasons and the sky. Overall this article calls for artists to engage and explore Stellarium and how it might be incorporated into their work to explore skyscapes.

Introduction

Astronomy, as in observing the stars and creating a cosmology, has been described as the oldest science.[1] But science is a rather modern term, and we should probably re-phrase the previous statement as: watching the skies has been with humanity since the moment we became aware of our surroundings and looked up. Clive Ruggles commented that:

The Sky is of universal importance. Cultural perceptions of the skies are vital in fulfilling humankind's most basic need to comprehend the universe it inhabits, both from a modern scientific perspective and from countless other cultural standpoints, extending right back into early prehistory.[2]

As a consequence, depictions of the stars and skies are potent expressions of cultural interactions with the sky phenomenon. This goes beyond capturing the landscape and ventures into the realm of engaging with the skyscape. As discussed by Fabio Silva, a skyscape is not only a single view that encompasses the sky framed by terrestrial objects, but rather the ensemble of interconnected meaningful objects therein.[3]

---

[1] Clive L. N. Ruggles, *Astronomy in Prehistoric Britain and Ireland* (New Haven: Yale University Press, 1999), p. 155.
[2] Clive L. N. Ruggles, 'Archaeoastronomy and Ethnoastronomy: Building Bridges between Cultures', Proceedings of the IAU Symposium No. 278 (2011): p. 6.
[3] Fabio Silva, 'The Role and Importance of Sky in Archaeology: An Introduction', in *Skyscapes: The Role and Importance of the Sky in Archaeology*, ed. Fabio Silva and Nicholas Campion (Oxford:

Contemporary professional astronomers, associated with such depictions, are challenged by the way in which they gather information on the objects in the sky. The land art artist Charles Ross mentioned in an interview with Saad-Cook,

> Astronomers have to spend so much time in front of computers now. I have heard them express concerns about how they are losing touch with a direct experience of the sky. It is a problem because being under the stars inspires a feeling that can lead to discovery. I am not dismissing the rational approach – it is necessary. But one can lose creative motivation by getting too separated from one's source.[4]

The author is a professional astronomer and can fully confirm this statement. But as another land art artist, Nancy Holt, said: 'I am not involved in astronomy: I am just looking at the world, that's all'.[5] As an artist she feels

> … that the need to look at the skies – at the moon and stars – is very basic and it is inside all of us. So when I say my work is an exteriorization of my own inner reality, I mean I am giving back to people through art what they already have in them.[6]

Therefore, to express the interconnectedness and depth of meaning encountered within a skyscape by a viewer requires the individual to immerse themselves in the landscape and skyscape experience, to feel that they are within the skyscape. Such a full experiential encounter with a skyscape will be transformative and allow viewers to carry out an intimate conversation between themselves and their environment. As one carries out such an elaborate engagement one becomes aware of the skyscape being a unity with the place experience. Here, place and its experience is understood in terms of Heidegger's exploration of *being* and Merleau-Ponty's body and sense oriented interaction.[7]

At this point we arrive at the main challenge when depicting a skyscape: How can something intrinsically reliant upon experience become purely a picture? Or phrased slightly differently: how can temporality be captured?

One possibility put forward within this paper is the use of planetarium software

Oxbow Books, 2015), p. 3.
[4] Janet Saad-Cook et al., 'Touching the Sky: Artworks Using Natural Phenomena, Earth, Sky and Connections to Astronomy', *Leonardo* (1988): pp. 125, 156.
[5] Saad-Cook, 'Touching the Sky', p. 128.
[6] Saad-Cook, 'Touching the Sky', p. 128.
[7] Graham Harman, *Heidegger explained: From phenomenon to thing* (Chicago and La Salle, IL: Open Court, 2013); Maurice Merleau-Ponty, *Phenomenology of Perception*, trans. Colin Smith (Delhi: Motilal Banarsidass, 1996).

called Stellarium, which allows the inclusion of custom made landscapes.[8] As a result, skyscapes can now be viewed for any date with the correct distribution of Sun, Moon, planets and stars; a necessity for researchers who are investigating the skyscapes of ancient peoples. Also the passage of time itself can be simulated in the sky, representing the Sun's spiral during the year and further cosmic rhythms. But we still remain removed from the site that the skyscape experience represents.

However, what we have created is an abstract representation of the skyscape. We acknowledge that it is not the real material *Site* that evoked feelings and emotions. What has been created can be described as a *Non-Site* as defined by Robert Smithson.[9] As such it allows us to use this interactive imagery full of information and encoded meaning to now negotiate our own understanding of the skyscape captured at the site. Smithson continues that such a *Site* versus *Non-Site* dialectic engagement by the viewer offers the possibility for a deeper interaction and self-reflection. Additionally, the importance of comprehending the components of the displayed skyscape in a single view is vital to allow a non-prescriptive exploration. In this way such a single view turns into a dialectic landscape, as described by my paper, 'The Experience of Watching', in which many emotions and meanings are displayed all at once forming a constellation of moments.[10] As they come together in a pattern and interconnectedness determined by the viewer, a skyscape experience can be explored. Similarly, Smithson describes the constellation of moments in his description of the dialectic landscape in the context of New York's Central Park as

> An endless maze of relations and interconnections, in which nothing remains what or where it is, as a-thing-itself, but the whole park changes like day and night, in and out, dark and light – a carefully designed clump of bushes can also be a mugger's hideout.[11]

The intentional non-linear way of presenting images in a single view mirrors the *Bild-Atlas* as used in Warburg's works.[12]

---

[8] Stellarium 0.14.2, stellarium.org.

[9] Robert Smithson, 'A Provisional Theory of Non-Sites', in *Robert Smithson: The Collected Writings*, ed. Jack D. Flam (Berkeley, CA: University of California Press, 1996), p. 364.

[10] Daniel Brown, 'The Experience of Watching: Place Defined by the Trinity of Land-, Sea, and Skyscape', *Culture and Cosmos* 17, no. 2 (2013): p. 11; Walter Benjamin, *On the Concept of History* (New York: Classic Books America, 2009).

[11] Robert Smithson, 'Frederick Law Olmsted and the Dialectic Landscape', in *Robert Smithson: The Collected Writings*, ed. Jack D. Flam (Berkeley, CA: University of California Press, 1996), p. 165.

[12] Christopher D. Johnson, *Memory, Metaphor, and Aby Warburg's Atlas of Images* (Ithaca, NY: Cornell University Press, 2012), p. 18.

It is worth noting that the temporality within the skyscape depiction is vital since it is the only way to include and feel emotions.[13] Hence, the creator of a skyscape will not carry out a quick snapshot of the location, but rather as Malcolm Andrews summarizes when talking about 'active passivity' used by landscape painters:

> Sustained direct contact with the chosen site, famously for example, in the case of Cezanne's 'Mont Sainte-Victoire' landscape, or Monet's series paintings of 1890s; and it meant bit by bit, in the face of the site, sloughing off the conventional 'picturing' habits. It meant saturating oneself in the site so that it ceases to be just a visual field, ceases perhaps to be 'landscape', but becomes a complex sensations, of light, colour, smell, sounds, tactile experience. It becomes an environment.[14]

Carrying out such a regular on-going engagement with the site will inevitably result in an awareness of cosmic rhythms and cycles. They will reveal the passing of time within the skyscape that to some extent is a result of cosmic rhythms. It will also cause an emotional attachment with the surroundings, mirrored in the objects and metaphors chosen to be included in the representation and the actual location of the skyscape depiction, as well as the time when the custom made landscape was gathered. As a result the creator of a skyscape experience in Stellarium has explored for themselves the essence of the place itself – at a given time.

The following will elaborate on why Stellarium especially is ideal to develop a skyscape experience and how it offers a creative canvas to capture emotions and include memories. Thereafter, possible actors the viewer might create within a skyscape narrative are introduced. Then, the challenges that Stellarium poses in skyscaping and the unique creative opportunities it can offer to artists will be outlined.

Exploring Skyscapes with Planetarium Software

Skyscapes require a correct depiction of objects visible in the sky. From an astronomer's point of view, that includes stars, Sun, Moon and planets in their ever-changing paths. Additionally, skyscapes have also been integrated into skyscape archaeology, an approach located in the disciplines of archaeoastronomy and cultural astronomy.[15] Here the skyscape is explored as seen, for example, by ancient societies many millennia ago. This supports the need to simulate the sky and allows access to what could have been observed at that time period or moment.

---

[13] Henry Bergson, *Creative Evolution* (New York: Henry Holt, 1911), p. 16.

[14] Malcolm Andrews, *Landscape and Western Art*, Vol. 10 (Oxford: Oxford University Press, (1999), p. 192.

[15] Fabio Silva and Nicholas Campion, eds., *Skyscapes: The Role and Importance of the Sky in Archaeology* (Oxford: Oxbow Books, 2015).

To achieve such a result planetarium programs are needed that model and represent the skies. These programs also allow for the visualisation of the temporality expressed by the paths of Sun, Moon, planets and stars. But skyscapes are not only depictions of sky. They need to contain landscape, therefore the horizon and terrestrial objects.

In the past, planetarium programs represented a very basic landscape only through the actual perfect mathematical horizon; this was essentially a line separating the upper and lower hemisphere into exactly equal halves. Further developments discussed by myself then allowed the display of a realistic horizon profile, changing the perfect level horizon into outlines of trees, church steeples or houses.[16] However, to evoke any skyscape representation the landscape needs to be shown in more detail. Current software such as Starry Nights and Stellarium changed this by integrating photo realistic panoramas to create a full 360 degrees landscape, above which the simulated sky extends. Given that Stellarium is a freeware and the controls are very easy to grasp, the following paper will only focus upon this software, since it allows anyone to engage with skyscape in the presented manner. Overall, the choice of Stellarium can be in some way seen to resemble a much freer and non-commercial way to engage with skyscape. Thornes described this notion when contrasting American land art to English environmental art represented by Richard Long:

> It was the antithesis of so-called American 'Land Art', where an artist needed money to be an artist, to buy real estate to claim possession of the land, and to wield machinery. True capitalist art.[17]

Therefore, the engagement of the skyscape with Stellarium, in the words of Thornes reflecting upon Long's work *A Line Made by Walking*, 'opened up the realisation that an artist can create art in the environment without any special skills or tools or canvas'.[18]

But the simple view of photorealistic panorama within Stellarium representing skyscapes has several important limitations. These constraints derive mostly from the temporal dimensionality required not only in the sky but also the land. It is clear that the use of preinstalled landscapes of unrelated sites to the skyscape experience is insufficient. I have outlined how the skyscape explorer can create their own land-

---

[16] Daniel Brown, 'Exploring Skyscape in Stellarium', *Journal of Skyscape Archaeology* 1, no. 1 (2015): pp. 93–112.

[17] Suzi Gablik, *Has Modernism Failed?* (New York: Thames and Hudson, 1984) , p. 44.

[18] John E. Thornes, 'A Rough Guide to Environmental Art', *Annual Review of Environment and Resources* 33 (2008): p. 403.

scape.[19] But following in the footsteps of the typical landscapes created for Stellarium results in a depopulated landscape.[20] The landscapes rarely contain any humans or wildlife. For the vast majority of skyscape experiences this is unrealistic. Additionally, temporality needs to be explored not only in the sky above but also needs to become visible in the landscape. However, there is no animation or otherwise simulated change included in the landscape.

Some changes have been included into Stellarium of late to allow for a more sophisticated change from a daytime to a nighttime landscape by including a light pollution layer. Its potential, for example, in motivating the need for dark skies has been described by Georg Zotti but I highlighted that they might be too simple and not capture the change in mood and atmosphere that occur during sunrise or set.[21]

Protagonists in the Skyscape Narrative

Working on skyscapes with Stellarium not only has challenges regarding the landscape. It also requires the skyscape explorer to become familiar with the movements of celestial bodies. This time not in a simulation and modelling fashion championed by scientists, but in a story-telling manner. Protagonists or arenas in which a narrative in the skyscape can unfold need to be understood and identified. Goethe and his 'delicate empiricism' approach stressed the need for scientific knowledge when engaging in creative activities within the landscape in the past.[22] Additionally, Constable, as described in Thornes and Goethe, saw it as an important perquisite for an accomplished landscape painter that they:[23]

> Should possess various sorts of knowledge. It is not enough for him to understand perspective, architecture, and the anatomy of men and animals; he must also have insights into botany and mineralogy, that he may know how to express

---

[19] Brown, 'Exploring Skyscape in Stellarium'.

[20] Landscapes (2015) http://www.stellarium.org/wiki/index.php/Landscapes [accessed 30 August 2016].

[21] Georg Zotti and Günther Wuchterl, 'Raising Awareness of Light Pollution by Simulation of Nocturnal Light of in Astronomical Cultural Heritage Sites', in *The Materiality of the Sky*, the Proceedings of the 22nd Annual Conference of SEAC, held in Malta, 22–26 September 2014, ed. Fabio Silva, Kim Malville, Tore Lomsdalen, and Frank Ventura (Ceredigion: Sophia Centre Press, 2016); Brown, 'Exploring Skyscape in Stellarium', p. 106.

[22] Arthur Zajonc, 'Goethe and the Science of his Time: An Historical Introduction', in *Goethe's Way of Science: a Phenomenology of Nature*, ed. Davis Seamon and Arthur Zajonc, (Albany, NY: State University of New York Press, 1998), p. 15.

[23] Thornes, 'A Rough Guide to Environmental Art', pp. 398–99.

properly the characteristics of trees, plants, and the character of the different sorts of mountains.[24]

Some of the possible arenas in which the skyscape narration can be enacted are part of the sky alone without any horizon or terrestrial interaction. The so-called circumpolar stars are a region of never-setting celestial objects that can be associated with eternal life. This has already been part of the land art installation *Star Axis* created by Charles Ross. Other stellar objects vary in their visibility from the circumpolar to a rising and setting behaviour, expressed by Bernadette Brady as Star Phases.[25] In the context of the Old Kingdom Egyptian culture Brady outlines how they offer an ideal canvas for metaphors.

Moving away from pure sky-related objects and regions, the horizon is a powerful and interesting skyscape component that can be incorporated in skyscape impressions. Such examples would be rising and setting events and horizon locations of the Sun, Moon or a star. Nancy Holt integrated this subject in her land art installation *Sun Tunnels*, displayed in FIGURE 4.1, as well as Olafur Eliasson in his play on perceiving skyscapes in his *Double Sunset* installation.

But there are also purely landscape related arenas influenced by the objects such as the Moon and the Sun. Such examples are found in the light and shadow displays with which the general public is most familiar. They stretch beyond, for example, the south facing aspect of a garden or property towards architectural oddities causing the unexpected focus of sunbeams from reflective surfaces, such as the *Walkie Talkie* building in London.[26]

The above outlined possible protagonists within a skyscape are only a few from a much larger ensemble that can be explored by the individual skyscape artist. They illustrate the multitude of subjects to work with in a skyscape, going beyond plain landscape motifs and metaphors, venturing into the sky relations above.

---

[24] Johann Peter Eckermann, *Conversations with Goethe*, ed. J. K. Moorhead, trans. John Oxenford (London: JM Dent and Sons, 1930), p. 562.

[25] Bernadette Brady, 'Star Phases: The Naked Eye Astronomy of the Old Kingdom Texts' , in *Skyscapes: The Role and Importance of the Sky in Archaeology*, ed. Fabio Silva and Nicholas Campion (Oxbow : Oxford, 2015), pp. 76–86.

[26] Mark Duell and Sam Webb, 'Now the Walkie Talkie building is melting BICYCLES: Dazzling light reflected from giant London skyscraper scorches bike seats (and you can even fry an egg)', *Daily Mail*, 3 September 2013, available at http://www.dailymail.co.uk/news/article-2409710/Walkie-Talkie-building-melting-bicycles-Light-reflected-construction-City-skyscraper-scorches-seat.html [accessed 30 August 2016].

FIGURE 4.1. The Sun Tunnels land art installation by Nancy Holt. Image taken by Retis, 2015, under the creative commons licence.

FIGURE 4.2. An extract from one skyscape experienced expressed through Stellarium and created for the Skyscapes Above Clifton project. Note the non-photorealistic landscape including birds and viewing windows. It represents the conflict experienced at this site through the fences and building site.

Challenges for Stellarium exploring Skyscapes

In 2015 I carried out a project, *Skyscapes above Clifton*, to capture a skyscape experience within Stellarium. This was an undergraduate research project funded by Nottingham Trent University and had at its core the strong collaborative work between astronomy and architecture students, fostering interconnectedness: both were using phenomenology as their main research methodology. Their project is discussed in detail in 'Clifton Skyscapes'; in brief, it consisted of detailed landscape and skyscape engagements with the Nottingham Trent University Clifton campus.[27] These interactions were self-exploratory as well as determined by semi-structured interviews. They also included exposure to the site at different times of the day.

As an output the project created Stellarium landscapes that were not only photorealistic: their emotional response, memories recorded and historic findings were the basis for an artistic modification of the landscapes displayed in FIGURE 4.2. Also the location of each skyscape representation was chosen so that the rising and setting of certain constellations relevant to the sites usage and history were aligned, for example, to an associated building.

This project illustrated the immense potential of Stellarium landscapes as a canvas to express a skyscape experience, and thereby capture emotions, memories and the temporality of the skyscape. However it also stressed challenges that still need addressing. Duncan Higgins, the artist in our *Skyscapes above Clifton* project, indicated that Stellarium itself as an interface or lens has its limitations but also opportunities.[28] Our project therefore could offer future projects a fruitful research question, or an artist a motif for self-exploration. Three main issues were identified:

1) The skyscape was still mostly explored during daytime and the sky component incorporated in the skyscape was a depiction of a nocturnal scenario. Clouds or contrails were not included. Especially when considering skyscape impressions as one aspect of environmental art the impressions of the atmosphere and clouds are vital.[29] They also represent the weather dimension an important aspect of the landscape and therefore skyscape.[30]

2) The horizon has been described by viewers of the *Skyscapes above Clifton* project as a too powerful magnet drawing the eye towards it. As a result the viewer tends

---

[27] William Ktorides, Rumbidzai Mukundu, and Daniel Brown, 'Clifton Skyscapes', in the Proceedings of SEAC 2015, *Mediterranean Archaeology and Archaeometry*, submitted.

[28] Duncan Higgins, informal communication, 19 January 2016.

[29] Thornes, 'A Rough Guide to Environmental Art', p. 407.

[30] Tim Ingold, 'Footprints through the Weather-World: Walking, Breathing, Knowing', *Journal of the Royal Anthropological Institute* 16, no. 1 (2010): pp. S121–39.

to be too distracted to further engage with the other components. Artists, such as Katherine Bash, have tried to break this overpowering feature, and artists, such as James Turrell in his *Ganzfeld* installation, have created light installations removing the horizon entirely, creating a whiteout experience.[31]

3) Even though the sky simulation can capture the temporal dimensionality, the project struggled in capturing it fully in the landscape implemented into Stellarium. Although animations were used to fade one landscape into another gathered during sunset, the passage of people and wildlife could not be represented. However other artists – especially photographers – have explored this, capturing the invisible (for example Tarja Trygg in her *Solargraphs*) by using long exposures.[32]

Looking beyond the specific *Skyscape above Clifton* project, skyscape experiences are very striking, deeply emotional, and cannot always be identified with specific motifs, metaphors or associated to memories. James Turrell's *Skyspaces*, described by Thornes, 'are works created by an opening in the ceiling of a chamber so that the sky appears as a solid object'.[33] Turrell said himself:

… what I am doing is making space that is sensitive to events that happen in the skies, taking light from those events and somehow making it work in a space. (…) Here attention is not directed to the event but to the space itself. It is space that responds in some manner when the event occurs. In other words, it has its own way of forming its response to this event.[34]

These statements capture a so far unaddressed challenge of exploring light and space within a skyscape in Stellarium. Photorealistic landscape impressions will not be able to capture this experience. It will be up to the artist to further develop the landscape and push the current landscape usage boundaries in Stellarium.

Creative opportunities with Stellarium
The previous section, although stating many challenges, also outlined the opportunities creative practitioners working with Stellarium can explore and tap into.

The author proposes that future engagement should try and address the following to further develop a holistic skyscape experiences:

[31] William L. Fox, 'Listening to Light: Katherine Bash and Observational Displacement', available at http://www.itinerantlaboratory.org/docs/WLFEssay.pdf [accessed 30 August 2016].
[32] Robert Fosbury and Tarja Trygg, 'Solargraphs of ESO', *The Messenger* 141 (2010): pp. 43–45.
[33] Thornes, 'A Rough Guide to Environmental Art', p. 406.
[34] Saad-Cook, 'Touching the Sky', pp. 129–30.

* Playing with the concept of horizon, up and down.
* Recreating and generating possibly a new space experience using light in the spirit of Turrell. Therefore, engendering a more transformative place experience and moving closer to a deeper skyscape experience.
* Embracing the *Site* versus *Non-Site* dialectic hailed by Smithson as a further way to negotiate the landscape dialectic. It uses the seemingly sterile non-immersive Stellarium skyscape experience, turning it into a positive by using it as an extra dimension to engage the viewer.
* Populating the skyscape with text as one way that humans can encode memories additionally to their imprints left in the skyscape.[35]

Conclusions

Skyscapes are informative and expressive ways in which different peoples have communicated their understanding of the skies, land, society and their own selves. Land, sky and people themselves form a tightly linked interconnected mesh that shapes and defines the skyscape experience. At its heart lays the place experience containing the engagement with time, emotions and memories unlocked within and through the surroundings or environment. Stellarium has been described as offering an ideal opportunity to capture a skyscape experience. It not only includes the temporal sky dimension but also offers a landscape that can be used as a canvas to capture memories and impressions for the viewer to engage with.

Using many examples drawn from land art installations and practitioners as well as a small project, *Skyscapes above Clifton*, it has been shown how much creative potential Stellarium offers beyond a mere scientific investigation. It mirrors the liberating move of the artist into the outdoors at the beginning of the nineteenth century by requiring the creator of Stellarium to venture into the skyscape to be captured. It offers tools accessible to all via its freeware character, which should make it as appealing as the method of phenomenology itself has been seen by Thornes and in tune with the way Richard Long carries out environmental art.[36] Stellarium is a true new land to be explored by artists and creative practitioners engaging with environmental art in a new manner that could be further heightened by Smithson's *Site* versus *Non-Site* dialectic. As such this paper should not only illustrate this potential but also be a call for creative practitioners to include Stellarium in their work to further the Skyscape agenda.

---

[35] Katherine Bash, 'Spatial Poetries: Heuristics for Experimental Poiesis' (PhD diss. University College London, University of London, 2011).
[36] Thornes, 'A Rough Guide to Environmental Art', pp. 394, 402.

Acknowledgements
I would like to acknowledge the financial support of the Sophia Centre for the Study of Cosmology in Culture in presenting his work at INSAPIX. I would also like to thank Duncan Higgins, Ana Souto and Katherine Bash for inspiring discussions, as well as Rumbidzai Mukundu and William Ktorides for their work on the *Skyscapes above Clifton* project.

THE OCULUS RIFT PLANETARIUM PROJECT: STARSIGHTVR

Alastair G. Bruce

ABSTRACT: The Oculus Rift Planetarium Project was a pilot-scheme, funded by a Science and Technology Facilities Council small-project public engagement award in 2015. The main aim was to explore the possibility of integrating planetarium software with the Oculus Rift virtual reality development headset (DK2). The initial project was deemed a success and work continues on the project under the name StarsightVR. The software allows multiple machines to connect to a stargazing host whilst allowing users an independent view of the night sky, with or without a VR headset. The software is open-source and has been designed with the capability for live internet-based stargazing talks thus providing a useful new tool for astronomical outreach.

Virtual reality headsets for use in the home have long been sought after but only in recent years have they begun to reach the level of sophistication or pricing required to be viable. One of the first companies to make waves in this market was Oculus.[1] The company's initial development work, crowd-funded through a kickstarter campaign, created a prototype headset that showed that a home-VR experience was within reach. The second iteration, the Oculus Rift DK2, improved on the original prototype and started shipping to developers in July 2015.

The idea behind StarsightVR was simple: adapt planetarium software to display with the Oculus Rift headset and so give the wearer the ability to look anywhere they liked on a simulated night sky.[2] Given that the Scottish weather and light summer skies frequently hamper live stargazing sessions at the Royal Observatory Edinburgh, such a tool would prove invaluable, both as a poor weather backup and as an educational tool, allowing full use of the planetarium software and its features.

In addition to a single-user VR stargazing experience, it was also desirable to include the capability for group experiences. Stargazing with the benefit of an expert can be very rewarding and this became a key component of the project. The planetarium software selected for the project was Stellarium, a popular open-source package and an ideal place to begin testing these ideas.[3]

With the basic idea in hand, a proposal was made to the STFC Public Engagement Small Awards Scheme. I (an astronomy PhD student) took the role of project leader with Professor Andy Lawrence as formal principal investigator. The project

---

[1] Oculus website: https://www.oculus.com/.
[2] StarsightVR website: http://starsightvr.org.uk/.
[3] Stellarium website: http://www.stellarium.org/.

was selected for full funding totalling £9,150, and work began towards the end of 2014. The software developer engaged to work on the project, Guillaume Chereau, is part of the Stellarium development team and was fortunately already very familiar with the base code.[4]

After purchase of three DK2 headsets and hardware for testing, the rest of the funds were used to buy the software development time required. Early tests were promising and allowed the addition of extra functionality to help improve the over-all experience. A key aspect of StarsightVR is that it does not make use of a fully 3D rendered environment. Given the distances involved in astronomical observations, 3D rendering provides little extra depth at the cost of more processing power re-quired for the headset, so the decision was made to keep the overhead as small as possible.

Integration of Stellarium with the DK2 headset is achieved by performing the required colour-correction and image distortion in Stellarium so that the image displayed by the headset is as desired. Updates on the position/orientation of the headset allow Stellarium to render the appropriate portion of sky. If no headset is present, the user will see a default Stellarium projection and can enable VR-mode with a simple toggle. Feedback from VR testing has proven very positive and as long as the hardware is providing a smooth frame rate the experience is very rewarding, particularly if there is someone on hand to guide the person wearing the headset around the sky. Simply turning off the ground for VR users is a memorable event indeed! One current caveat is that the colour correction is not always ideal, partly exacerbated by stars being rendered as point-like white objects using r/g/b pixels.

The remote control aspect in StarsightVR is achieved using a handler for key events triggered on the host machine, such as turning on constellation lines, which are then broadcast to connected clients for replication. This ensures that connected machines are synchronised with the host but are also running independently, allow-ing each user the freedom to look wherever they wish. This works regardless of the presence of a VR headset. Users can enable the remote control through a standard Stellarium plugin interface.

In order to help provide a live, multi-user connected stargazing tour (with or without a headset), additional features have been added. The first is the addition of an arrow alerting users as to the direction of the host's selected target. This ensures that navigating around the sky is very straightforward. A second feature is the ability to replace the Moon with any of the solar system planets. This was done for the ben-efit of headset users as it was felt that many people would find attempting to 'zoom

---

[4] Developer website: https://noctua-software.com/about .

FIGURE 5.1. Post-doctoral astronomer Chelsea MacLeod with DK2 headset. StarsightVR press release; photo credit Paul Sutherland 2015.

in' to an object disorienting. The added benefit of these features is that they provide a memorable view of these objects. Similarly, some deep-sky objects have had a toggle added to make them larger on the sky. Not yet included is built-in audio support for the host astronomer, allowing talks to be heard by connected users. However, there are many suitable third-party apps available, so this extension could easily be added in future.

With the original project funding spent, the proof-of-concept build of Starsight-VR has managed to achieve all of the goals set out in the project proposal. In order to continue development, two further avenues of funding have been donations, via the ROE Trust, together with a modest crowdfunder campaign. Monies raised are being directed towards improving and maintaining the current build; for more information see the StarsightVR website.[5] For future work, we hope to begin scheduling live stargazing shows and, when funding permits, integrating different operating systems and the latest VR headsets. This market is evolving quickly, as is the software, and it remains to be seen which vendor(s) will stand the test of time.

The current generation of VR headsets were designed primarily for gaming but they are clearly well suited as educational tools. StarsightVR is not a full-blown 3D package as its main purpose was to be an immersive stargazing experience only. Consequently, it should prove less demanding than some other VR applications.

---

[5] ROE Trust website: http://roetrust.org.uk.

Also, it is freely available. For those that have no easy access to a full dome planetarium, StarsightVR fills the void between these and the clouds...

ACKNOWLEDGEMENTS: Many thanks go to the STFC Public Engagement Small Awards Scheme and panel (project ref: ST/M002322/1) and also to Andy Lawrence for supporting the project from the outset. Thanks also to the University of Edinburgh and the Principals Career Development Scholarship (focus: public engagement), of which Alastair is a recipient.

ADVENTURES IN SPACE: HARMONY, SUSTAINABILITY AND ENVIRONMENTAL ETHICS

Nicholas Campion

ABSTRACT: The space programme is driven by many imperatives, of which the most overt are economic and geopolitical. There is evidence that we are at the beginning of an acceleration of the space programme, which will raise issues of industrialisation, pollution and militarisation, with attendant ethical considerations and legal issues. In this paper I consider the current issues, relate them to questions of human rights and sustainability and, this being an astronomical publication, I argue that astronomy itself provides one foundation for notions of human rights and sustainability. Astronomers should have the confidence to bring their history to the global discussions now taking place.

Introduction: Ethics and Law

To assert that there is a moral good on which we can all agree is one thing. The difficulty of deciding exactly what this good is was addressed by Aristotle in the *Nicomachean Ethics*. He pointed out that if there is no such thing as a universal good which is written into the fabric of the universe, which is what Plato had claimed, then the value of any action can only be measured its context.[1] If so, then we have an inevitable conflict of interest between the interests of scientists, for whom the ethical good consists of exploring the nature of the universe, industrialists who see opportunities for profit and politicians who plan an extension of terrestrial power.

Legal considerations concerning space date back to 1959 when the Committee on the Peaceful Uses of Outer Space (COPUOS), administered by the United Nations Office for Outer Space Affairs, was set up by the United Nations General Assembly even earlier in order 'to govern the exploration and use of space for the benefit of all humanity: for peace, security and development'.[2] This was followed in 1960 by the foundation of the International Institute of Space Law.[3] The Outer Space Treaty, which came into force in October 1967, established the following principles: space should be used for peaceful purposes for the benefit of all; the placing of weapons in space is prohibited; and states are responsible for any consequences arising from

---

[1] Aristotle, *The Nicomachean Ethics*, trans. J.A.K Thomson, revised Hugh Tredennick (London: Penguin 2004), 1094a, 1–3, 1096a, 1–7. Plato, *Timaeus*, trans. R.G.Bury (Cambridge MA: Harvard University Press 1931), for example 28C, 41B, 42E. See also David Bostock, *Aristotle's Ethics* (Oxford: Oxford University Press, 2000), p. 8; J.O. Urmson, *Aristotle's Ethics* (Oxford: Blackwell: 1988), p. 23.
[2] The Committee on the Peaceful Uses of Outer Space, available at http://www.unoosa.org/oosa/en/ourwork/copuos/index.html [accessed 4 March 2016].
[3] International Institute of Space Law, http://www.iislweb.org/ [accessed 4 March 2016].

their activities.[4] These principles were then developed in a second treaty ratified in 1984.[5] Significantly, these treaties account only for the activities of states, not the actions of private corporations, which were not envisaged at the time: corporations such as SpaceX or Virgin Galactic have obligations only to their shareholders. Despite denials to the contrary, the United States appears to have now unilaterally assumed the right to license private exploitation of space.[6] Some commentators regard this act as sabotaging the 1967 treaty, abandoning any attempt at a global space policy.[7] In addition to privatisation there is concern that existing agreements focus on peaceful uses of space and ignore the possibility of space war.[8] The actual problems therefore may be few at the moment but are potentially huge.

The Current Context

The space programme enjoyed just under twelve glorious years of adventure, innovation and extraordinary achievement between the launch of Sputnik, the first satellite, on 4 October 1957 and Neil Armstrong's Moon walk on 20 July 1969. After the first Moon landing the Americans found it difficult to maintain public interest in space exploration and after the end of the Apollo Programme as a whole in 1972, human activity in space was confined to Low Earth orbit.[9] In order to procure continued public support the benefits of space travel were often presented in the most banal terms: at one time the non-stick frying pan had been held up as the principal

---

[4] UNOOSA (United Nations Office for Outer Space Affairs 'Treaty on Principles Governing the Activities of States in the Exploration and Use of Outer Space, including the Moon and Other Celestial Bodies', http://www.unoosa.org/oosa/en/ourwork/spacelaw/treaties/introouterspacetreaty.html [accessed 18 October 2016].

[5] UNOOSA (United Nations Office for Outer Space Affairs) 'Agreement Governing the Activities of States on the Moon and Other Celestial Bodies', 1979, http://www.unoosa.org/pdf/publications/ST_SPACE_061Rev01E.pdf; see also UNOOSA (United Nations Office for Outer Space Affairs), 'Status of International Agreements relating to Activities in Outer Space', http://www.unoosa.org/oosa/en/ourwork/spacelaw/treaties/status/index.html [accessed 16 October 2016].

[6] 'H.R.2262 – U.S. Commercial Space Launch Competitiveness Act', Congress.Gov, https://www.congress.gov/bill/114th-congress/house-bill/2262 [accessed 14 July2017].

[7] James Rathz, 'Law Provides New Regulatory Framework for Space Commerce', *The Regulatory Review*, 31 December 2015, https://www.theregreview.org/2015/12/31/rathz-space-commerce-regulation/ [accessed 14 July2017].

[8] Jamie Doward, 'The new rulebook for real-life star wars', *The Guardian*, 21 May 2017, https://www.theguardian.com/science/2017/may/20/star-wars-first-rules-manual-military-in-space-laws [accessed 14 July 2017].

[9] Kendrick Oliver, *To Touch the Face of God: The Sacred, the Profane, and the American Space Program, 1957–1975* (Baltimore, MD: Johns Hopkins University Press, 2013), pp. 52–54.

benefit of the space programme, a claim which is widely accepted but also challenged.[10] The most prominent beneficiary of the space programme in recent years has arguably been the smart phone, which depends on satellite additions to terrestrial communications systems: life has been revolutionised, but with little or no public attention paid to the space programme on which the phone often depends.

The evidence suggests that we are now at the beginning of a period of renewed acceleration of space activity in the solar system, and according to some commentators, such as Dan Brown, this is a welcome development.[11] We are now faced with three principal issues in space: geopolitical struggle, industrialisation and tourism (given, that as Martin Rees says in his introduction to this volume, tourism may not be a good description for such a hazardous enterprise). With the Chinese planning a Moon base, private individuals contemplating travel to Mars, and the latest technology making mining on the Moon possible, by the middle of the century space travel in the solar system may be a commonplace. The sky, already 'over-exploited' according to Jacques Arnould, is about to become more congested.[12]

The Problem

Such developments will bring consequences in their wake. For example, unfettered private commercialisation and industrialisation, combined with the biological hazards of terraforming, present potentially serious pollution.[13] There is now a recognition of this problem, including legal ramifications.[14]

---

[10] Anon., 'Moon landing brought non-stick frying pans and evolution insights', *The Daily Telegraph*, 16 July 2009, available at http://www.telegraph.co.uk/news/science/space/5843595/Moon-landing-brought-non-stick-frying-pans-and-evolution-insights.html [accessed 23 July 2016]; John Emsley, 'Molecule of the Month: Teflon: The non-stick myth that stuck: Did you think that your hi-tech frying pan was a spin-off from the space race? John Emsley explains that the truth is the other way around', *The Independent*, 7 July 1994, available at http://www.independent.co.uk/news/science/molecule-of-the-month-teflon-the-non-stick-myth-that-stuck-did-you-think-that-your-hi-tech-frying-1414648.html [accessed 23 July 2016].

[11] Daniel Brown, 'From comets to planets near and far, space probes reveal the universe', 15 July 2015 6.22AM BST, available at http://theconversation.com/from-comets-to-planets-near-and-far-space-probes-reveal-the-universe-44158 [accessed 1 August 2016].

[12] Jacques Arnould, *Icarus' Second Chance: the Basis and Perspectives of Space Ethics* (New York and Vienna: Springer, 2011), p. 84.

[13] Paris Arnopoulos, *Cosmopolitics: Public Policy of Outer Space* (Toronto: Guernica Editions, 1998), pp. 189–212, 258; James Clay Moltz, *Crowded Orbits: Conflict and Cooperation in Space* (New York: Columbia University Press, 2014), pp. 121–45; E.C. Hargrove, *Beyond Spaceship Earth: Environmental Ethics and the Solar System* (San Francisco, CA: Sierra Club Books, 1986), pp. xii–xiii.

[14] Lotta Viikari, *The Environmental Element in Space Law: Assessing the Present and Charting the*

Whereas in some terrestrial cultures it has now become unacceptable to drop one's litter or pollute the environment at will, in effect, all countries can act in space exactly as they wish: there is no equivalent in space of the United Kingdom's 'Keep Britain Tidy' campaign.[15] Can we keep space tidy? The problem was described by Nola Taylor Redd in 2013:

> There are also millions of pieces of debris smaller than a third of an inch (1 cm). In Low Earth-orbit, objects travel at 4 miles (7 kilometers) per second. At that speed, a tiny fleck of paint packs the same punch of a 550 pound object traveling at 60 miles per hour. Not only can such an impact damage critical components such as pressurized items, solar cells, or tethers, they can also create new pieces of potentially threatening debris.[16]

This debris is an explicit hazard to activity in Low Earth orbit. And it could get a lot worse: imagine the unregulated open-cast mining of the Moon and the resulting discharge of huge amounts of rock into the lunar environment. High-tech fixes are one option.[17] For example, the European Space Agency has plans to use cleaning satellites to remove debris by lassooing it, sweeping it up in large nets.[18]

Problematically, some space debris is there not just as a result of carelessness but as a result of calculated action, and that in clear breach of the 1967 Treaty. The worst single instance of intentional pollution was the destruction of a weather satellite in 2007 by the Chinese, as part of the country's general military build-up. As the UK *Daily Telegraph* reported, 'The successful anti-satellite test, the first by any country for two decades, drew attention to China's military build-up and raised fears of a news arms race in space'.[19] China is indeed continuing its military build-up in space,

---

Future (Leiden: Martinus Nijhoff, 2008).

[15] Keep Britain Tidy, available at http://www.keepbritaintidy.org/home/481 [accessed 23 July 2016].

[16] Nola Taylor Redd, 'Space Junk: Tracking and Removing Orbital Debris', 8 March 2013, available at http://www.space.com/16518-space-junk.html [accessed 30 April 2016].

[17] Anon., 'Moon Express Becomes First private Company in History to Initiate a Commercial Lunar Mission Approval Process with the U.S. Government', 8 April 2016, http://www.moonexpress.com/news/moon-express-becomes-first-private-company-history-initiate-commercial-lunar-mission-approval-process-u-s-government/ [accessed 14 July 2017]; Sophie Curtis, 'Billionaire entrepreneur to start mining the Moon for gold and platinum by end of 2017', Mirror, 3 February 2017, http://www.mirror.co.uk/science/billionaire-entrepreneur-start-mining-moon-9746422, [accessed 14 July 2017].

[18] European Space Agency, 'E.Deorbit: Pushing the Boundaries of Space Technologies', 11 July 2016, available at http://www.esa.int/Our_Activities/Space_Engineering_Technology/Clean_Space/e.Deorbit_Pushing_the_Boundaries_of_Space_Technologies [accessed 17 July 2016].

[19] Richard Spencer, 'Chinese missile destroys satellite in space', *The Daily Telegraph*, 19 January

as reported occasionally by the western media. This is from the online *Washington Free Beacon* in 2015: 'China recently conducted a flight test of a new missile capable of knocking out US satellites as part of Beijing's growing space warfare arsenal'.[20] As Christopher Impey wrote,

> … perhaps a new geopolitical space race is brewing between the United States and China… in China space activity is twinned with military activity. It is a very secretive activity where we have to guess what their true intentions are. There clearly is concern about the 'weaponization' of space and I think that concern is justified. [21]

Elsewhere he concluded,

> For those who worry about the militarization of space, this all brings to mind the purported Chinese curse: "May you live in interesting times."[22]

The US's militarising aspirations, meanwhile, were unambiguously incorporated into government policy in Air Force Space Command's 'Strategic Master Plan' of 2003:

> The Strategic Master Plan (SMP) describes how we will transform the command into a space combat command. It outlines how we will sustain, modernize, divest and transform our forces in order to maximise our warfighting capabilities. This plan is the command's roadmap to ensure our military remains dominant in space, in the air, on the ground and on the sea.[23]

As Arnould noted, the 1967 Treaty prohibited the weaponisation of space, but not its militarisation.[24]

2007, available at http://www.telegraph.co.uk/news/worldnews/1539948/Chinese-missile-destroys-satellite-in-space.html [accessed 30 April 2016].

[20] Bill Gertz, 'China Tests Anti-Satellite Missile: New ASAT interceptor threatens U.S. spy satellites', *The Washington Free Beacon*, 9 November 2015, available at http://freebeacon.com/national-security/china-tests-anti-satellite-missile/ [accessed 30 April 2016].

[21] Christopher Impey, 'Our Future Off Earth', Gresham College lecture at the Museum of London, 26 April 2016, available at http://www.gresham.ac.uk/lectures-and-events/our-future-off-earth [accessed 1 May 2016]. See also for China, Christopher Impey, *Beyond: Our Future in Space* (New York: W.W. Norton and Company, 2016), pp. 139–44.

[22] Impey, *Beyond*, p. 122.

[23] Air Force Space Command, 'Strategic Master Plan', 1 October 2003, available at http://www.wslf-web.org/docs/final%2006%20smp--signed!v1.pdf [accessed 1 August 2015].

[24] Arnould, *Icarus*, p. 87.

Ethics and Morality: An Ethical Policy for Space

The question is, can anything be done to restrict military conflict in space, or prevent worse pollution than we already have? As Henry and Taylor argue, the need to develop an ethical policy is certainly one solution, one which has already been discussed in detail by Reiman.[25] I have also earlier considered the need for an extension of debates on the ethics of terrestrial environmentalism into the space environment.[26] In that paper I questioned the COMEST report's conclusion that the construction of an ethical policy for space should not take emotional appeals into account.[27]

To compose an ethical policy for space, though, is simpler said than done. For example, Mark Williamson argued that, compared to other issues 'space demands a somewhat different [ethical] philosophy, based on detailed knowledge of the space environment'.[28] Williamson also argued that to be of any practical use a space ethics policy and must progress from the 'philosophical and academic' and 'be targeted towards the design of an ethical code or policy'.[29] He added that we have to consider 'who is asking [ethical] questions and who is answering them?': no question is ever asked without a prior context, or a set of assumptions about the way the world works.[30] Williamson is correct that a policy is needed, but issues in the space environment are not necessarily different in principle to those of a terrestrial nature, even if they differ in detail.

Williamson himself had in mind a number of groups whose views may be diametrically opposed; for example, the interests of space scientists who value the pristine environment are opposed to those commercial and political interests who see it as no more than a material resource to be exploited. If competing positions are to be reconciled it is necessary to formulate an ethical policy to which rivals can all consent. Only then, Williamson argues, can a legal framework for space protection be constructed.[31]

---

[25] Holly Henry and Amanda Taylor, 'Re-thinking Apollo: envisioning environmentalism in space', in *Space Travel and Culture: From Apollo to Space Tourism*, ed. David Bell and Martin Parker (Oxford: Blackwell Publishing, 2009), pp. 191–203; S. Reiman, 'On Sustainable Exploration of Space and Extraterrestrial Life', *Journal of Cosmology* 12 (2010): pp. 3894–3903.

[26] Nicholas Campion, 'The Moral Philosophy of Space Travel: A Historical Review', in *Commercial Space Exploration: Ethics, Policy, Governance*, ed. Jai Galliot, (Abingdon: Ashgate, 2015), pp. 9–22.

[27] COMEST (World Commission on the Ethics of Scientific Knowledge and Technology), Sub-Commission Report on 'The Ethics of Outer Space' (New York: UNESCO Headquarters, 2000) available at http://unesdoc.unesco.org/images/0012/001220/122048E.pdf [accessed 4 October 2015].

[28] Mark Williamson, 'Space ethics and protection of the space environment', *Space Policy* 19 (2003): pp. 47–52, here p. 52.

[29] Williamson, 'Space ethics', p. 52.

[30] Williamson, 'Space ethics', p. 49.

[31] Williamson, 'Space ethics', pp. 149–80.

The existence of multiple questions and contexts invites comparisons with descriptive morality, which recognises the variation of morality and ethics from one group to another. Under such circumstances we cannot arrive at an absolute set of moral values, only describe, compare and contrast the values of different groups.

Descriptive morality is inevitably relativistic: if different moralities are described and not judged then all must be morally equivalent. The *Stanford Encyclopaedia of Philosophy* offers a concise summary of the problem: 'if one is not a member of the relevant society or group, or is not the relevant individual, then accepting a certain account of the content of morality, in the descriptive sense, has no implications for how one should behave.'[32] Descriptive morality in this sense is the morality of individual cultures and so is necessarily culturally contextualised: a culture which prohibits certain practices cannot condemn their use in other cultures The implications for the industrialisation and militarisation of space are clear: if ethics are designed to minimise the harm in space resulting from human activities, then what constitutes harm in the first place is not always clear and what is of benefit to one person or government may be harmful to another.[33] Such debates are rehearsed endlessly in relation to the terrestrial environment: for example, the OECD has summarised the claims and counter-claims concerning whether the harm caused by global warming is greater than the measures taken to avert it.[34]

The first step in the construction of any public ethical policy is the definition of key terms. At first sight this would seem simple: an ethical code is the application of morality to conduct. As one general definition has it, ethics is both 'the philosophical study of the moral value of human conduct, and of the rules or principles that ought to govern it', and 'a code of behaviour considered correct, especially that of a particular group, profession or individual.'[35] However, this is far from adequate, for if ethics is an application of morality, we are trapped in a circular argument dominated by the insoluble question of what version of morality exactly should be the basis of an ethical code: as descriptive morality recognises, what is moral in one culture may be immoral in another. Defining any universal theory of morality is therefore complex, if not impossible.

The utilitarian philosopher John Stuart Mill provided one solution. He defined

---

[32] 'The Definition of Morality', first published Wednesday 17 April 2002; substantive revision Monday 8 February 2016, *The Stanford Encyclopaedia of Philosophy*, http://plato.stanford.edu/entries/morality-definition/ [accessed 30 April 2016].

[33] Moltz, 'Crowded orbits, p. 192; See also Reiman, 'Sustainable Space Exploration'.

[34] OECD, 'Climate Change Mitigation: WHAT DO WE DO?' 2008, available at http://www.oecd.org/env/cc/41751042.pdf [accessed 25 July 2016].

[35] *Collins Concise English Dictionary* (London: Collins, 1992), p. 439.

morality as 'the rules and precepts for human conduct, by the observance of which [a happy existence] might be, to the greatest extent possible, secured'.[36] There is a rejection of the absolute here, the words 'the greatest extent possible' suggesting the compromise necessary to formulate a globally binding agreement. Even when ethical arguments are based on a notion of the absolute, realpolitik dictates that practical solutions can only be reached through negotiation and compromise.

The relativity of descriptive morality contrasts with normative morality, which is absolute in the sense that 'all rational persons, under certain specified conditions, would endorse it'.[37] In theory, then, normative morality should be accepted by all rational, right-thinking people, the world over. This, in turn, begs the question of what is rational? The word 'rational' implicitly contrasts with its opposite, the 'irrational', as if the two modes of mental cognition exist in a binary struggle.

This kind of binary distinction was recognised by Bertrand Russell. Writing in 1943, at the height of the Second World War, Russell produced the following unambiguous statement:

> Man is a rational animal – so at least I have been told. Throughout a long life, I have looked diligently for evidence in favour of this statement, but so far I have not had the good fortune to come across it, though I have searched in many countries spread over three continents. On the contrary, I have seen the world plunging continually further into madness. I have seen great nations, formerly leaders of civilization, led astray by preachers of bombastic nonsense. I have seen cruelty, persecution, and superstition increasing by leaps and bounds, until we have almost reached the point where praise of rationality is held to mark a man as an old fogy regrettably surviving from a bygone age.[38]

Russell considered that humanity is inherently rational but too easily slides into irrationality, a condition which he regarded as necessarily dangerous and therefore harmful to the general good. All around him he saw European rationality subverted by the irrationality of Nazism and war. For Russell, then, the militarisation of space, whether by the Chinese or Americans, would be irrational.

A strong current of thought in sociology pioneered by Max Weber challenges the rational-irrational dichotomy and argues that it is in the very nature of being human to make rational choices.[39] Human beings cannot, therefore, be irrational and what

[36] John Stuart Mill, *Utilitarianism*, ed. G. Sher (1861; Indianapolis, IN: Hackett, 2002), p. 12.
[37] 'The Definition of Morality'.
[38] Bertrand Russell, 'An Outline of Intellectual Rubbish', pp. 69–108, in *Unpopular Essays* (1943; Oxford: George Allen and Unwin, 2009), p. 69.
[39] Stephen Kalberg, 'Max Weber's Types of Rationality: Cornerstones for the Analysis of Rationalisa-

may at first appear to be irrational may actually be the product of rational considera-tions. For example, in the Weberian sense, the choices made by religious people may be considered rational if they contribute to a healthier, more prosperous and fulfilled life, an argument taken up by theorists of religion such as Bainbridge and Stark.[40] Yet, Weberian typology, in which even irrational choices are rational, breaks down. For example, is there a difference between a rational choice made on the basis of, say, the best available empirical evidence on the one hand, and the apparently evidence-free belief in supernatural powers or alien visitation, on the other? However, the politics of compromise dictates that, even if we believe that our opponents are irrational, we must assume that they are making rational choices. Otherwise there can be no conversation about ethical problems, and no solution. It may seem irrational to blow up a satellite and create a huge amount of potentially lethal debris, but the Chinese made a perfectly rational choice to test their weapons systems. What for Russell is irrational is therefore served by rational considerations.

Telling Stories about Space

How can we persuade competing parties to agree to a single ethical system? Should we stick to the facts or appeal to emotions? Williamson highlighted the core problem of a successful space ethics policy: if success is defined by practical implementation: how is it possible to persuade people who see space as either empty or dangerous that it needs protecting? In this respect the COMEST report of 2000 explicitly ruled out emotional appeals, arguing that scientific evidence alone should be adequate to for-mulate an ethical policy for outer space.[41] But is scientific evidence alone sufficient?

The problem parallels that faced by climate-change scientists in the US, who find that scientific data and calls for regulation of polluting activity, are dismissed by emotional appeals to unregulated economic liberty as fundamental to American identity, and by the claim that those who oppose exploitation of the environment are un-American.[42]

Scientists need a positive counter-narrative of their own: as Donald Spence wrote, 'Interpretations are persuasive not because of their evidential value but because of

---

tion Processes in History', *The American Journal of Sociology* 88, no. 5 (1980): pp. 1145–79.

[40] Rodney Stark and William Simms Bainbridge, *The Future of Religion: Secularization. Revival and Cult Formation* (Berkeley, CA: University of California Press, 1985); Rodney Stark and William Simms Bainbridge, *A Theory of Religion* (New Brunswick, NJ: Rutgers University Press, 1987).

[41] COMEST, 'The Ethics of Outer Space'.

[42] M. Lynas, 'We must reclaim the climate change debate from the political extremes', *The Guard-ian*, 12 March 2015, available at http://www.theguardian.com/commentisfree/2015/mar/12/climate-change-reclaim-debate-political-extremes [accessed 12 March 2015].

their rhetorical appeal'.[43] The questions, then, are 'what value do we put on space exploration?', and 'how does it express our values?' Above all, what are the stories that we construct about it? I will propose two stories.

Space as Special

My first story draws on suggestions of 'place' as special, as Tilley wrote, and 'storied', as Lane said, can be brought to the debate.[44] The concept of the Earth and sky and space as a single system is fundamental to classical cosmology. Plato argued that God moulded the Cosmos '...to be One single Whole, compounded of all wholes, perfect and ageless and unailing'.[45] Aristotle wrote that ourania (the sky) is 'the body continuous with the extreme circumference which contains the moon, the sun, and some of the stars; these we say are "in the heaven"'.[46]

The ancient formulation of the universe as a single organism found one of its most enthusiastic recent interpreters in Carl Sagan. In his classic work, *Cosmos* he argued that

> The Cosmos is all that is or ever was or ever will be. Our feeblest contemplations of the Cosmos stir us – there is a tingling in the spine, a catch in the voice, a faint sensation, as if a distant memory, of falling from a height. We know we are approaching the greatest of mysteries.[47]

He also wrote,

> As the ancient myth makers knew, we are children equally of the earth and sky. In our tenure of this planet, we have accumulated dangerous, evolutionary baggage – propensities for aggression and ritual, submission to leaders, hostility to outsiders, all of which puts our survival in some doubt. We have also acquired compassion for others, love for our children, a desire to learn from history and experience, and a great, soaring passionate intelligence – the clear tools for our continued survival and prosperity.[48]

[43] Donald Spence, *Narrative Truth and Historical Truth: Meaning and Interpretation in Psychoanalysis* (London: W.W. Norton & Co., 1982), p. 32.

[44] Belden Lane, *Landscapes of the Sacred: Geography and Narrative in American Spirituality*, expanded edition (Baltimore, MD: John Hopkins University Press, 2001), p. 59; Christopher Tilley, *A Phenomenology of Landscape*, (Oxford: Berg, 1994), pp. 10–11.

[45] Plato, *Timaeus*, trans. R. G. Bury (Cambridge, MA: Harvard University Press, 1931), 33a.

[46] Aristotle, *On the Heavens*, trans. J. L. Stocks, available at http://classics.mit.edu/Aristotle/heavens.1.i.html [accessed 8 June 2013].

[47] Carl Sagan, *Cosmos: The story of Cosmic Evolution, Science and Civilisation* (London: Futura Publications, 1983) p. 20.

[48] Carl Sagan, 'Who Speaks for Earth', The School of Cooperative Individualism Library, available at

In this sense outer space is a place of value and a potential source of wisdom.

The Astronomical Basis of Human Rights Theory

My second story is one which astronomers themselves have contributed to the debate: the origin of human rights theory in the notion that the law of the universe requires human equality in all things. The greatest statement of natural rights occurs in the opening lines of Rousseau's *Social Contract*, first published in 1762: 'Man is born free, and everywhere he is in chains'.[49] Humanity is naturally free and all constraints are artificial. The origins of human rights theory are diverse and have a strong basis both in Scripture and Natural Rights Theory, which was developed throughout the sixteenth and seventeenth centuries, notably by John Locke (1632-1704), who argued that the right to life, liberty and property are innate, not culturally derived.[50] Significantly Locke was a close friend of his younger contemporary, Isaac Newton, and the two men shared and exchanged ideas.[51]

The astronomical basis of human rights theory is one of the political consequences of what I term Political Newtonianism.[52] The simple argument familiar to eighteenth century radicals, was that as the entire universe was governed by a single law, so all people, from kings to commoners, must be subject to one law under which they are all equal. In 1728, the year after Newton's death, the natural philosopher John Theophilus Desaguliers wrote the earliest known manifesto of Newtonian political theory, applying the new astronomy to the management of the state, including the idealistic lines,

---

http://www.cooperative-individualism.org/sagan-carl_who-speaks-for-earth-1980.htm [accessed 7 October 2016].

[49] Jean-Jacques Rousseau, *The Social Contract*, trans. Maurice Cranston (London: Penguin, 2006), I.1.

[50] John Locke, *An Essay Concerning Human Understanding* (London: Penguin, 1997); John Locke, *Two Treatises of Government* (Cambridge: Cambridge University Press, 1988); James T. Axtell, 'Locke, Newton and the Elements of Natural Philosophy', in Richard Ashcraft, *Revolutionary Politics and Locke's "Two Treatises of Government"* (Princeton, NJ: Princeton University Press, 1986), pp. 419–29.

[51] James L. Axtell, ' Locke, Newton, and "The Elements of Natural Philosophy"', *Paedagogica Europaea*, Vol. 1 (1965), pp. 235-245; G. A. J. Rogers, 'Locke's Essay and Newton's *Principia*', *Journal of the History of Ideas* 39, No. 2 (Apr.-Jun., 1978): pp. 217–32.

[52] Nicholas Campion, 'Astronomy and Political Theory', in *The Role of Astronomy in Society and Culture*, International Astronomical Union Symposium 260, UNESCO, Paris, 19–23 January 2008, ed, David Valls-Gabaud and Alec Boksenberg (Cambridge: Cambridge University Press), pp. 595–602.

> What made the Planets in such Order move,
> He said, was Harmony and mutual Love.[53]

For Desaguliers harmony, equality and progress were facts of life revealed by Newton and embedded in the fabric of the universe. Desaguliers would also have been well aware of Johannes Kepler's recent restatement of the Pythagorean and Platonic view of the harmonious cosmos.[54] The consequences were profound and touched every aspect of philosophy. A prime example was 1750 Anne-Robert-Jacques Turgot's (1727–1781) articulation of the theory of progress in 1750: it was Newton, Turgot insisted, who had shown the way:

> One man, Newton, has subjected the infinite to the calculus, has revealed the properties of light which in illuminating everything seemed to conceal itself, and has put into balance the stars, the earth and all the forces of nature.[55]

Turgot's theory of progress, was elaborated by another Newtonian, the Marquis de Condorcet in the 1790s.

> Our hopes for the future state of the human species may be summed up in three important points: the elimination of the inequality between nations; progress in equality within the same peoples; and finally the real perfection of mankind. [56]

And for Thomas Paine, who did so much to persuade the Americans to break with Britain, the perfect order of the planets was a profound demonstration of the truth of God's natural creation, and, quoting the French radical, the Marquis de Lafayette, he wrote that the truths which Nature had engraved on the heart of every citizen carried an innate love of liberty.[57] Natural rights belong to people through the fact of

---

[53] John Theophilus Desaguliers, *The Newtonian System of the World: the Best Model of Government, an Allegorical Poem*, II.17–18, cited in Porter, Roy, *Enlightenment: Britain and the Creation of the Modern World* (London: Penguin 2000), p. 137.

[54] Johannes Kepler, *The Harmony of the World*, trans. E.J. Aiton, A.M. Duncan, J.V. Field, (American Philosophical Society, Philadelphia, 1997); Kenneth Guthrie, *The Pythagorean Sourcebook and Library* (Grand Rapids, MI: Phanes, 1987); Plato, *Republic*, 2 Vols, trans. Paul Shorey (Cambridge MA: Harvard University Press 1937), X.617B–C.

[55] Anne-Robert-Jacques Turgot, 'A Philosophical Review of the Successive Advances of the Human Mind', in Ronald L. Meek (ed.), *Turgot on Progress, Sociology and Economics* (Cambridge: Cambridge University Press, 1991), p. 59; see also p. 56.

[56] Antoine-Nicolas de Condorcet, *Sketch for a Historical Picture of the Progress of the Human Mind*, trans. June Barraclough (New York: Noon Day Press, 1955), p.173.

[57] Thomas Paine, *The Age of Reason* (Mineola, NY: Dover, 2003), pp. 191–92.; Thomas Paine, 'The Rights of Man;' in Michael Foot and Isaac Kramnick (eds), *The Thomas Paine Reader* (London:

their existence and freedom is the default position of the Newtonian universe.[58] And the force which manages society for the best, Paine declared in 1776, paraphrasing Newton in the very year that American independence was declared, is like a 'gravitating power'.[59] It is irresistible: it must succeed.

Rousseau, who gave the American and French revolutionaries the notion of humanity as inherently free, was no admirer of Newton and had attacked astronomy in his *Discours sur les Sciences et les Arts* (*Discourse on the Sciences and Arts*), written in 1750.[60] But Newton's achievement was to provide a rationale for human equality which was written into the fabric of the universe. John Locke, himself, whose theory that good government required the consent of the people was fundamental to the ideology of the American independence movement, was influenced by Newton.[61] Where we find human rights, then, we always find the legacy of Isaac Newton.

Human Rights Theory

The United States Declaration of Independence in 1776 paraphrased and expanded Rousseau as follows:

> We hold these truths to be self-evident, that all men are created equal, that they are endowed by their Creator with certain unalienable Rights, that among these are Life, Liberty and the pursuit of Happiness.[62]

In this view, God, the supreme lawmaker, endowed humanity with its basic rights, which are therefore inherent within each human being. The same religious tendency was characteristic of those who composed the French Declaration of the Rights of Man in 1789 which, like the American declaration was profoundly influenced by Rousseau. The crucial sections are these:

> Therefore the National Assembly recognizes and proclaims, in the presence and under the auspices of the Supreme Being, the following rights of man and of the citizen:

Penguin 1987, 2003,) p. 207.

[58] Thomas Paine, 'The Rights of Man', p. 217.

[59] Paine, Thomas, *Common Sense* (London: Penguin 1986), p. 66.

[60] See Jeff S. Black, *Rousseau's Critique of Science: A Commentary on the Discourse on the Sciences and the Arts* (Lanham, MD.: Lexington Books, 2008), pp. 74–76.

[61] See Rogers, 'Locke's Essay and Newton's Principia'.

[62] The Declaration of Independence: A Transcription, 4 July 1776, http://www.archives.gov/exhibits/charters/declaration_transcript.html [accessed 30 April 2016].

Articles:
1. Men are born and remain free and equal in rights…
2. The aim of all political association is the preservation of the natural and impre-
scriptible rights of man. These rights are liberty, property, security, and resistance
to oppression...
4. Liberty consists in the freedom to do everything which injures no one else…
5. Law can only prohibit such actions as are hurtful to society.[63]

There are two key features in this passage. First, the founding principle that all men
are equal and, second, the consequence of this that nobody has the right to infringe
anyone else's freedom. Further, by virtue of being human, individuals have the right
to property and security.

The 1776 declaration was then updated in the United Nations Universal Declaration of Human Rights of 1948. Following the American and French declarations,
the United Nations' Article 1 also paraphrases the opening lines of Rousseau's *Social
Contract*. It reads

All human beings are born free and equal in dignity and rights. They are endowed
with reason and conscience and should act towards one another in a spirit of
brotherhood.[64]

The UN declaration removes the Creator, and so emphasises that rights are indeed
held by the very act of being human:

Whereas recognition of the inherent dignity and of the equal and inalienable rights
of all members of the human family is the foundation of freedom, justice and peace
in the world.[65]

The removal of the Creator is not really that big a step, as the religion of the founding
fathers of the United States was already drifting steadily towards deism, the substitu-
tion of the personal God by a remote creating power.[66]

While Newton did not create Natural Rights theory, his work was widely

---

[63] Declaration of the Rights of Man and Citizen Approved by the National Assembly of France, August 26, 1789, available at http://www1.curriculum.edu.au/ddunits/downloads/pdf/dec_of_rights.
pdf [accessed 7 October 2016].

[64] United Nations Universal Declaration of Human Rights in English, available at http://www.ohchr.
org/EN/UDHR/Pages/Language.aspx?LangID=eng [accessed 7 October 2016].

[65] United Nations Universal Declaration of Human Rights in English.

[66] Catherine L. Albanese, *A Republic of Mind and Spirit: A Cultural History of American Metaphysical
Religion* (New Haven: Yale University Press, 2007).

perceived to have made it a consequence of the physical laws of the universe and therefore incontrovertible. The legacy of Political Newtonianism then connects 1776 to 1789 and 1948.

Sustainability

In August 2012 the United Nations extended the 1948 Declaration and appointed an Independent Expert (now Special Rapporteur), whose task was to extend human rights theory to environmental concerns. The resulting report no longer needed to paraphrase Rousseau, but it did develop the consequences of his cosmology in the context of modern environmentalism.

> All human beings depend on the environment in which we live. A safe, clean, healthy and sustainable environment is integral to the full enjoyment of a wide range of human rights, including the rights to life, health, food, water and sanitation. Without a healthy environment, we are unable to fulfil our aspirations or even live at a level commensurate with minimum standards of human dignity. At the same time, protecting human rights helps to protect the environment. When people are able to learn about, and participate in, the decisions that affect them, they can help to ensure that those decisions respect their need for a sustainable environment.[67]

The 2012 statement had progressed in recognisable steps from the understanding of human dignity as a fundamental starting point in the 1776 and 1789 declarations and the 1946 Charter, to a utilitarian assessment of how such dignity may be guaranteed through assertion of the need for a sustainable environment. The inherent freedom which is a quality of being human in Rousseau's cosmology now required a sustainable environment.

Reiman also argued that there should be an emphasis on sustainable development, but what constitutes sustainability?[68] Fortunately the United Nations has worked this out in successive summits, the 2000 Millennium Summit and the 2005 World Summit, resulting in the adoption in 2016 of the 2030 Agenda for Sustainable Development.[69]

The agenda sets out some very clear principles. It takes from the American model

---

[67] 'Special Rapporteur on human rights and the environment (former Independent Expert on human rights and the environment)', available at http://www.ohchr.org/EN/Issues/Environment/SREnvironment/Pages/SRenvironmentIndex.aspx [accessed 16 April 2016].

[68] Reiman, 'On Sustainable Exploration'.

[69] United Nations, 'Sustainable Development Goals', available at http://www.un.org/sustainabledevelopment/ [accessed 18 April 2016].

the need for prosperity, but relates this to the fact that we all live on a single planet, that the interests of the poor and vulnerable are vital to the welfare of all, that current political and economic interests are secondary to the concerns of future generations, and that economic, social and technological progress must be framed within sustainable principles. The paragraph on Prosperity reads,

> We are determined to ensure that all human beings can enjoy prosperous and fulfilling lives and that economic, social and technological progress occurs in harmony with nature.[70]

Sustainability has now morphed into harmony: sustainability is only possible if we recognise the need to live in harmony with the entire natural world. Harmony, though, is not defined. It is assumed that we all know what it means. Just as sustainability has been defined by the UN, so, perhaps, we can expect that harmony will be defined. Until it is defined, it may be up to us to do so.

The 2030 agenda also recognises that we live on a single planet, the ultimate lesson of the Apollo spacecraft photographs of the whole Earth which helped galvanise the environmental movement in the 1960s.[71] Additionally, if there is no boundary between terrestrial and celestial nature, then the UN's injunction to live in harmony means to live in harmony with the sky. And this is where astronomy comes in: to apply Turgot's view, the entire Newtonian cosmos is harmonious.

Conclusion

At a moment when human engagement with near space is set to increase, we need wider narratives concerning the value of space if we are to construct an ethical policy which protects it for all. Elsewhere I have argued that this can be achieved by the projection of stories about the value of the celestial environment onto the celestial. One potent story is derived from natural rights theory as filtered through Newtonianism: that human beings have natural rights because they are encoded into the fab-

[70] United Nations General Assembly, 'Follow-up to the outcome of the Millennium Summit Draft resolution submitted by the President of the General Assembly. Draft outcome document of the United Nations summit for the adoption of the post-2015 development agenda', p 2.35, 'Annex: Transforming our world: the 2030 Agenda for Sustainable Development', 12 August 2015, available at http://www.un.org/ga/search/view_doc.asp?symbol=A/69/L.85&Lang=E [accessed 16 April 2016].
[71] Denis Cosgrove, 'Contested Global Visions: One-World, Whole-Earth, and the Apollo Space Photographs', *Annals of the Association of American Geographers* 84 (1994): pp. 270–94; Denis Cosgrove, *Apollo's Eye: A Cartographic Genealogy of the Earth in the Western Imagination* (Baltimore, MD: Johns Hopkins University Press, 2001).

ric of the universe. Newtonian astronomy contributed to a world view in which the notion of a single mathematical law governing the universe, and to which all parts of the universe complied?, generated the idea that all people should be subject to a corresponding single political and legal system. Natural rights theory was codified in the 1948 United Declaration and applied to the natural environment in the 2015 Agenda for Sustainable Development. Peace, prosperity and freedom can now, the United Nations decrees, be possible only if we live in harmony with nature. The same principle can be applied to the space environment, reinforced by the argument that Newtonian physics proclaims the equal rights of all human beings, and therefore their equal right to have a say in human activity in space. Sustainable development is therefore only possible if we live in harmony with the entire natural environment, including space.

Newton's laws may have been modified, adapted and challenged by Relativity and quantum mechanics, but the legacy of eighteenth century Political Newtonianism continues to shape political discourse. It provides the foundation for a normative morality, which must apply everywhere, across the entire universe, at all times. Its egalitarianism provides a rationale for the guarantees of security and property rights in the 1946 UN charter. Its argument that there is a universal order which necessarily connects all parts of the universe, including humanity and nature, provides a rationale for sustainable economics. It can also provide a narrative for developing sustainable activity in space. And if the United Nations insists that harmony is vital for prosperity in earth, so it must be in space. This should provide a story around which astronomers, scientists, environmentalists and human rights activists can all unite.

CONDENSING FROM A FLUID HAZE: JOHN PRINGLE NICHOL, THE NEBULAR HYPOTHESIS AND NINETEENTH-CENTURY COSMOGONY

Howard Carlton

ABSTRACT: A significant astronomical debate of the mid-nineteenth century revolved around competing hypotheses of cosmogony. Traditional explanations held that God had created a static cosmos nearly six thousand years ago. Two credible astronomers, however, had begun to believe that traces of the developmental processes which led to the birth of stars and planets still persisted in the form of the unresolvable nebulae. An evolutionary cosmogony seemed to sit well with emergent long geological timescales and therefore proved attractive to those thinkers who were not persuaded of the correctness of Mosaic chronology and traditional creation stories. A key figure in this debate was the Scottish political economist and radical polemicist John Pringle Nichol (1804–1859), who was a proponent of what had come to be known as the 'nebular hypothesis'. He was opposed by astronomers and theologians of a more conservative religious and political mien, for whom an evolving universe conflicted with their belief in final causes and also threatened to undermine social stability. Some commentators believed that the claimed resolution of the Orion nebula by Lord Rosse's giant telescope in the mid-1840s had destroyed Nichol's argument for ongoing stellar development. This paper will show how Nichol repurposed the evidence produced by his opponents and proceeded to promote his alternative views to a generally receptive audience. He thus paved the way for an absolute expansion of universal time and distance, the acceptability of evolution, and a general move towards rationalist religious views during the later nineteenth century.

One of the most significant astronomical debates of the nineteenth century revolved around conflicting explanations of the origins of the stars and planets. Traditional explanations held that God had created the cosmos in one short burst of activity approximately six thousand years ago.[1] In the eighteenth century, however, some philosophers had suggested that the universe had been formed by some kind of developmental process, traces of which might still be visible in such artefacts as the 'unresolvable nebulae', and that such changes might still be ongoing in the universe at large.[2] Such an evolutionary view of cosmogony seemed to sit better with emerging geological narratives and also to allow time for the possible development of biological transmutations.

---

[1] James Ussher, *The Annals of the World* (London: J. Crook, 1658), pp. iv–vi.

[2] For example Emmanuel Swedenborg, Thomas Wright of Durham, Johann Lambert and Immanuel Kant. See Marilyn B. Ogilvie, 'Robert Chambers and the Nebular Hypothesis', *The British Journal for the History of Science* 8, no. 3 (1975): pp. 214–32, p. 216. See also Michael J. Crowe, *The Extraterrestrial Life Debate 1750–1900: The Idea of Plurality of Worlds from Kant to Lowell* (1986; repr., Mineola, NY: Dover, 1999), pp. 97–98, for a discussion as to whether or not Swedenborg had described an early form of the nebular hypothesis.

A key figure in this debate was the Scottish political economist and lapsed cleric John Pringle Nichol (1804–1859) who, after being awarded the chair in Astronomy at Glasgow University in 1836,[3] began to reduce his output of radical polemics and transferred his attention to the promotion of astronomical discoveries, including the idea of the 'nebular hypothesis' as a valid explanation for the universe as we see it today.[4] He was opposed by astronomers and theologians of a more conservative religious and political bent. In the mid-1840s his antagonists believed that a key observation of the great nebula in the constellation of Orion had destroyed his argument for an older, larger and still developing universe. This paper will show how Nichol managed to evade, or rather repurpose, the apparently clinching evidence produced by his opponents, thus paving the way for the adoption of long chronologies which would allow sufficient elapsed time for events like the shaping of the earth's surface and the development of a diversity of living species to unfold and deliver the 'modern world'.

One key idea which subsequently became a component of the broader nebular hypothesis sought specifically to explain the origins of the solar system. Pierre-Simon Laplace (1749–1827) built on the views of several of his predecessors by pointing out that the shared orbital directions and similar inclinations of most of the objects in our solar system are highly improbable per se, but fit well with the idea of condensation from a rotating nebula.[5] Laplace was, however, a potentially controversial authority on matters philosophical, being identified in Britain as a rationalist, empiricist and possibly atheistic 'fellow traveller' of the French revolution.[6]

An alternative nebular hypothesis came about as a result of observations made by William Herschel (1738–1822) who had classified the nebulae in a fashion analogous to the typologies employed by contemporary naturalists.[7] Herschel came to the conclusion that the unresolvable nebulae, of which perhaps the best-known was the great nebula in Orion, are formed from a 'shining fluid'. These objects, he believed, were future sidereal systems in their nascent stage.[8]

---

[3] James Coutts, *A History of the University of Glasgow: From its Foundation in 1451 to 1909* (Glasgow: James MacLehose and Sons, 1909), p. 388.

[4] John Pringle Nichol, 'Art. VI. State of Discovery and Speculation Concerning the Nebulae', *London and Westminster Review* 25 (July 1836): pp. 390–409.

[5] Gerald J. Whitrow, 'The Role of Time in Cosmology', in *Cosmology, History and Theology*, ed. Wolfgang Yourgrau and Allen D. Breck (New York: Plenum Press, 1977), pp. 159–77, pp. 167–68.

[6] Roger Hahn, *Pierre Simon Laplace, 1749–1827: A Determined Scientist* (Harvard: Harvard University Press, 2005), pp. 202–4.

[7] Simon Schaffer, 'Herschel in Bedlam: Natural History and Stellar Astronomy', *The British Journal for the History of Science* 13, no. 3 (1980): pp. 211–39.

[8] Stephen G. Brush, 'The Nebular Hypothesis and the Evolutionary World View', *History of Science*

Nebular cosmogony was given fresh impetus and a wider audience in the 1830s through the writings of the Scottish political economist and lapsed cleric John Pringle Nichol (1804–1859) who had, as a young man, been '… licensed as a preacher', but '… he only appeared two or three times in the pulpit, and at one of these […] he completely broke down. Whatever was the cause, he soon abandoned theological study, and devoted himself to that of astronomy'.[9]

Nichol wrote a lengthy essay in the *London and Westminster Review* of 1836 in which he laid out a comprehensive summary of the nebular hypothesis to date.[10] Opponents of developmental hypotheses, in their enthusiasm for re-establishing the pre-eminence of a natural theology in which final causes were seen to dominate, attempted to dismiss the evidence as presented by Nichol and subsequently by the anonymous, at the time, publication of *Vestiges of the Natural History of Creation* in 1844.[11] It is also evident that these antagonists were reacting in part to the political views of '… "those washed & unwashed Radicals" who had kept the country in a state of alarm since the Reform Bill'.[12] They viewed the ideas of 'progressive' cosmology as being in some way bound up with an agenda derived in large part from Ricardian political economics and promulgated in the articles which Nichol had contributed to various liberal journals.[13]

It would have come as a relief to them when an apparently major blow to the nebular hypothesis was struck by the use of the 'Leviathan of Parsonstown', a six foot reflecting telescope built by William Parsons (1800–1867), the third Earl of Rosse, at Birr Castle in the midlands of Ireland;[14] an interestingly unsuitable location in

---

25, no. 3 (1987): pp. 245–78, pp. 251–52.

[9] James MacLehose, *Memoirs and Portraits of One Hundred Glasgow Men* (Glasgow: James MacLehose and Sons, 1886), p. 249; and J. H. Parker, ed., 'Professor Nichol', *The Gentleman's Magazine: and Historical Review, July 1856–May 1868*, Nov 1859): p. 207, British Periodicals p. 536.

[10] Nichol, 'Nebulae'.

[11] See, for example, Nichol, 'Nebulae', pp. 406–7, and Simon Schaffer, 'The Nebular Hypothesis and the Science of Progress', in *History, Humanity and Evolution: Essays for John C. Greene*, ed. James R. Moore (Cambridge: Cambridge University Press, 1989), pp. 131–64, pp. 153–56.

[12] James A. Secord, *Victorian Sensation: The Extraordinary Publication, Reception, and Secret Authorship of Vestiges of the Natural History of Creation* (Chicago: University of Chicago Press, 2000), p. 227, which includes this quotation from a letter written by W. Smyth to W. Whewell on 29 January 1845.

[13] For example, *Tait's Edinburgh Magazine*, and John Stuart Mill's *The London and Westminster Review*. See Schaffer, 'Nebular', p. 145.

[14] Simon Schaffer, 'Astronomers Mark Time: Discipline and the Personal Equation', *Science in Context* 2, no. 01 (1988): 115–45, p. 126. Parsonstown subsequently reverted to the more traditional name 'Birr' and King's County is now known as County Offaly.

view of the restricted opportunities for observation it provided due to the prevailing climate.[15] The context in which these observations took place included the contemporary disruption which arose as a result of the great potato famine, which significantly reduced the usage of the telescope between 'first light' in 1845 and the end of the famine in 1848.[16] The parish of Birr was also the location of one of the few nineteenth-century Catholic schisms in Ireland, a breakaway movement led by the Crotty cousins, Michael and William, 'who became respectively, a Church of Ireland parson and a Presbyterian minister after having [...] first attempted to organise a dissident group of Catholics at Parsonstown'.[17] In 1846 there was even a report of a religiously motivated attack on the giant telescope 'upon <u>Religious Grounds</u>'.[18]

One of the key objectives pursued by Rosse in constructing the Leviathan of Parsonstown was to re-examine the objects previously identified by the Herschels as unresolvable nebulae with a view to potentially reclassifying them as stellar clusters.[19] He was certainly prompted in his endeavours by Thomas Romney Robinson (1792–1882), the former 'infant bard of Belfast', a Church of Ireland minister and Director of the Armagh Observatory whose religious enthusiasm led him to become Rosse's closest supporter and a frequent co-observer.[20] It can also be inferred that another of Rosse's objectives in the whole enterprise of the Leviathan was to demonstrate his mastery of the requisite engineering techniques, his benevolence as a landlord and the ability of the relatively unskilled workers from his demesne, under his expert supervision, to carry out the necessary tasks of casting and polishing speculum metal.[21]

Robinson was so determined to undermine the concept of ongoing stellar development that as early as November 1840 he knew what the outcome of future nebular

---

[15] Michael Hoskin, 'Rosse Robinson and the Resolution of the Nebulae', *Journal for the History of Astronomy* 21, no. 4 (1990): pp. 331–44, pp. 333, 341.
[16] Margaret Hogan, 'William Parsons' influence on the town and community of Birr', in *William Parsons 3rd Earl of Rosse: Astronomy and the Castle in Nineteenth-century Ireland*, ed. R. Charles Mollan (Manchester: Manchester University Press, 2014), pp. 103–11.
[17] K. Theodore Hoppen, *Ireland since 1800: Conflict and Conformity* (Harlow: Longman, 1999), p. 79.
[18] Humphrey Lloyd to Sir John Herschel, 17 October 1846, Royal Society Herschel MSS HS 11.285. Emphasis in the original. This supposed attack was also reported in the weekly gossip column of *The Athenaeum* dated 17 Oct. 1846. This article, however, also suggested that the original report may have been spurious.
[19] Wolfgang Steinicke, *Observing and Cataloguing Nebulae and Star Clusters: From Herschel to Dreyer's New General Catalogue* (Cambridge: Cambridge University Press, 2010), pp. 105–6.
[20] J. A. Bennett, *Church, State and Astronomy in Ireland* (Armagh: The Armagh Observatory, 1990), pp. 59–84.
[21] Simon Schaffer, 'On Astronomical Drawing' in *Picturing Science, Producing Art*, ed. Caroline A Jones and Peter Galison (New York: Routledge, 1998), pp. 441–74, here pp. 457–59.

observations would be. Following a series of observations made with the three-foot reflector (a precursor to the Leviathan), he expressed himself in very confident terms to an audience at the Royal Irish Academy to the effect that 'all Nebula observations must be reformed'.[22] In 1845 Robinson was similarly satisfied with the outcomes of his initial observations of nebulae through the eyepiece of the Leviathan: 'We have however worked to some purpose, for of the 43 nebulae which we have examined All have been resolved'.[23] Rosse, whilst publicly more cautious than Robinson about claims of nebular resolvability,[24] when conversing with visitors at dinner, repeated Robinson's claims and stated that 'he altogether rejects the nebular hypothesis'.[25]

It is rather ironic that Nichol was the first person to have had an opportunity to view the Orion nebula with the Leviathan. Following his night of observation in December 1845, Nichol declared that he was unable to resolve the nebula into individual stars.[26] Successful resolution was subsequently claimed by Rosse in a letter to Nichol in early 1846 and then reported in *The Times* and the *Athenaeum*.[27] These accounts were not supported by any graphical representations at the time and, surprisingly, it was not until 1867 that a drawing of the nebula, as seen via the Leviathan, was made available to the public press. Given the meteorological and atmospherical uncertainties arising from the location of the telescope and also the frequent deterioration of the main mirror, together with the subjective difficulty of deciding whether or not one is seeing very fine stars scattered like 'grains of sand' or a nebulous haze, it is perhaps not surprising that detailed illustrations of this object were few and far between.[28]

Nichol had a surprisingly amicable relationship with Lord Rosse, to the extent that he had previously courted controversy within the University of Glasgow by agreeing to loan some of its observatory equipment to Rosse.[29] The unintended consequences of this action included, in due course, Nichol's own bankruptcy.[30] This generosity

---

[22] Sir James South to Lord Rosse, 11 November 1840, Birr Castle MSS K 1.2.

[23] Thomas Romney Robinson, *Astronomical Diary*, 12 March 1845, Birr Castle MSS L2.1.

[24] See, for example, Hoskin, 'Rosse Robinson', p. 338, and Simon Schaffer, 'The Leviathan of Parsonstown: literary technology and scientific representation', in *Inscribing Science: Scientific Texts and the Materiality of Communication*, ed. T. Lenoir (Stanford, CA: Stanford University Press, 1998), p. 214.

[25] Frances Power Cobbe to Louise Cobbe, dated to the 1840s, Birr Castle MSS J 7. Also see Frances Power Cobbe, *Life of Frances Power Cobbe by Herself*, 2 vols (New York: Houghton, Mifflin and Co., 1895), Vol. 1, p. 177. NB: Rosse seemed to be slightly misquoting Robinson.

[26] John Pringle Nichol, *The Architecture of the Heavens* (London: John W. Parker, 1850), pp. 113–14.

[27] Schaffer, 'Leviathan', pp. 214, 216.

[28] Schaffer, 'Drawing', p. 462.

[29] Coutts, *History*, pp. 388–89.

[30] Schaffer, 'Nebular', p. 141.

seems surprising in the light of his earlier quite trenchant views on the iniquitous behaviour of some members of the Protestant ruling class and the use of tithes to support the Church of Ireland, which would also have placed him in opposition to Robinson.[31] Perhaps he was seduced by the impressive and progressive nature of the technology to be deployed in Ireland and he may also have been convinced that the planned observations would provide confirmatory evidence in favour of his preferred cosmogony. On receiving Lord Rosse's claim of the Orion nebula's potential resolvability, Nichol publicly accepted the word of a gentleman; an acceptance which presumably reflected a disinclination to contradict a peer of the realm.[32] A second factor which must have played a role in Nichol's acquiescence to the Rosse and Robinson interpretation of the observations of Orion was his earlier material support for the project. Thirdly, he may have been anxious to avoid being seen as a supporter of the anonymous author of *Vestiges*; a controversial tract which had woven Laplace's planetary formation hypothesis into a narrative of evolution that appeared to leave only marginal room for the activities of a creator.[33]

The reported resolvability of the Orion nebula was a set-back for Nichol and the nebular hypothesis. One contemporary commentator observed that 'Without doubt then, the nebular hypothesis must be abandoned', relying in part on the news that Nichol had himself written to the editor of the *Glasgow Argus* to report that the concept 'is no longer tenable'.[34] Despite his apparent concurrence with the Irish observers, however, Nichol was unwilling to give up the idea of a developing cosmos and in due course began to promote variants of his previously stated views. He published a new book in 1846 which, whilst containing fulsome praise for the achievements of Lord Rosse, subtly undermined the inference drawn by Robinson and Rosse that the nebular hypothesis had been totally disproved. He challenged the solidity of their conclusions by means of such contrary evidence as Herschel senior's planetary nebulae, the persistent un-resolvability of the Andromeda nebula, and the continuing applicability of Laplace's idea of planetary formation.[35] But the key argument which Nichol deployed in the book and reiterated in his subsequent lecture series was an ingenious re-framing of the Orion nebula observations to support his views regard-

---

[31] Bennett, *Ireland*, pp. 80, 87.

[32] Schaffer, 'Nebular', pp. 141–42. Also see Secord, *Sensation*, pp. 403–10 for a discussion of the role of gentlemen in science.

[33] Secord, *Sensation*, pp. 466–67 documents Nichol's negative reaction to the publication of Vestiges and also provides a reference to the announcement of his bankruptcy in *The Times* of 30/3/1842.

[34] Anon., 'Journal of Science, Inventions, and improvements', *The Critic*, 11 April 1846, pp. 3, 67.

[35] John Pringle Nichol, *Thoughts on Some Important Points Relating to the System of the World* (Edinburgh: William Tait, 1846), pp. 39–47, 56–68.

ing the scale and age of the universe.[36] To summarise his argument succinctly, Nichol held that if the nebulae consist ultimately of clusters of stars which are only just resolved by the Leviathan, then the distance of the stars (and hence the time taken for light to travel from them to the earth) is considerably greater than previously supposed, to the extent that the Orion nebula must be at least 60,000 light years away and that the furthest visible objects in the sky are perhaps thirty million light years away; a point which did not escape the reporter from the *Manchester Guardian*; 'About 5,000 of these clusters had been observed, from some of which the light occupied 30 million years in coming to us.'[37]

Thus Nichol neatly contrived to turn the evidence which Robinson, in particular, felt had destroyed the nebular hypothesis into yet another leg of support for astronomical and geological developmental hypotheses which could also encompass ideas of change and progression in the spheres of biology and society.

In 1848 Nichol published a populist tract entitled *The Stellar Universe* which was aimed particularly at 'younger Intelligences.'[38] To emphasize the lengthy chronologies implied by his speculations he invited the reader to understand that the arrangement of constellations in the sky has not been fixed since the dawn of time, but must have themselves changed over time due to the solar system's proper motion:

> ... we infer the Sun [...] could reach that remotest distance to which Lord Rosse's telescope can pierce, in about TWO HUNDRED AND FIFTY MILLIONS OF YEARS; and [...] that many of the mountains of our Earth may, through its whole duration, have been in being, rearing their peaks towards different constellations,[39]

The subtle implications of this statement were no doubt welcome to geologists and suitably impressive to his audiences.

Space, ironically, does not permit a closer examination of several discussion points which arise from this narrative. One such is the question of who had author-

---

[36] The lecture series at the Athenaeum was reported by *The Manchester Guardian* editions of 12, 15, 22 and 26 May 1847. The lectures in Manchester were similarly covered by the same journal in its 10 and 24 September 1851 editions. The series delivered in New York was published in book form as: John Pringle Nichol, *Views of Astronomy: Seven Lectures delivered before the Mercantile Library Association of New York in the Months of January and February 1848* (New York: Greeley & McElrath, 1848).

[37] Anon., 'Professor Nichol's Lectures on Recent Astronomical Discoveries', *The Manchester Guardian*, 12 May 1847.

[38] John Pringle Nichol, *The Stellar Universe: Views of its Arrangements, Motions, and Evolutions* (Edinburgh: John Johnstone, 1848), p. viii.

[39] Nichol, *Stellar*, pp. v–ix, 196–97. Emphasis in original.

ity to make statements about astronomical observations in the nineteenth century. Similarly, we ought to consider the interactions which occurred between astronomers and the various technologies they employed for observing and recording. Another key question would be why there was so little overt challenge to Nichol's restatement of the problematic resolution of Orion in his own terms. A partial explanation may lie in the variety of views to which readers were exposed at the time. The reported outcome of the observations at Birr Castle depended on which periodical they subscribed to, ranging in outlook from conservative organs such as *The Times* and *The Athenaeum* through to *Chambers' Edinburgh Journal* and other more progressive imprints. Perhaps the most surprising reaction was that of the Catholic *Dublin Review* which noted that '… the observations hitherto made upon nebulae are too imperfect to form a safe foundation for any hypothesis.'[40]

Another key observation which can be drawn from this episode is the elusiveness of objectivity in the making of cosmogonies. Rosse, Robinson and Nichol managed the gathering and reporting of data in such a way as to bolster their preferred metaphysical and theological positions. A charitable observer might say that they had succumbed to confirmation bias. Despite Robinson's certainty and his aristocratic partner's assertion of resolvability, Nichol managed to snatch a limited form of victory from the jaws of apparent defeat. Whether he had planned his response in advance cannot be known, but he managed in practice to execute a classic Morton's fork; if the Orion nebula is unresolvable then Herschel's developmental hypothesis stood confirmed but, if not, the universe was in fact older and more extensive than had previously been thought.[41] Nichol had succeeded in distancing humanity from a personal God in the way that theologians had feared and had himself become a preacher of the new 'religion' of astronomy. His earlier political inclinations were perhaps sated by the recent successes of other would-be reformers in Britain. These could be contrasted with the repetition of the dangers of revolutionary fervour represented by recent uprisings in France and elsewhere. Radical and progressive astronomy, which quietly demolished existing intellectual structures without unwanted physical repercussions, was safer, more popular and rather more lucrative than any inclinations of a more activist nature.

---

[40] Schaffer 'Leviathan', pp. 215–17.

[41] A Morton's Fork is a specious piece of reasoning in which contradictory arguments lead to the same (unpleasant) conclusion. It is said to originate with the collecting of taxes by John Morton, Archbishop of Canterbury in the late fifteenth century, who held that a man living modestly must be saving money and could therefore afford taxes, whereas if he was living extravagantly then he was obviously rich and could still afford them. John Ayto, ed., *Oxford Dictionary of English Idioms*, 3rd edition, (Oxford: Oxford University Press, 2009), pp. 232–33.

GALILEO GALILEI'S MEMORIAL TOMB IN SANTA CROCE: AN HONORIFIC MONUMENT TO A FLORENTINE GENIUS

Liana De Girolami Cheney

'Galileo Galilei is the greatest light of our time'
—Ferdinand II de' Medici, Grand Duke of Tuscany

ABSTRACT: Galileo Galilei (Pisa 1564–Arcetri 1642), the Father of Modern Science, was a Tuscan artist, an astronomer, a mathematician, a physicist and a philosopher. His renowned innovations and discoveries paved the way for the scientific understanding of the universe in the modern era, and included a navigational compass, improvement of the telescope, the four moons surrounding Jupiter, the nature of the Milky Way, the rings around Saturn, the phases of the planet Venus, and sunspots on the Sun. Moreover, Galileo's legacy in the area of his scientific writings derived from his numerous publications such as *The Little Balance* (1598), *On Motion* (1592), *The Operations of the Geometrical and Military Compass* (1604), *Discourse on Bodies in Water* (1602) and *Dialogue Concerning the Two Chief World Systems* (1632). After his death in 1642, Galileo's body could not be buried on consecrated ground because the Inquisition of the Catholic Church condemned him for his support of Copernicus' heliocentric notion and his theory on the motions of the earth. Challenging the Church's refusal, Ferdinando de' Medici, Grand Duke of Tuscany and a devoted patron of Galileo, was able to obtain a humble burial in Santa Croce for Galileo's remains. Many years later, in 1703, Galileo's pupil, Vincenzo Viviani, bequeathed a substantial endowment to build a monumental memorial for both Galileo and himself in Santa Croce. In 1737, the commemorative structure was designed and built by the Foggini family, the sculptor Girolamo Ticciati and the architect Giovan Battista Nelli. The sepulcher is a remarkable funerary monument celebrating the accomplishment of a *genius locus*, Galileo Galilei. This paper will analyse the originality of the tomb in terms of its Renaissance composition and humanistic symbolism.

Introduction

In Santa Croce the sepulcher of Galileo Galilei is a remarkable funerary monument celebrating the accomplishment of a Florentine *genius*. This chapter analyses the originality of the tomb in terms of its Renaissance composition and humanistic symbolism. Three points will be considered: 1) the history of the commission for Galileo Galilei's tomb; 2) the visual iconographic sources for the tomb; and 3) the symbolism of Galileo Galilei's tomb.

Galileo Galilei (Pisa 1564–Arcetri 1642), the Father of Modern Science, was a Tuscan artist, an astronomer, a mathematician, a physicist and a philosopher.[1]

---

[1] See Erwin Panofsky, 'Galileo as a Critic of the Arts: Aesthetic Attitude and Scientific Thought', *Isis* 47, no. 1 (March 1956): pp. 3–15; Erwin Panofsky, *Galileo as a Critic of the Arts: Aesthetic Attitude*

His renowned innovations and discoveries paved the way for the scientific understanding of the universe in the modern era, which included a navigational compass, improvement of the telescope, the four moons surrounding Jupiter, the nature of the Milky Way, the rings around Saturn, the phases in the planet Venus,
and sunspots on the Sun. Moreover, Galileo's legacy in the area of his scientific
literary also derived from his numerous publications such as *The Little Balance*
(1598), *On Motion* (1592), The *Operations of the Geometrical and Military Compass* (1604), *Discourse on Bodies in Water* (1602), *The Starry Messenger* (1610)
Discourses and Mathematical Demonstrations Concerning The Two New Sciences
(1619) and *Dialogue on the Two Chief World Systems: Ptolemaic and Copernican*
(1632).[2]

I. History of the commission for Galileo Galilei's tomb

After his death in 1642, Galileo's body could not be buried on consecrated ground
because the Inquisition of the Catholic Church condemned him for his support of
Copernicus' heliocentric notion and his theory on the motions of the earth. Chal-
lenging the Church's refusal, Ferdinand II de' Medici, Grand Duke of Tuscany
(1610–1670) and a devoted patron of Galileo, was able to obtain a humble burial for
Galileo's remains below the campanile in a very small chapel, the Novices Chapel,
Cappella del Noviziato, dedicated to the Medici patrons Cosmas and Damian, in the
Franciscan basilica of Santa Croce.[3]

---

and Scientific Thought (The Hague: Narinus Nijhoff, 1954); Eileen Reeves, *Painting the Heavens: Art
and Science in the Age of Galileo* (Princeton, NJ: Princeton University Press, 1997); Horst Bredekamp, 'Gazing Hands and Blind Spots: Galileo as Draughtsman', in *Galileo in Context,* ed. Jürgen Renn
(Cambridge: Cambridge University Press, 2001); and Filippo Camerota, *Linear Perspective in the
Age of Galileo: Ludovico Cigoli's Prospettiva practica* (Firenze: Olschki, 2010). See J.L. Heilbron, *Galileo* (Oxford: Oxford University Press, 2010); and Paolo Galluzzi, ed., *Galileo: Images of the Universe
from Antiquity to the Present* (Florence: Giunti, 2009), exhibition catalogue at the Strozzi Palace in
Florence; Galileo Galilei, *Dialogue Concerning the Two Chief World Systems,* trans. Stillman Drake
(New York: Modern Library, 2001); and Maurice A. Finocchiaro, *The Routledge Guidebook to the
Galileo's Dialogue* (New York: Routledge, 2013).
[2] See Galluzzi, *Galileo: Images of the Universe.*
[3] See Paolo Galluzzi, 'The Sepulchers of Galileo: The 'Living' Remains of a Hero of Science', in *The
Cambridge Companion to Galileo,* ed. Peter Machamer (Cambridge: Cambridge University Press,
1998), Ch. 12, p. 424; Paolo Galluzzi, 'I sepolchri di Galileo. Le spoglie 'vive' di un eroe della scienza', in *Pantheon di Santa Croce a Firenze,* ed. Luigi Berti (Florence: Giunti, 1993), pp. 145–82; and
A. Favaro, 'Studi e ricerche per una iconografia galileiana', *Atti del Reale Istituto Veneto di Scienze,
Lettere ed Arti, A, A,* 72 (1912–13): second part, pp. 1035–47; M. Gregory, 'Le tombe di Galileo e il
palazzo di Vincenzo Viviani', in *La città degli Uffizi,* exhibition catalogue (Florence: Sasoni, 1983),

From the start, the Grand Duke wanted to honour Galileo's accomplishments by arranging an elaborate funeral with a public oration and the construction of a marble mausoleum to be a burial site of Galileo with his family in the centre of the Basilica of Santa Croce. He requested assistance from Vincenzo Viviani (1622–1703), a Florentine mathematician, astronomer and engineer who had trained with Galileo in his earlier years and remained his devotee throughout his life. He was also the president of the Accademia dell' Arte del Disegno and a member of the Accademia del Cimento. In 1656, after Galileo's death, Viviani published a collected edition of Galileo's works, excluding the *Dialogue*. The Grand Duke, impressed by Viviani's fame, continued to encourage him to pursue plans for Galileo's memorial tomb in the Basilica. Although the Grand Duke and Viviani made numerous attempts following the death of Galileo, they were unsuccessful. Galileo's funerary plan was prohibited by Urban the VIII, the Barberini Pope, who vehemently objected to this honorific memorialization.

Viviani's hopes began to wane when his most important advocate, the Grand Duke, died in 1670; his son and successor, Cosimo III de' Medici, was apathetic in this pursuit. Moreover, in Rome the new pope, Innocent the X, was opposed to involving the church in this controversial request. Unable to obtain monetary support from Galileo's family – his son Vincenzo (Vincenzo Gamba 1606–1649) had died in 1649, his second daughter Suor Maria Celeste (Virginia 1600–1634) had died earlier in 1634, and his first daughter, Suor Arcangela (Livia, 1601–?) had no financial means – Viviani took on the task to visually memorialize his master. He commissioned several sculptors to represent honorific images of Galileo. Pietro Antonio Novelli (1600–1662) was responsible for creating a clay model of Galileo's death mask and a bust sculpture, while Ludovico Salvetti was to cast a bronze bust of Galileo based on Giovanni Caccini's carve marble head of 1612. Regrettably, Salvetti's bronze bust of Galileo remained unfinished.

Not discouraged by lack of Medicean support, in 1674 Viviani organized a simple ceremony to honour his master in the Cappella del Noviziato where a bronze bust of Galileo was placed in front of a tondo decorated with a scallop shell (FIG. 8.1). This classical and Christian symbol is an allusion to prosperity and the regeneration of life as well as a reference to Christian baptism and pilgrimage to an eternal life.[4] Below the bust, there is cartouche with carefully selected honorific references to Galileo's accomplishments. Below the cartouche, there is a wall marble plaque eulogizing

---

pp. 113–18; and Frank Büttner, 'Die ältesten Monumente für Galileo Galilei in Florenz', *Kunst des Barock in der Toskana* (Munich: no publ. listed, 1976), pp. 1013–37.

[4] See James Hall, *Dictionary of Subjects and Symbols in Art* (New York: Harper & Row, 1974), p. 280.

FIGURE 8.1. Galileo Galilei's First Tomb, 1642–74, Cappella del Noviziato, Basilica of Santa Croce, Florence. Photo credit: Liana De Girolami Cheney.

Galileo. The Latin epitaph was carefully composed by Simone di Bindo Peruzzi, Member of the Accademia Columbaria and Reader of the Tuscan Language.[5]

Meanwhile, between 1686 and 1697, Viviani acquired several adjoining homes in via Sant'Antonino 11 in Florence, and wished to combine them and renovate them as a palace (FIG. 8.2). He requested the assistance of the architect Giovanni Battista de' Nelli to achieve this project. Today, the complex is known as Palazzo dei Cartelloni

[5] See Galluzzi, 'The Sepulchers of Galileo', p. 436; Bredekamp, 'Gazing Hands and Blind Spots', p. 155.

Figure 8.2. Vincenzo Viviani, Palazzo dei Cartelloni, 1686–1697, Florence. Photo credit: Liana De Girolami Cheney.

(The Palace of the Billboards).[6] Viviani also commissioned the sculptor Giovanni Battista Foggini to compose a bronze bust of Galileo, based on Giovanni Caccini's terracotta model of 1610, to be placed at the centre of the façade, above the main entrance door. Behind the bust, a large, marble plaque eulogizes in Latin Galileo's scientific accomplishments and honours his noble and just persona. The bust is laterally framed by two large stone-scrolls with two vignettes in the life of Galileo. On

---

[6] See Gregory, 'Le tombe di Galileo e il palazzo di Vincenzo Viviani', pp. 113–18.

99

the left, on the balcony of his home, Galileo is viewing the Medicean planets with a telescope, while on the right two young figures are testing the motion of projectiles with canons. Framing the entrance door of the façade are two large cartouches decorated with festoons. They contain lengthy descriptions (in Latin) of Galileo's astronomical and scientific discoveries, namely, the telescope, the Medicean planets, sunspots, longitudinal sea measurement and projectile motion. Viviani has visually honoured his palace with the memory of his teacher and mentor.

In 1703, Viviani died leaving all his possessions to his nephew, with the obligation of relocating Galileo's grave in the Basilica of Santa Croce. Viviani's request went unfulfilled until his inheritance passed to a Florentine senator, Giovanni Battista Clemente de' Nelli, who abided by Viviani's wish and, upon receiving papal approval from the new Medicean pope, Clement XII, provided a proper burial for Galileo. The Florentine senator and historian, along with his predecessors, the Grand Duke and Viviani, understood the scientific and humanistic genius of Galileo, and they all wanted to honour him as a companion of famous learned Florentines such as Leonardo Bruni, Carlo Marsuppini and the 'divine' Michelangelo, who were all buried in the Basilica of Santa Croce.

Finally, on 12 March 1737, the cadaver of Galileo was exhumed from the first interment made of bricks in the Cappella del Noviziato and moved to the new marble tomb in the Basilica of Santa Croce. Also, the cadaver of Viviani was collected and united with his master in the new site. In his will, Viviani requested to be buried next to Galileo, hence he was initially buried in a brick tomb in the Cappella del Noviziato, then his body was exhumed and moved with Galileo's body to their new resting place in the Basilica of Santa Croce.[7] A private and civic ceremony took place with torches and candles during the transport of the bodies to the new burial site in Santa Croce. 'Some ritual removal of a single vertebra, three fingers from the right hand and a tooth where kept for posterity'.[8] Finally, the wishes of Galileo, Viviani and the Grand Duke were realized: on 12 March 1737, almost a century after his death in 1642, Galileo received a proper burial and a memorial monument honouring his humanist and scientific achievements in the consecrated ground of the Basilica of Santa Croce.

---

[7] There is a claim that a third body was found in the Cappella del Noviziato and transported to the new burial site in the basilica. It is believed that it was the body of Galileo's oldest daughter, Suor Arcangela. See Dava Sobel, *Galileo's Daughter: Historical Memoir of Science, Faith and Love* (New York: Walker, 1997), p. 366.

[8] See Sobel, *Galileo's Daughter*, p. 366.

II. Visual iconographic sources for the tomb

In 1294, the Florentine sculptor and architect, Arnolfo di Cambio, under the aus-
pices of the Florentine Republic, built the Basilica of Santa Croce, the largest Italian
Franciscan church. A humanist and religious centre, the church is also known as
a funerary pantheon of *uomini e donne famosi*, and includes cenotaphs for Dante
Alighieri, Niccolò Machiavelli, Gioacchino Rossini, Vittorio Alfieri and Ugo Fos-
colo, as well as the well known humanist tomb of the Cinquecento artistic genius,
Michelangelo Buonarroti, and the Quattrocento historians and chancellors of the
Florentine Republic, Leonardo Bruni and Carlo Marsuppini.

During the Florentine Quattrocento a type of humanistic tomb for *uomini famosi*
developed.[9] It consisted of an altarpiece or niche decorated with an architectural frame
constructed of a classical triumphal arch sheltering a funeral composition. The classical niche rests an on elaborate base decorated with a relief of garland motifs. Four
levels of compositional designs are incorporated in the niche. In descending order
from top to bottom, on the first level there is at top of the tomb the inclusion of the
coat of arms of the deceased or his family. The second level consists of a lunette
containing the religious depiction of the Madonna and Christ Child with adoring
angels. The third level is comprised of an effigy of the deceased reclining on a bier.
Corresponding attributes of the profession are held by the deceased, e.g., the historian and statesman, Leonardo Bruni, is crowned with laurel and holds a book. The
last level in the tomb is a large plaque or epitaph with a Latin inscription eulogizing

---

[9] For studies on Renaissance Funerary monuments, see Philipp Fehl, 'Death and the Sculptor's Fame:
Artists' Signatures on Renaissance Tombs in Rome', *Biuletyn Historii Sztuki* 59 (1997): pp. 196–
217; Philipp Fehl, *Monuments and the Art of Mourning* (Rome: Unione Internazionale degli Istituti
di Archeologia Storia e Storia dell'Arte in Roma, 2007); Philipp Fehl, 'On the Representation of
Character in Renaissance Sculpture', *The Journal of Aesthetics and Art Criticism* 31, no. 3 (1973):
pp. 291–307; Giulio Ferrari, *La Tomba nell'arte italiana* (Milan: Ulrico Hoepli Editore, 1916); Rab
Hatfield, 'Giovanni Tornabuoni, i fratelli Ghirlandaio e la cappella maggiore di Santa Maria Novel-
la', in *Domenico Ghirlandaio 1449–1494*. Atti del Convegno Internazionale Firenze, 16–18 Ottobre
1994, ed. Wolfram Prinz and Max Seidel (Florence: Centro Di, 1996), pp. 112–17; Linda A. Koch
'The Early Christian Revival at S. Minato al Monte: The Cardinal of Portugal Chapel', *Art Bulletin*
78, no. 3 (1996): pp. 527–55; Sarah Blake McHam, ed., *Looking at Italian Renaissance Sculpture*
(Cambridge: Cambridge University Press, 1998); Erwin Panofsky, *Tomb Sculpture* (New York: Harry N. Abrams Publisher, 1992); John Pope-Hennessy *Italian Renaissance Sculpture*, 4th ed., Vol.
2 (London: Phaidon, 1996); Michel Ragon, *The Space of Death: A Study of Funerary Architecture,
Decoration, and Urbanism*, trans. Alan Sheridan (Charlottesville, VA: University Press of Virginia,
1983); Sharon T. Strocchia, *Death and Ritual in Renaissance Florence* (Baltimore: The Johns Hopkins
University Press, 1992); and Shelley E. Zuraw, 'The Public Commemorative Monument: Mino da
Fiesole's Tombs in the Florentine Badia', *Art Bulletin* 80, no. 3 (1998): pp. 452–77.

the accomplishments of the deceased. Some times the epitaph also includes a dedicatory inscription from a family or a civic organization. The composition of the altar also alluded to the natural separation of the body from the spirit (soul): 1) from the coffin, actual burial and death; 2) to the effigy, symbolic death and separation of the soul; 3) soul ascending to heaven through the intercession of Mary and Christ; and 4) the angels honour and protect the soul of the deceased, symbolized emblematically in the family coat of arms.

Cinquecento humanist tombs elaborate on the accomplishment of the deceased by transforming the third level: removing the reclining figure on the bier and adding a bust or a seated sculpture representing the deceased. This innovation is referred to as a *statua parlante* (speaking statue) or *scultura in testa* (awakening sculpture), and was appropriated from classical art by Donatello in his sculptures, for example Saint Mark at Orsanmichele, and followed by Michelangelo's portrayal of Duke Lorenzo and Duke Guiliano in the Medici Tombs in San Lorenzo.

Furthermore, another level is added to the funerary composition: the effigy holding the attributes is substituted with a female personification of virtues carrying the effigy's attributes or identifying with the attribute, for example a female figure may hold a palette and a paintbrush, thus personifying Painting. This new funerary style was assimilated by Giorgio Vasari in the tomb for his artistic mentor, Michelangelo, where in front of the coffin three personifications of the Fine Arts are mourning and guarding his tomb. They represent the arts of Architecture, Painting and Sculpture. Michelangelo excelled in all the Fine Arts. Cinquecento funerary art with the inclusion of Mannerist conceits became an apotheosis of beauty on the memory of the deceased.

III. Symbolism of Galileo Galilei's tomb

The studies of several scholars have focused on the historical development of the Galileo's burial site, namely, Paolo Galluzzi, Dava Sobel, J. L. Heilbron, Frank Büttner, Horst Brederkam, Mina Gregory and others, but none have thoroughly addressed the symbolism of the tomb, hence my interest in the topic.[10]

In 1737, the commemorative structure was designed and built by the architect

---

[10] See also Jünger Renn, *Galileo in Context* (Cambridge: Cambridge Press, 2001); Stillman Drake, trans. and ed., *Discoveries and Opinions of Galileo* (New York: Anchor Books, 1957); Mark A. Peterson, *Galileo's Muse* (Cambridge, MA: Harvard University Press, 2011); Evelyn Lincoln, *Brilliant Discourse: Pictures and Readers in Early Modern Rome* (New Haven: Yale University Press, 2014); Elizabeth A. Spiller, 'Reading through Galileo's Telescope: Margaret Cavendish; the Experience on Reading', *Renaissance Quarterly* 53, no. 1 (2000): pp. 192–221; and Patricia Likos Ricci, 'The Celestial Paragone Galileo and the Rivalry of the Astronomer', paper presented INSAP, New York, 2013.

FIGURE 8.3. Giovanni Battista de' Nelli and Foggini family, Galileo Galilei's Second Tomb, 1737, Basilica of Santa Croce, Florence. Photo credit: Liana De Girolami Cheney.

and mathematician Giovanni Battista de' Nelli and the Foggini family. The overall designed was composed by Giulio Foggini. The portrait bust of Galileo was created by the sculptor Giovanni Battista Foggini. The figure of *Astronomy* was carved by sculptor Vincenzo Foggini and *Geometry* was carved by the sculptor Girolamo Ticciati (Fig. 8.3)

The original design for Galileo's second tomb was likely inspired by Stefano della Bella's drawings and etchings for Galileo's *Dialogue* for the Landini edition of 1632. The della Bella drawings at the Uffizi (n. 7991 F and 8042 F) differ in the position of the three personifications. Three female personifications depict Astronomy, crowned with stars and holding a globe, Optics (or Philosophy) rests a book on her lap, and Mathematics or Geometry holds a geometrical sector or compass.

Galileo, standing in front of them, provides them with the most updated technology, the telescope, to see the beauty of the cosmos, the Medicean planets, the moons and the sunspots. In the background to the right, a cannon and ships allude to Galileo's military and maritime discoveries. Della Bella's etching is based on the Uffizi drawing (Inv. 7991 F). The allusion of the three personifications or muses is similar to Vasari's Tomb of Michelangelo where three muses, which are personifications of the Fine Arts of Architecture, Painting, and Sculpture, are protecting Michelangelo's remains. Furthermore, inspired by Vasari's Tomb of Michelangelo, della Bella was likely alluding as well to the complex nature of Galileo, a humanist, musician, artist, philosopher, writer, scientist, mathematician, astronomer and engineer. In the same manner, Vasari honoured his fellow artists in the editions of the *Vite* (*The Lives*) of 1550 and 1568, including the celebratory Fine Arts and allegory of Fame as their personifications or muses.

Due to the papal discord with Galileo's supportive view of Copernicus, as noted in the astral depiction in the Third Dialogue, Foggini opted to discard the della Bella design, and suggested a second design for Galileo's tomb, as seen in the engraving of 1737. Foggini reduced the number of figures to two, Astronomy and Geometry, and omitted the personification of Optics or Philosophy by instruction of the papacy, fearing repercussions from past denunciations.

The architectural composition reveals the appropriations from Vasari's Tomb of Michelangelo, in particular the colossal pilasters framing the altar and the flaming urns. After the damage of the 1966 floods, a transformation occurred in the restoration, including the removal of the triumphal arch motif as seen in the photograph of the Florentine artist, Giacomo Brocci.

Foggini composed a powerful pyramidal composition that projects Galileo's tomb into the space of the viewer, connecting the past (Galileo) with the present (the viewer). Visually analysing Galileo's tomb, one observes the funerary composition of the Quattrocento and Cinquecento *uomini famosi*. The horizontal four levels of honorific recognition are seen from top to bottom. Framing the coat of arms of Galileo's family are flaming urns (Latin for *uro*, meaning burning), symbols of eternal life or immortality. Galileo's coat of arms contains, in a cartouche, a red ladder with three horizontal rungs, alluding to his ancestors' family name, Galileo Buonaiuti, and a rebus meaning 'good help'.[11] The symbolism of the three-step ladder signifies the connection between the natural realm of matter and the spiritual realm. The three steps are a symbol of fearlessness, or a suggestion to be on guard against spiritual and corporeal enemies. In the Bible, *Exodus*, lines 461–65, cites this verse: 'Is sculptured on his shield: "a man in arms, His ladder fix'd against the enemies' walls"'.[12]

The second level depicts, in a scalloped niche, the bust of Galileo resting on a platform and holding the attributes of his profession and inventions – books, a globe or cosmos, a compass and a telescope. Foggini has captured the energy and vitality of an adult Galileo, based on earlier portraits and likely using the clay models of Caccini and the numerous portraits of painters such as Domenico Robusti and Ottavio Leoni, discarding the images of an aging Galileo like those of Justus Sustermann and others. The second level is united with the third level by the platform whose elaborate cartouche contains the four Medicean planets or moons of Jupiter, discovered in his numerous telescopic observations and recorded in his notebooks.

The third level consists of a ceremonial coffin framed by the personifications of Astronomy and Mathematics or Geometry. Their contrapposto stance balances the composition. Astronomy holds a scroll with the celestial disc, while Geometry or Mathematics holds a table (notebook) with studies and calculations on gravity. At the fourth level is the second epitaph eulogizing Galileo, craftily composed by senator Clemente de' Nelli.[13]

The vertical movement from the bottom of the tomb, where the epitaph is written, to the top of the monument where the coat of arms is portrayed, alludes to the Christian transformation of the body from a physical realm, the deceased interment, to the elevation of the soul into a metaphysical realm, where the soul of the deceased

---

[11] Sobel, *Galileo's Daughter*, p. 16.

[12] The full stanza reads:

No man devise
is sculptured on his shield: a man in arms,
His ladder fix'd against the enemies' walls,
Mounts, resolute, to rend their rampages down;
And cries aloud (the letters plainly mark'd),
Not Mars himself shall bent me from the Towers.

[13] The full discussion of the transportation of the bodies is disclosed in Nelli's *Instrumento*, presently in the Florentine Archivio di Stato. See Gregory, 'Le tombe di Galileo e il palazzo di Vincenzo Viviani', pp. 113–18.

has left the body in order to reach the divine or heaven. Galileo as a spiritual and religious man would have understood this transformation and likely would have considered a parallel between the mechanical use of an instrument as telescope, a physical object, to view a natural realm, and the visualization of heaven, a metaphysical realm of God.

Considering that Galileo's memorial tomb was not completed until the early eighteenth century (1737), I would like to suggest that an additional source of inspiration was employed for its execution, since the architect and sculptors of Galileo's tomb were then familiar with the renowned Florentine Baroque architect and sculptor Lorenzo Bernini, who resided in Rome. Perhaps also as a scoff to Pope Urban VIII, who condemned Galileo's writings, and with typical Florentine audacity, Foggini appropriates sculptural elements from Bernini's Tomb of Urban VIII of 1627–47 in Saint Peter's at the Vatican.[14] For example, Bernini's composition for the virtue of *Charity* in Urban's tomb transforms into Foggini's personification of *Astronomy* in Galileo's tomb, and Bernini's *Justice* in Urban's tomb resembles Ticciati's personification of *Geometry.* Thus, artistically, years later, Foggini and Ticciati vindicated Galileo with a sculptural pun by appropriating papal alleged virtues of charity and justice to represent Galileo's scientific accomplishments.

Another interesting question is posed in Foggini's tomb composition: what about the statues' gaze.[15] What are Galileo and the personifications looking at? From the location of the tomb and observing Galileo's gaze, Foggini cleverly has composed an axial line from Galileo's eyes to the oculus or window above in the main altar of Santa Croce, hence he is looking at light and at the stars. The same occurs with the placement of the personifications, who both look in the opposite direction to Galileo's gaze; however, they are also looking at the light that permeates from the stained glass window around the oculus or rose window at the entrance of the basilica.

The visual trajectories are not accidental. Metaphorically with this type of design, Foggini has triumphantly liberated Galileo from the constriction of the Inquisition, allowing him to observe the cosmos for eternity. Viviani and the Grand Duke would have applauded the creation of this honorific memorial and Galileo has thankfully found a place to rest and contemplate the universe that he loved.

---

[14] See Howard Hibbard, *Bernini* (Baltimore, MD: Penguin, 1966), pp. 124–28.
[15] See Samuel Y. Edgerton, *The Mirror, the Window and the Telescope* (Ithaca: Cornell University Press, 2009); and Camerota, *Linear Perspective in the Age of Galileo.*

Mars and the Mediums

Clive Davenhall

ABSTRACT: Spiritualism in its modern guise began in the mid-nineteenth century and its heyday was during the late nineteenth century and the first part of the twentieth, when it was a worldwide movement with several million adherents. The Martian 'canal craze' occurred during much the same period. Mars was reported to be criss-crossed with a network of fine lines, the famous 'canals', and these were popularly regarded as artificial waterways, constructed by intelligent beings to stave off extinction by husbanding their scarce and dwindling water supplies. These ideas were widely disseminated and popular with the public, though they remained, at best, problematic amongst astronomers. There was some common ground between these two developments, with some spiritualists reporting communication with Martians or making astral journeys to the planet (or, less often, elsewhere in the solar system or further afield). This chapter discusses this unlikely conjunction of ideas.

The second half of the nineteenth century and the first half of the twentieth were the heyday of both spiritualism and the Martian 'canal craze'.[1] The canal craze imagined the planet as a dying, desert world where an advanced civilisation had hoarded its dwindling water resources. A surprising confluence of these ideas led to reports of occult communication with Mars. These reports are widely scattered and, unsurprisingly, rarely feature in the astronomical literature, though there have been a few modern surveys.[2] Mediums reported travelling to Mars, and less frequently elsewhere, by astral projection and similar techniques. They reported communication with intelligent Martians by thought transference, recollected past lives on Mars and anticipated reincarnation there. This paper outlines this unexpected corner of the interaction between astronomy and wider culture. Subsequent sections summarise the relevant background, both spiritualist and astronomical, present examples of actual reports by

---

[1] For a comprehensive discussion of spiritualism see Janet Oppenheim, *The Other World: Spiritualism and Psychical Research in England 1850–1914* (Cambridge: Cambridge Univ. Press, 1985). For a more popular account see Ronald Pearsall, *The Table Rappers* (1972; repr. Stroud: Sutton, 2004). The Martian canal craze has been well-studied. Michael Crowe gives a succinct account in 'The Battle over the Planet of War', Chapter 9 of his *The Extraterrestrial Life Debate 1750–1900* (New York: Dover, 1999), 480–546. There is also much relevant material in: O. Morton, *Mapping Mars* (London: Fourth Estate/HarperCollins, 2003); R. Markley, *Dying Planet* (Durham, NC: Duke University Press, 2005); Robert Crossley, *Imagining Mars* (Middletown, CT: Wesleyan University Press, 2011); Maria D. Lane, *Geographies of Mars* (Chicago: University of Chicago Press, 2011); and H.V. Hendrix, G. Slusser and E.S. Rabkin, eds., *Visions of Mars* (Jefferson, NC: McFarland, 2011).

[2] The most comprehensive recent survey of Mars and spiritualism is Chapter 7 of Crossley, *Imagining Mars*, pp. 129–48. There is also useful material in Chapter 8 of Jerome Clark, *Hidden Realms, Lost Civilizations and Beings from Other Worlds* (Detroit: Visible Ink, 2010), pp. 129–68 and Gareth Medway, 'Mediums, Mystics and Martians', *Magonia* 99 (2009): pp. 3–9.

mediums and similar ideas employed in fiction, and finally discuss the phenomenon.

Background

Mars was seen to display a disk soon after the telescope was applied to astronomy in the early years of the seventeenth century and the first permanent markings on it were identified later in the same century.[3] Thereafter it was the subject of careful if sporadic observation. By the mid-nineteenth century a consensus understanding of the planet had been achieved. The rotation period differed by only a few minutes from that of the Earth. There were permanent markings on the surface, though their appearance varied somewhat. The darker, greenish patches were usually thought to be seas and the lighter red-brown regions to be continents. The polar caps waxed and wained with the seasons and were considered to be made of water ice. A substantial atmosphere supported occasional clouds. There was already speculation about whether the planet was inhabited, in an established tradition of 'the plurality of worlds'.[4]

The 1877 perihelic opposition brought the planet unusually close, thus positioning it favourably for observation, and it was extensively studied by a large number of observers. The most unexpected results were obtained by Giovanni Schiaparelli (1853–1910), a Professor of astronomy from Milan, who had previously shown little interest in the planet.[5] He produced maps showing unprecedented detail and introduced a new, elegant system of names, based on classical mythology and geography, which still forms the basis of Martian nomenclature. His maps were notable for showing, particularly in the northern hemisphere, a grid of hitherto unsuspected dark lines criss-crossing the lighter 'continents' and linking the darker 'seas'(FIGURE 9.1).[6] Schiaparelli called these features *canali*, Italian for 'channels', but the term was

[3] The definitive account of the early history of observation of Mars is Camille Flammarion, *La planète Mars et ses conditions d'habitabilité* (Paris: Gauthier-Villars, 1892). An English edition has recently been published: William Sheehan, ed., *Camille Flammarion's the Planet Mars*, trans. Patrick Moore (New York: Springer, 2015). See also Patrick Moore, 'The Mapping of Mars', *J. Brit. Aston. Assoc*, 94 no. 2 (1984), 45-54.
[4] Michael Crowe, *Extraterrestrial Life Debate*, 480-546.
[5] See M. Beech, 'Schiaparelli, Giovanni' in the *Biographical Encyclopaedia of Astronomers* (henceforth BEA), T. Hockey (ed.), 2007 (New York: Springer, 2007a), 1020-1021 and also obituaries Anon., *Astrophys J*. 32 (1910), 313-319 and 'E.B.K.' (probably E.B. Knobel), *Mon. Not. R. Astron Soc*. 71 (1911), 282–87.
[6] Schiaparelli produced maps for the 1877 opposition and several subsequent ones. They were originally published as memoirs of the Milan Observatory, of which he was Director, and were reproduced in Flammarion's *La planète Mars* and have appeared many times since. See Sheehan, *Flammarion's Mars*, 248, 258, 280–81, 288, 302–3. On-line versions are available at, for example http://

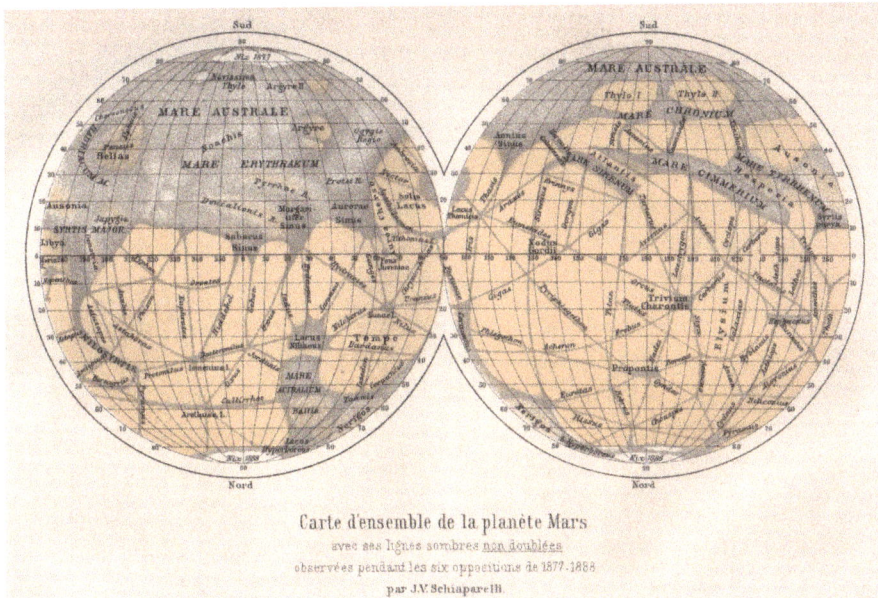

Carte d'ensemble de la planète Mars
avec ses lignes sombres non doublées
observées pendant les six oppositions de 1877-1888
par J.V. Schiaparelli

FIGURE 9.1. A composite map of Mars drawn by Schiaparelli from his observations made during the six oppositions between 1877–88 (reproduced from Camille Flammarion, *La planète Mars et ses conditions d'habitabilité* (Paris: Gauthier-Villars), 1892 ed., betw. pp. 440–41.

mis-translated into English as 'canals', suggesting an artificial origin, though Schiaparelli always remained non-comital about their nature. At successive oppositions Schiaparelli produced further maps showing an increasingly complex canal network. Initially only he saw the canals, but soon other observers were reporting them and by around 1890 their existence, if not their nature, was considered well-established.

The nature of the canals remained a matter of debate (and in the event they turned out to be optical illusions). The idea that they were artificial and constructed by intelligent beings was never accepted by most astronomers. However it had two powerful and effective advocates, Camille Flammarion and Percival Lowell, who popularised it amongst the wider public.[7] Camille Flammarion (1842–1925) was a prolific,

---

bibliodyssey.blogspot.com/2008/04/channelling-Martian-maps.html [accessed 21 August 2016].

[7] Flammarion was a prolific author and wrote many popular astronomy books, including *La Pluralité des Mondes Habités* (Paris: Mallet-Bachelier, 1861), *Les terres du ciel* (Paris: Didier, 1877) and *Astronomie populaire* (Paris: C. Marpon et E. Flammarion, 1880). For an English-language account of his ideas about Mars see, for example, book IV, chapter IV of the English translation of *Astronomie populaire*: Camille Flammarion and J. Ellard Gore, *Popular Astronomy: a General Description of the Heavens* (Chatto and Windus: London, 1894), 373–98, especially 382 onwards. For biographical details see R. Baum, 'Flammarion, Camille' in BEA, 934–35 and the obituary by A.F. Miller, 'Camille

self-educated French astronomer and populariser of astronomy. He was widely-read in France and discussed his ideas in numerous popular astronomy books, articles and also in works of fiction. Percival Lowell (1855-1916) was a wealthy American businessman, traveller and orientalist. He became fascinated by Mars in the early 1890s (initially by reading one of Flammarion's books) and founded the observatory that bears his name in Flagstaff, Arizona principally to study the planet. He was an eloquent and persuasive author who wrote several books expounding his ideas about the planet and his observatory issued a blizzard of press releases.

Flammarion and Lowell's ideas about Mars differed somewhat. Flammarion adhered to the earlier idea that the dark patches were seas and the lighter ones continents.[8] These continents supported lush vegetation of a riot of colours, but predominantly red and orange rather than green (H.G. Wells borrowed this idea for the 'red weed' that briefly overran South East England in *The War of the Worlds*). Percival Lowell's conception of the planet was significantly different. For him the orange areas were true desert and the darker ones arid semi-desert supporting sparse vegetation. The seas that the planet once possessed had long-since vanished. Both, however, saw the canals as the artificial constructs of an advanced civilisation.

Spiritualism in its modern form emerged in 1848 with the Fox sisters from Hydesville in Wayne County, New York State, with table-rapping and other phenomena.[9] It quickly became popular throughout the US and Europe and by the turn of the century had become a mass-movement with about eight million followers. The central tenet of spiritualism is that the spirit or soul enjoys a continued existence after death and, further, it is possible for the living to communicate with the spirits of the deceased. Some people were particularly adept at such communication and would act as 'mediums', mediating communications for others by relaying questions and answers. The movement quickly developed its now-familiar trappings: groups of communicants sitting in a darkened room with a medium, speaking in altered voices, automatic writing, sound effects such as bells and trumpets, and even the manifestation of 'aports' or physical objects.

Though spiritualism was a mass-movement it had little formal organisation.

Flammarion: his Life and his Work', *J. R. Astron. Soc. Canada* 19 (1925), 265–85. Lowell wrote three books setting out his ideas about Mars: *Mars* (Boston: Houghton Mifflin: 1895), *Mars and its Canals* (New York: Macmillan, 1906) and *Mars as the Abode of Life* (New York: Macmillan, 1908). His life and career have been well-studied. His most comprehensive biography is David Strauss' *Percival Lowell* (Cambridge, MA: Harvard University Press, 2001). More concisely, Strauss also contributed Lowell's entry in BEA, 710–11.

[8] See Flammarion and Gore, *Popular Astronomy*, or Sheehan, *Flammarion's Mars*, 435–41.

[9] See Janet Oppenheim, *Other World* or Ronald Pearsall, *Table Rappers*.

Rather, communication was through periodicals, lectures and practitioners. The later nineteenth century also saw the development of similar movements, such as theosophy and the founding of societies to investigate and assess the validity of these ideas, such as the Society for Psychical Research (in 1882) and the American Society for Psychical Research (in 1885).

Examples

Writing in 1959 the psychoanalyst and journalist Nandor Fodor recalled that 'at least a dozen well-known mediums have been involved with the planet Mars' and it seems likely that there were others.[10] Only a few examples will be mentioned here.

William Denton (1823-83), originally from Darlington, County Durham, settled in Ohio, where he became a geologist and political activist, advocating women's rights, the abolition of slavery and temperance.[11] He and his family also practiced 'psychometry', a form of divination that purports to discern the nature and history of objects merely by handling them. Between 1863 and 1874 William and his wife Elizabeth wrote the three volumes of *The Soul of Things* which included reports of trips that several family members had made to Mars while in a trance state.[12] Together they found several more-or-less human races (each member encountered a somewhat different species), all with civilisations broadly similar to the terrestrial one. Anne Denton Cridge, William's sister, reported a dense life-sustaining atmosphere, great mountains and lush valleys that were home to large crocodile-like reptiles.

Daniel Paul Schreber (1842-1911), originally from Leipzig, was a successful and highly respected judge, presiding over the Saxon High Court of Appeals, until in early middle age he suffered the onset of the mental illness that would periodically plague him for the rest of his life. Its underlying cause may have been an unusually strict upbringing. He suffered three periods of illness: 1884–85, 1893–1902 and 1907–11, separated by periods of recovery. After the second he wrote a book about

---

[10] Nandor Fodor, *The Haunted Mind: a Psychoanalyst Looks at the Supernatural* (New York: Helix Press, 1959), 262.

[11] For William Denton see Jerome Clark, *Hidden Realms*, 142-143, 151. For additional details see the biographical note prepared by the Wellesley Historical Society and available at http://www.wellesleyhistoricalsociety.org/documents/Denton%20Family%20Papers%20-%20Research%20Guide.pdf [accessed 21 August 2016].

[12] William and Elizabeth M.F. Denton, *The Soul of Things: or Psychometric Researches and Discoveries* (Boston: Walker Wise and Co, 1863); William and Elizabeth M.F. Denton, *The Soul of Things: or Psychometric Researches and Discoveries II* (Wellesley, Mass: E.M.F. Denton, 1873) and William and Elizabeth M.F. Denton, *The Soul of Things: or Psychometric Researches and Discoveries III* (Wellesley, Mass: E.M.F. Denton, 1874).

his condition, *Memoirs of My Nervous Illness* (1903).[13] It scandalised his family, but was quickly adopted by psychiatrists; a rare example of a textbook written by the patient rather than the doctor. It remains well-known, perhaps because Freud subsequently wrote a book re-assessing it in terms of psychoanalysis.[14]

Schreber's illness took several forms, including memories of previous incarnations. He wrote:

> ...I had lived for years in doubt as to whether I was really still on earth or whether on some other celestial body. Even in the year 1895 I still considered the possibility of my being on Phobos, a satellite of the planet Mars mentioned by the voices in some other context, and wondered whether the moon, which I sometimes saw in the sky, was not the main planet Mars.[15]

Interplanetary journeys were peripheral to Schreber's condition, but played an important part in the case of Hélène Smith, one of the better-known examples of spirit voyages to Mars. She was studied by Théodore Flournoy (1854–1920), the first Professor of psychology at the University of Geneva.[16] Flournoy, a contemporary of Freud and early mentor to Jung, was an important figure in the development of psychology, particularly in Switzerland. He reported Mlle Smith's case in *From India to the Planet Mars* which on publication in 1899 became something of a best-seller, running through several editions in a few months. An English translation appeared in the same year.[17]

Hélène Smith (or Helen Smith in the English edition, real name Catherine-Elise Müller; Flournoy adopted the pseudonym to protect her identity) came from a respectable Geneva family and worked as a secretary. She was also a well-known medium who gave seances to a circle of friends. She would enter a trance state in which

---

[13] Daniel Paul Schreber, *Denkwürdigkeiten eines Nervenkranken* (1903); edition consulted: Ida Macalpine and Richard Hunter (eds, trans.), *Memoirs of my Nervous Illness* (Cambridge, MA: Harvard University Press, 1988). Alex Pheby's recent novel *Playthings* (Norwich: Galley Beggar, 2015) is a fictional account of Schreber's case.

[14] Sigmund Freud, *The Schreber Case*, trans. Andrew Webber (London: Penguin, 2002).

[15] Schreber, *Memoirs*, p. 88.

[16] See Ronald Earl Goldsmith, *The Life and Work of Theodore Flournoy, 1854–1920*, 1979, PhD Thesis, Michigan State University, Dept. of History; and James Witzig, 'Theodore Flournoy, A Friend Indeed', *J. Ann. Psychology*, 27 (1982), pp. 131-48.

[17] Théodore Flournoy, *Des Indes à la planète Mars: étude sur un cas de somnambulisme avec glossolalie* (Paris: F. Alcan ; Geneva: Ch. Eggimann, 1900). English edition (somewhat abridged): *From India to the planet Mars: a Study of a Case of Somnambulism with Glossolalia*, trans. Daniel B. Vermilye (New York; London: Harper & Bros, 1900). See https://archive.org/details/cu31924022985299 [accessed 21 August 2016].

she spoke in the persona of spirits she encountered, and produce automatic writing and drawings. During a trance she would recall previous lives, mostly under the tutelage of 'Leopold', a spirit guide. Most of Mlle Smith's previous lives were terrestrial (and included an Indian princess; hence the title of Flournoy's book). However, one sequence involved communication with Mars. Amazingly she produced an entire Martian language, complete with script, vocabulary and grammar, and also drawings of the Martian scenes she witnessed. The language was later discovered to be a version of her native French, but it had been produced entirely automatically.

Flournoy thought Mlle Smith entirely sincere, but did not believe her experiences were real. He describes them as products of cryptomnesia and as a 'romance of the subliminal imagination'. Mlle Smith considered Flournoy's skepticism a betrayal and ended cooperation when the book was published.While Hélène Smith is well-documented, little is known of Sara Weiss other than that she was the American author of two books reporting astral journeys to Mars (or Ento in the Martian tongue): *Journeys to the Planet Mars* (1903), and *Decimon Huydas: A Romance of Mars* (1906).[18] Both journeys were made with the assistance of spirit guides. Her principal guide was one Carl De L'Ester but others included such luminaries as Giordano Bruno and Alexander von Humboldt. The journeys had occurred during 1893–94 though the books were not published until a decade later.

The first book, *Journeys*, is a rambling, digressive travelogue with detours into spiritualist doctrine. There are sections on the Martian language and flora and several well-executed drawings of the latter. *Decimon Huydas* is similar but has a narrative, relating a domestic tragedy that had occurred a few hundred years earlier. Both books are informed by a contemporary understanding of Mars. Giant irrigation projects, embankments and other civil engineering projects are mentioned and there are explicit references to Schiaparelli and Flammarion.

Finally, Hugh Mansfield Robinson was a practicing solicitor who had been the Town Clerk of Shoreditch (1900–11).[19] He was also a psychic and medium; his first astral visit to Mars was in 1918. His Mars was home to an advanced civilisation and on his first visit he arrived at a radio station. The Martians had a highly developed radio technology and generated hydro-electric power using waterfalls in the mountains and canals. The Martians themselves were 6–8 feet tall, with oriental features

---

[18] Sara Weiss, *Journeys to the Planet Mars; Or, 'Our mission to Ento'* (New York: The Bradford Press, 1903); *Decimon Hûýdas: A Romance of Mars* (Rochester, NY: The Austin Publishing Co., 1906). A review of the former appeared in the *New York Times*, 19 December 1903, available at http://www.nytimes.com/ref/membercenter/nytarchive.html [accessed 21 August 2016].

[19] Nandor Fodor, *Haunted Mind*, 'Interplanetary Fantasies', pp. 259–69.

and large ears. Robinson's guide was the Lady Oomaruru, who later turned out to be a reincarnation of Cleopatra.

During Mars' 1926 opposition Robinson tried to validate his astral communication by also establishing radio contact with the planet. He tried sending messages using standard commercial services and equipment then available to the public. No convincing replies were received, though his efforts were widely reported in national and local newspapers. He made further attempts in 1928 and on subsequent occasions before sinking in obscurity. The idea of communicating with Mars by radio had been discussed intermittently since the turn of the century and the idea resurfaced in the early 1920s, which may have sparked Robinson's interest. During the 1924 opposition the astronomer David Todd had organised a coordinated and widely-reported attempt to listen for signals.[20]

Examples from Fiction

The idea of astral communication with Martians or astral travel to the planet was not unusual in proto-science fiction of the later nineteenth and early twentieth centuries. The best-remembered of these stories is undoubtedly Edgar Rice Burroughs' *A Princess of Mars*, first published in 1912, and its many sequels.[21] In the first story the hero projects himself to Mars basically by force of will and strength of desire. There he finds a vaguely Lowellian desert world, criss-crossed by canals and inhabited by various advanced (and not so advanced) civilisations.

George du Maurier (1834–96) was a book illustrator and cartoonist best remembered for his work for *Punch*.[22] He also wrote three novels: *Peter Ibltson* (1891), *Trilby* (1894) and *The Martian* (published posthumously in 1897).[23] Both of the two earlier novels include some supernatural elements. *The Martian* is largely a conventional story that tells a fictionalised version of its author's own career. However, towards the end it is unexpectedly revealed that throughout his life the author has been unwittingly directed by a telepathic Martian. She had arrived on Earth about a hundred years earlier in a meteor shower and in the interim had tutored several earthlings.

[20] Steven J. Dick, *The Biological Universe: the Twentieth-Century Extraterrestrial Life Debate and the Limits of Science* (Cambridge: Cambridge University Press, 1996), pp. 401–10.

[21] Edgar Rice Burroughs, *A Princess of Mars* (Chicago: A.C. McClurg, 1917). Originally serialised under the pseudonym Norman Bean and title 'Under The Moons of Mars' in *The All-Story magazine*, beginning February 1912.

[22] Leonée Ormond, 'Du Maurier, George Louis Palmella Busson (1834–1896)', *Oxford Dictionary of National Biography*, Oxford University Press, 2004; online edn, October 2008, http://www.oxforddnb.com/view/article/8194, [accessed 21 August 2016].

[23] George Du Maurier, *The Martian* (London: Harper & Bros., 1897).

Latterly some of the Martians have taken an interest in directing suitably inclined earthlings towards an interest in and appreciation of higher things: aesthetics, philosophy *etc*. Du Maurier's Mars is an aged world that is nearing the end of the period during which it can support life. The amphibious seal-like Martians live near the equator, the only part of the planet still habitable.

Richard Ganthony's *A Message from Mars* (1899) was a popular and widely-performed play. Though it was first performed in 1899 the earliest reference to a printed version appears to be a revision from 1923.[24] Ganthony, collaborating with the novelist Mabel Knowles, later adapted it into a novel.[25] It also inspired both a spin-off by Owen Hall and a spoof by Mostyn T. Pigott, and was filmed three times, in: 1903 (New Zealand), 1913 (UK) and 1921 (US).[26]

The story concerns one Horace Parker, the 'most selfish man on earth'. He is visited by a messenger from Mars, who has come to show him the error of his ways, rather in the manner of Dickens' *A Christmas Carol*. The visitor has considerable psychic powers, and they are the means by which he journeyed to earth. The stage production was a notable success for the celebrated actor Sir Charles Hawtrey (1858-1923) who also starred in the 1913 film.

Finally *To Mars via the Moon* (1911) by Mark Wicks is an early space adventure.[27] The narrator designs and builds a spaceship, the *Areonal*, in which he and two companions travel to Mars. The book was partly intended as an introduction to astronomy for younger readers and early chapters present astronomically-accurate descriptions of the Sun, Moon and Mars. Once the explorers arrive on Mars its geography is entirely Lowellian; indeed the book is dedicated to Lowell. The Martian civilisation is also similarly Lowellian: peaceful and technologically and spiritually advanced, with all the usual attributes of a literary utopia. So far, there has been no connection with spiritualism, but it turns out that the Martians, or at least some of them, are reincarnated earth-humans, and the narrator meets his dead son.

Little is known of Wicks: his only other published work is on the entirely different

---

[24] Richard Ganthony *A Message from Mars* (New York: Samuel French, 1923).

[25] Richard Ganthony and Lester Lurgan, *Message from Mars* (London: Greening & Co, 1912). Lester Lurgan was a pseudonym for the novelist Mabel Knowles (1875–1949).

[26] Owen Hall, *The Silver Slipper: a Modern Extravaganza* (London: Francis, Day & Hunter, 1901); Mostyn T. Pigott, 'The Messenger from Mars', *The World* Christmas number, 1900 (London) 5-17; The British Film Institute has recently released a full restoration of the 1913 version: *A Message from Mars*, dir. J. Wallett Waller. See: http://player.bfi.org.uk/film/watch-a-message-from-mars-1913/ [accessed 21 August 2016].

[27] Mark Wicks, *To Mars via the Moon: an Astronomical Story* (London: Seeley & Co, 1911).

topic of making organs.[28] However, the astronomical descriptions and familiarity with Lowell's ideas clearly demonstrate a sound general knowledge of astronomy and the dedication and preface make it clear that he was familiar with Lowell's books.

Discussion

Ideas of interplanetary spirit travel and reincarnation on other planets seem odd, if not ridiculous, now. However, they have a long pedigree. Following the Copernican revolution of the sixteenth and early seventeenth centuries the Earth and the other planets became essentially the same sort of body, all orbiting the Sun. The question of whether the other planets were inhabited then arose naturally and a tradition of the 'plurality of worlds' was widely, if not universally, accepted: as all the planets were God's creation they were thought likely to be inhabited. Bernard de Fontenelle's *Entretiens sur la pluralité des mondes* (1686) in which a philosopher discusses astronomy with a *marquise* as they walk in the latter's garden under the stars effectively and eloquently spread the idea. [29]

The modern idea of communicating with the inhabitants of other planets begins with the Swedish mystic Emanuel Swedenborg (1688–1772) who amongst his voluminous writings reported conversations with spirits from all the then-known planets and described the nature of each of them.[30] Swedenborg's writings remained known and he was certainly an influence on some of the authors of astral journeys. A more direct influence, however, seems likely to have been Camille Flammarion. In addition to being an influential populariser of astronomy whose books were widely read he was a firm believer in spiritualism. In this respect he differed from Percival Lowell who consistently adopted a skeptical attitude to spiritualism and an entirely materialist approach to his Martian studies. Flammarion joined the *Societé Parisienne des Études Spirites* as a young man and later held office in it. Towards the end of his life he was also President of the Society for Psychical Research. He regularly

---

[28] Mark Wicks, *Organ Building for Amateurs. A Practical Guide for Home-workers* (London: Ward, Lock and Co. Ltd, 1887).

[29] Bernard le Bovier de Fontenelle, *Entretiens sur la pluralité des mondes* (*Conversations on the Plurality of Worlds*, 1686). Three English translations appeared within a couple of years of the French original. The second of these, by playwright and author Aphra Behn, has recently been republished: Bernard de Fontenelle, *A Discovery of New Worlds*, trans. Aphra Behn (London: Hesperus, 2012).

[30] Swedenborg's extra-terrestrial visitations were originally collected as *De Telluribus in Mondo Nostro Solari* (1758), which was largely extracted from his earlier *Arcana Caelestia* (1749–56). For a modern English translation see E. Swedenborg, *The Worlds in Space*, trans. J. Chadwick (London: The Swedenborg Society, 1997). For a general introduction to Swedenborg see, for example, G. Lachman, *Into the Interior: Discovering Swedenborg* (London: The Swedenborg Society, 2006).

contributed to spiritualist journals and pamphlets. These publications discussed re-incarnation on other planets and the idea features in his novels *Lumen* and *Urania*.[31]

The influence of Flammarion is obvious, for example, in Gustave Le Rouge's early space adventure *Prisoner of the Vampires of Mars* and its sequel.[32] The version of Mars that features in this story, particularly the description at the start of Chapter 8, might have been taken from one of Flammarion's astronomy books. Le Rouge does not use reincarnation or astral travel but does introduce his own variation: his hero travels to Mars in a physical spacecraft, but it is propelled by the psychic power of Indian mystics. There are also instances of telepathy in the novels.

Hélène Smith and Theodore Flournoy were both French-speaking Swiss. Flournoy was certainly familiar with Flammarion's work; he quotes from one of his books in *From India to the Planet Mars*.[33] He notes that Hélène Smith moved in spiritualist circles and that, though he did not know if she had ever read an astronomy book, she was certainly familiar with Flammarion's ideas about both astronomy and spiritualism, including reincarnation on other worlds. The question of communication with Mars was first discussed amongst Mlle Smith's group in 1892, but there was then a gap of two years before it resurfaced.

During 1894 the circle to which Mlle Smith belonged began holding seances at the house of one M. Lemaître, who expressed an interest in knowing 'what is happening upon other planets'. Mlle Smith was also introduced to a Mme Mirbel, a widow seeking to contact her recently deceased son, Alexis. During a seance held in October 1894 Hélène Smith relayed messages to Mme Mirbel from Alexis and a companion, Prof. Raspail. Both Alexis and Prof. Raspail were communicating from an unspecified location that was not discussed. The next seance was held the following month on 25 October. Alexis and Prof. Raspail reappear, but are now explicitly located as resident on Mars. The 1894 opposition of Mars occurred on 20 October, between these two seances. This opposition was well-placed for observation and was both anticipated and widely reported. Mlle Smith may well have read reports about it in either newspapers or periodicals and, remembering M. Lemaître's remark earlier in the year, inserted Martian elements into her communication for Mme Mirbel.

---

[31] Camille Flammarion, 'Lumen' in *Recits de L'Infini* (*Stories of Infinity*, 1872). A modern English edition is available: *Lumen*, trans. Brian Stableford (Middletown, CT: Wesleyan University Press, 2002). Camille Flammarion, *Uranie* (1889), trans. Mary J. Serrano (New York: Cassell, 1890).

[32] Gustave Le Rouge, *Le Prisonnier de la Planête Mars* (Paris: Méricault, 1908) and *La Guerre des vampires* (Paris: Méricault, 1909). A recent English translation of both novels is available: *Prisoner of the Vampire of Mars*, trans. David Beus and Brian Evenson (Lincoln, NE: University of Nebraska Press, 2015).

[33] Théodore Flournoy, *From India to the planet Mars*, pp. 139–53.

Following this episode Mlle Smith's Martian narrative does not reappear until February 1896 when it is fully developed.

The various accounts of spiritual journeys to Mars, both purportedly real and avowedly fictional, were influenced by the contemporary understanding of the planet to varying degrees, ranging from none to re-capitualisations of astronomical texts, complete with the reproduction of maps. However, where there is astronomical influence it always followed the popular accounts of Lowell or Flammarion. I am not aware any accounts that describe the rather more inhospitable planet that had become the consensus amongst astronomers by the early twentieth century, let alone the considerably more inimical world later revealed by robotic spacecraft. This failure to find the real Mars is further convincing evidence, if such is needed, against spiritualism.

While Mars was the most common destination for spiritual interplanetary travel it was not the only one. Other examples include Flammarion's own *Lumen*, where the recently deceased narrator travels to several external star systems, and the early German film *Algol: Tragedy of Power* (1920) which sees a visit from an extra-terrestrial, presumably from a planet orbiting the eponymous star.[34]

Astral journeys were not the only engagement between astronomy and spiritualism; astronomers feature amongst the membership of the Society for Psychical Research and its American counterpart, and some have served terms as President. A few have been more directly involved. For example Farie MacGeorge, Chief Observer at the Great Melbourne Telescope from 1870–73, resigned in order to be free to pursue his spiritualist interests.[35] Sir William Peck (1862–1925), the respected Director of the Edinburgh City Observatory on Calton Hill, was a member of the Golden Dawn who reportedly undertook astral journeys, as did other members of the Golden Dawn.[36]

Finally, this discussion stops at the end of the inter-war period, but interplanetary spiritual journeys continued post-war, merging into the UFO and 'new age' movements. As robotic spacecraft revealed an inhospitable solar system the destinations receded to planets orbiting other stars. The Aetherius Society will serve as an example.[37]

---

[34] Hans Werckmeister, *Algol. Tragödie der Macht* (*Algol: Tragedy of Power*, 1920).

[35] Richard Gillespie, *The Great Melbourne Telescope* (Melbourne: Museum Victoria, 2011).

[36] For brief biographical details of Sir William Peck see *Former Fellows of the Royal Society of Edinburgh 1783–2002 Biographical Index Part Two*, available online: https://www.royalsoced.org.uk/cms/files/fellows/biographical_index/fells_indexp2.pdf [accessed 21 August 2016]. For Peck and the Golden Dawn see Murphy Pizza and James R. Lewis, eds., *Handbook of Contemporary Paganism* (Leiden: Brill), p. 33. For the Golden Dawn and astral travel see Alex Owen, *The Place of Enchantment: British Occultism and the Culture of the Modern* (Chicago: University of Chicago Press, 2004).

[37] The Aetherius Society, http://www.aetherius.org/ [accessed 21 August 2016].

A Cosmic End and its Anthropological and Theological Implications

José G. Funes, S.J.

ABSTRACT: Motivated by curiosity about the end of the universe, I consider the cosmic end at different scales. The question of the end of the universe is one of the big questions which cut across human cultures. The scientific method is not the only approach for addressing such questions, but it is certainly an important one. We have quite a good picture of the early universe, but it remains a bit uncertain to predict its future scientifically. Our predictions depend on the different scales of time and space that we consider. Eventually, in the very distant future, the universe will be shredded. It is moving toward a final state in which it will be cold and dark. Obviously, this long-term scenario would be hostile to the existence of life. This prospect poses many questions at different levels: anthropological, philosophical and religious. The search for answers is what humanity has been doing for centuries, inspired by astronomical phenomena.

Human curiosity is the driving force for the scientific development, in which belief systems and philosophy still have their valid place. —Werner Arber[1]

To the scientist, and especially to the Christian scientist, corresponds the attitude to examine the future of humanity and the Earth. —Pope Francis[2]

Attracted by curiosity about the end of the universe, and encouraged by the words of Pope Francis about the importance of examining the future, I have considered the cosmic end at different scales, as far as this enormous task can be done for this paper.

There is a painting by Paul Gauguin with the title: *Where are we? Where did we come from? Where are we going?*[3] These are the big questions which cut across every human culture. The subject of the end of the universe is one of those big questions.

To address these questions the scientific method is not the only approach but certainly an important one. Scientific knowledge seeks the explanation of the observed phenomena based on natural causes. It should be able to explain the observed data

---

[1] Werner Arber, Preface to *Evolving Concepts of Nature*, Plenary Session of the Pontifical Academy of Sciences, 24-28 October 2014, available at http://www.casinapioiv.va/content/accademia/en/events/2014/nature.html [accessed 10 October 2016].

[2] Address of Pope Francis to the Plenary Session of the Pontifical Academy of Sciences on the occasion of the inauguration of the bust in honor of Pope Benedict XVI, 27 October 2014, available at https://w2.vatican.va/content/francesco/en/speeches/2014/october/documents/papa-frances-co_20141027_plenaria-accademia-scienze.html [accessed 10 October 2016].

[3] Museum of Fine Arts, Boston, MA, Artwork, Paul Gauguin, *Where Do we Come From? What are We? Where are We Going?*, available at http://www.mfa.org/collections/object/where-do-we-come-from-what-are-we-where-are-we-going-32558 [accessed 10 October 2016].

postulating a new model to predict results that must be verified with further observations. As Chris Impey has written:

> Science mostly answers the question of how things got to be the way they are. Yet if we stop at the present day, the job is only half done, as every good story needs an ending.[4]

We can only think about the past and the future of the universe from its present, i.e., from the local universe and from the data we have collected and interpreted in a theoretical framework. We verify our ideas about the beginning and the end of the universe with a reality check, i.e., confronting them with the experimental data. To survey the local universe, we have used telescopes for centuries. I like to think that we are a species with long eyes and that Galileo Galilei is the forefather of this people with long eyes.[5]

We have a good picture of the early universe. Let us follow T. S. Eliot's words 'In my beginning is my end'; in the initial conditions of the universe is written somehow its end. Though there are many unknowns, our current understanding of physics allows us to reconstruct the history of the universe since the universe was 10^{-43} seconds old.[6] We cannot go beyond this limit in our look-back time; there we arrive at the present limit of human knowledge.

It is uncertain to scientifically predict the future. Our predictions will depend on the different time scales that we take into account. Thus we can consider the end of Earth, of the Sun, of our galaxy and of the whole universe. In this paper I won't discuss any biological evolution or technology development that would exceed the goal of these considerations.

The beginning

If we ponder the cosmic end from the Earth to the universe, we should discuss the beginning and evolution of planets, stars and galaxies. Clearly this would be a gigantic effort, much beyond my abilities. Allow me to say that we have a good comprehension of the formation and evolution of planets, stars and galaxies.

In the last decades we have achieved a very solid foundation for the Big Bang the-

---

[4] Chris Impey, *How it Ends: From You to the Universe* (New York: W.W. Norton, 2010), p. 11.

[5] José G. Funes, Preface to *ASTRUM 2009, Astronomy and Instruments, Italian Heritage Four Hundred Years After Galileo*, ed. Ileana Chinnici (Livorno: Edizioni Musei Vaticani and Sillabe, 2009), p. 19.

[6] J.O. Bennett, M.O. Donahue, N. Schneider, and M. Voit, *The Cosmic Perspective*, 8th ed. (San Francisco: Pearson Education, 2016), p. 651.

ory, gathering observational, experimental and theoretical evidence to support the case for the standard Big Bang model, which is the best up-to-date explanation of the origin, evolution and current state of the universe. A key component of the standard Big Bang model is the hypothesis of inflation, an extremely rapid expansion of the early universe, introduced by Alan Guth and Andrei Linde, that is supported by observational evidence.[7]

A good amount of experimental data has been collected to confirm the case for the standard hot Big Bang that I briefly summarize here:[8]

1. *The expansion of the universe.* All galaxies are receding from us with a velocity which is proportional to their distance. This observational evidence was discovered by Edwin Hubble and it is known as Hubble's law.[9] Saul Perlmutter (Supernova Cosmology Project), Brian P. Schmidt and Adam G. Riess (High-Z Supernova Search Team) in 1998 discovered the accelerating expansion of the universe through observations of distant supernovae.[10]

2. *The Cosmic Microwave Background Radiation.* This is the radiation released when the universe was about 380,000 years old. It was detected by Arno Penzias and Robert Wilson in 1965 and observed by NASA's satellites COBE and WMAP, and recently by Planck, an ESA mission.[11]

3. *Abundances of elements.* The Big Bang model predicts the ratio of protons and neutrons during the period of nucleosynthesis. The chemical composition of the universe should be 75% hydrogen, 25% helium and a trace amount of other light elements. This prediction corresponds to observations of cosmic abundances.[12]

According to current observational evidence the universe is composed of 27% dark matter, 5% ordinary matter, and 68.3% dark energy, and has an age of 13.8 billion years.[13]

Taking into account current knowledge of the beginning and the present of the universe, we expect that, in a very distant future, the universe will continue to expand, going toward a final state of cold and darkness. But before discussing the future of the universe, I briefly discuss the cosmic future at different scales.[14]

---

[7] M.Y. Marov, *The Fundamentals of Modern Astrophysics* (New York: Springer, 2015), Chapter 11.

[8] Bennett et al., *The Cosmic Perspective*, p. 653.

[9] Bennett et al., *The Cosmic Perspective*, pp. 610–17.

[10] Bennett et al., *The Cosmic Perspective*, p. 685.

[11] Bennett et al., *The Cosmic Perspective*, pp. 653–57.

[12] Bennett et al., *The Cosmic Perspective*, p. 659.

[13] Bennett et al., *The Cosmic Perspective*, p. 687.

[14] In the presentation of the future at different scales I follow A. J. Meadows, *The Future of the Universe* (London: Springer, 2007); Impey, *How it Ends*; Bennett et al., *The Cosmic Perspective*.

The Future Earth: A Declining World

It is very challenging to predict the future of Earth since there are many factors that play an important role in its evolution. The Earth has been and will be affected by geological processes that determine the shape of the surface: impact cratering, volcanism, tectonics and erosion.

Geological activity depends on the fundamental properties of the planet, especially the size. Earthquakes release energy from the earth's crust through seismic waves. The core and mantle are changing continuously. In its final state, about 3 billion years from now, the Earth will be a cold solid body. At this point, there will be no earthquakes on the surface.[15]

Another factor to be taken into account is the change in the atmosphere and in the oceans. The Earth's surface has gone through periods of warmer and colder climates. The timescale on which major ice ages have occurred corresponds with the timescale for continental drift (about 400 million years).[16]

We also need to consider the dynamical evolution of the Earth-Moon system. Some calculations suggest that, within the next 4.5 billion years, the Sun would be overhead at the poles rather than the equator. Changes of this sort will alter considerably the distribution of temperature across the terrestrial surface.[17]

Prolonged periods of high volcanic activity can act as a trigger for climatic change. Also, the world will be warmer or colder depending on how freely equatorial currents can flow as the final layout of continents is approached. The winds too will adapt to the final continental pattern.

If the level of carbon dioxide in the atmosphere continues to rise, as at present, the average temperature in 2100 could be between 2 and 5°C above its current value. Cities around the globe that currently have moderate summer temperatures would be substantially hotter. At the same time, the higher temperatures could well lead to more violent hurricanes. As a consequence, in 2-3 billion years the Earth would be a declining world.[18]

In this timescale the Sun will considerably grow in size, turning into a red giant (about 100 times bigger than its current size). The Sun will expand out as far as the Earth's present orbit. This will happen some 8 billion years from now.[19] In its expansion, the Sun will blow off a considerable part of its mass into space. As a result, the

---

[15] Meadows, *The Future of the Universe*, Chapter 2.

[16] Meadows, *The Future of the Universe*, Chapter 3.

[17] Meadows, *The Future of the Universe*, Chapter 3.

[18] Meadows, *The Future of the Universe*, Chapter 3.

[19] Meadows, *The Future of the Universe*, Chapter 3 and Bennett et al., *The Cosmic Perspective*, p. 542.

Earth will move into a more distant orbit. By the time the Sun expands to the Earth's present orbit, the Earth itself will have moved out to nearly twice its current distance from the Sun.

Impacts on the Earth

Geological data shows that large impacts, which happened in the past, have caused mass extinctions. More recently, Earth experienced the smaller but well-documented entrance of the Chelyabinsk meteor in Earth's atmosphere over Russia on 15 February 2013.[20] This kind of event may occur every couple of years though the probability of major impact is very low in our life times.

The range of the impactor size could go from small bodies to large objects like asteroids and comets. At the moment no one knows for sure if with the current technology we would be able to destroy or divert a potential large object such as an asteroid or comet.

The Solar Neighbourhood

The next step is to consider the nearby stars and how they could affect life on Earth.[21] Taking into account the solar neighbourhood, a star would wander to within a distance of 3 light-years from the Sun every 100,000 years on average, and a nearby star could pass by about a light-year from the Sun. This is far enough away to have little effect on the Sun and planets. The closest approach to the Sun by any other star over the next few billion years is likely to be at a distance of about 10,000 AU. The gravitational pull of the star would disturb comets, especially the outermost ones. The shower of comets that would result may well increase the impact rate on the planets for a prolonged period after the encounter with the passing star.

A galaxy is a system of stars, gas, dust and dark matter gravitationally bound together with a total mass ranging from 10 million to 1000 billion times that of the sun.[22] Gas and dust are the material between stars, and it is called interstellar medium. This is the material from which new stars form.

Collisions between stars may be unlikely in the solar neighbourhood but collisions between stars and interstellar clouds are possible. The Sun is likely to encounter a molecular cloud once or twice every billion years.[23] The gravitational pull of the

---

[20] *New York Times*, available at http://www.nytimes.com/2013/02/16/world/europe/meteorite-fragments-are-said-to-rain-down-on-siberia.html?_r=0 [accessed 10 October 2016].

[21] Meadows, *The Future of the Universe*, Chapter 7.

[22] Marov, *The Fundamentals of Modern Astrophysics*, Chapter 10.

[23] Meadows, *The Future of the Universe*, Chapter 7.

cloud as the Sun moves through it would disturb comets, causing a major shower of comets into the inner regions of the solar system.

In addition, within a molecular cloud, the solar wind will be almost entirely suppressed, and the Earth will be surrounded by interstellar material. This may affect the upper atmosphere, and also what happens at the Earth's surface. It could affect the ozone layer and the amount of sunlight that the Earth receives. Encounters with dense clouds could affect longer-term future of the Earth.

Any supernova that occurs within 150–200 light-years of the Sun is likely to have a noticeable effect on the solar system.[24] The particles and radiation from a nearby supernova could produce significant, though temporary, changes in the Earth's atmosphere. Examination of the radioactivity in meteorites suggests that cosmic-ray bombardment of the solar system originated by a supernova becomes stronger every 100–200 million years.

At present, the atmosphere and the magnetic field of the Earth protect us from the direct effects of cosmic rays. Cosmic rays can interact with the Earth's upper atmosphere, altering the ozone layer and allowing through a flood of ultraviolet light. It also seems that an abundance of cosmic rays could affect the greenhouse effect lowering temperatures at the terrestrial surface.

A supernova also produces gamma rays that would cause cellular damage to all land creatures within thirty light-years and inflict destruction on the food web. An even more dramatic stellar cataclysm could disrupt genetic material at a distance of a thousand light-years. Luckily, violent star death is rare.

Galaxy Future

Galaxy formation and evolution is a complex combination of hierarchical clustering, gas dissipation, merging, and secular evolution.[25] Galaxies are tracers of cosmic evolution over the last 13 billion years. Galactic time scale is the combination of two clocks. One clock is the cosmological time scale (the Hubble time, basically the age of the universe) and the other time scale is related to stellar evolution. The combination of both gives rise to galaxy evolution.

Our knowledge of galaxy evolution showcases our understanding of cosmology, stellar evolution and galaxy dynamics. It is an excellent example of how scientific knowledge, achieved independently, can be put together to shed light on a complex process that involves other physical processes at different scales.

[24] Meadows, *The Future of the Universe*, Chapter 7.

[25] J.G. Funes, 'Galaxy Evolution', in *Proceedings of Scientific Insights into Evolution of the Universe and Life, Pontificia Academia Scientiarum Acta* 20 (2009): pp. 91–98, here p. 91.

As we know from Hubble's law, galaxies move farther apart over time. Meanwhile stars are born and die within the galaxies. They are born from the gravitational collapse of gas clumps in molecular clouds. Massive stars, when they die, enrich the gaseous environment with new and heavier elements. Eventually this gas cools off in molecular clouds completing the star-gas-star cycle. Looking to the far future, in a trillion years, the cycle will be broken as more gas is trapped in stellar corpses (white dwarfs, neutron stars and black holes) and less is left over to form new stars.

In each and every galaxy, the lights will gradually go out. In 10^{100} years (10^{90} times the age of the universe) clusters of galaxies will become clusters of black holes; finally black holes will evaporate. The Milky Way will be one of those galaxies fading into darkness.[26]

The Fate of the Universe

According to our current comprehension of the universe, dark energy seems to be the driving force for the accelerated expansion of it. If this is the case and dark energy does not change with time and there are no other factors, in the very distant future the universe eventually will be shredded. This final stage of the universe is known as the *Big Rip*.

Thus the universe is going toward a final state of cold and darkness, thermal death, which says that the universe will go toward a state of maximum entropy (*Big Freeze*). The long-term scenario, with everything in the universe gradually dying, is obviously hostile to life.

Emptiness and Questions: Anthropological Implications

The scientific study of the cosmic end raises the question of the meaning of human existence and the purpose of the universe. Using the poetic language of T. S. Eliot, the end of the universe can be metaphorically described as it follows:

> This is the way the world ends
> This is the way the world ends
> This is the way the world ends
> Not with a bang but a whimper.

The process of scientific research on the temporal end of the universe is also a spiritual journey to the final frontier of our existence. Looking at a not very bright perspective for life, we may experience what Friedrich Nietzsche sums up effectively in few words:

---

[26] Meadows, *The Future of the Universe*, Chapter 9; and Bennett et al., pp. 687–90.

When you look long into an abyss, the abyss looks into you.

Similarly, we could feel emptiness in front of the vastness of the cold and dark universe in its final stage, as the sacred author of the book of Ecclesiastes sees the fragility and contingency of this world:

A vast emptiness – Qoheleth says – an immense void, everything is empty.[27]

Some issues that once were considered reserved for philosophical or religious speculation today can be resolved through scientific knowledge. As Louis Caruana points out, one of the crucial tasks of philosophy is to build conceptual bridges between the scientific image and the everyday image, i.e., the more immediate image of the world.[28] Moreover, as Caruana affirms, science regarding the remote future is extremely anti-anthropocentric because it exposes the utter insignificance of human beings in the vastness of the universe.

If this is one of the tasks of philosophical research, we can raise the question about the place of humans in cosmic eschatology and about the purpose of the universe.[29] In this context it would be very useful to examine the anthropic principle, with differences in its strong and weak formulation. It states that any valid theory of the universe must be consistent with the existence of human beings or life.[30]

Given the current understanding of the early and present universe, i.e., that we live in an accelerated expanding universe, it is expected that in the very remote future of trillions and trillions of years, the universe will continue to expand to a final state of cold and darkness.

While the anthropic principle does not refer explicitly to the eschatological discussion, we can investigate the relationship between the future of the Earth and humanity on one side, and the future of the universe on the other hand. With regard to the connection between the fate of humanity and the future of life, it is illuminating what Martin Rees says:

---

[27] Ecclesiastes 1:2, a free translation according to a note to Eccl. 1:2 of the *New American Bible*, revised edition (Iowa Falls, IA: World Bible Publishers, 2011).

[28] L. Caruana, *Il fine e la fine dell'universo: scienza, interpretazione, simbolo, e analogia*, public lecture at the Gregorian University, Rome, 31 January 2014, private communication.

[29] Eschatology (from Greek *eskhatos*, last) is the branch of theology or biblical studies concerned with the end of the world. Here I use the word in a broad meaning referring to science, culture and philosophy.

[30] José G. Funes, 'La imagen actual del universo y algunas de sus implicaciones filosóficas', *Revista Portuguesa de Filosofia* 58 (200): p. 953.

The most crucial location in space and time (apart from the big bang itself) could
be here and now.[31]
The wider cosmos has a potential future that could even be infinite. But will these
vast expanses of time be filled with life, or as empty as the Earth's first sterile seas?
The choice may depend on us, this century.[32]

It is very difficult to make any statements or assumptions about the future of life in
the universe. Though we know from direct observation on Earth that life is resilient;
it has a remarkable ability to adapt and evolve in hostile conditions. Life could have
spread elsewhere in the universe or in other universes, if the hypothesis of the existence
of multiverse is confirmed. The end of the universe predicted by cosmology
raises other questions about life:

1. If our location in the universe is crucial for life, will life end with Earth?
2. Is life a common phenomenon?
3. What about life in trillions and trillions of years when the universe fades?
4. If there are other universes, could there be life in those places?

Because of the nature of scientific concepts addressed in this paper, it is also necessary
to point out some issues that are at the basis of the discussion of the human role
in the future of the universe.

1. Can science, cosmology in our case, make 'predictions'?
2. What are the epistemological limits of these 'predictions'?
3. How to interpret these 'predictions' in the framework of scientific realism?
4. If we consider the temporal end of the universe, does this have anything to do
with its purpose?

Speaking of the future we are assuming a concept of time. What is the connection
between the cosmological time and anthropological time?[33] Since the dawn of humanity
cosmological time and cultural time have been closely linked.

Does our understanding of the cosmic future have any impact on our conception
of humanity? What impact has there been and is there in human thought due to the
scientific and religious cosmic-end stories?

These are some of the open questions that we would like to address in future
studies.

---

[31] Sir Martin Rees, *Our Final Hour: A Scientists's Warning* (New York: Basic Books, 2004), Chapter 1.
[32] Rees, *Our Final Hour*, Chapter 14.
[33] A. Frank, *About Time: Cosmology and Culture at the Twilight of the Big Bang* (New York: Free
Press, 2011).

Theological Implications

The question of the future of the universe is inextricably linked to the question of God, as Gonzalo Zarazaga points out:

> There is no field of knowledge or reality, whose ultimate reality and rationality does not imply in any way the question of the origin, purpose and rationale, i.e., the question of God as the ultimate foundation that determines all reality in history.[34]

I have mentioned that one of the tasks of philosophical research is to build bridges between the scientific image and the everyday image of the world. A similar task for theological studies should be to build bridges between the image of God that comes from faith and the everyday image of God.

We may wonder if our understanding of God and his relation to the world can benefit. Is cosmology somehow relevant to the religious eschatology? What and how can theological research learn from scientific eschatology?

We would also like to explore the connection between the scientific interpretation of the end of the universe and its philosophical and theological interpretation. A good scientist in the search path of truth must remain open to the interpretation of reality being aware that scientific knowledge is incomplete just as the philosophical and theological thoughts are, as Pope Francis has said addressing theologians and philosophers:

> The theologian who is satisfied with his complete and conclusive thought is mediocre. The good theologian and philosopher has an open, that is, an incomplete, thought, always open to the maius of God and of the truth, always in development...[35]

There are many issues in science that are 'incomplete' in many respects, and this paper is also incomplete; however I intend to address the cosmic future also from a Christian approach as a key to understanding the history of the universe. Joseph Ratzinger's words are enlightening in this regard:

> If the cosmos is history and if matter represents a moment in the history of the spirit, then matter and spirit are not forever next to each other in a neutral manner,

---

[34] G. Zarazaga, 'Hablar de Dios en el nuevo escenario científico y cultural', *Teología* 52, no. 118 (2015).

[35] Address of Pope Francis to the Community of The Pontifical Gregorian University, 10 April 2014, available at http://m.vatican.va/content/francescomobile/en/speeches/2014/april/documents/papa-francesco_20140410_universita-consortium-gregorianum.html [accessed 10 October 2016].

it is necessary to consider one last 'complexity' in which the world finds its Omega and unity. Then there is one last link between matter and spirit in which the destiny of man and the world finds compliance, even if today we cannot define the type of such a connection. Then the 'last day' is one in which the fate of each man will be fulfilled because the fate of humanity has found fulfillment.[36]

I have discussed the 'Last Day' from the scientific point of view pointing out its anthropological and theological implications. More than answers there are many open questions still ahead. I believe the search for answers makes us humans. This is what humanity has been doing for centuries, inspired by astronomical phenomena.

---

[36] Joseph Ratzinger, *Introduzione al Cristianesimo* (Brescia: Queriniana, 2010), pp. 347–48.

THE PHOTOGRAPHIC PLATE ARCHIVE AS AN INSPIRATION FOR ART PROJECTS

Michael Geffert

ABSTRACT: This article describes art projects with historical photographic plates which are preserved in the Argelander-Institut für Astronomie der Universität Bonn. The plates are the scientific observations of astronomers from Bonn taken over nearly a hundred years. Since these observations were done in the last century, their scientific value today is limited. Non-scientific activities based on the old observations are presented. Due to the interest of artists, some years ago the idea was born to use these plates also for projects on the border between modern art and astronomy. Here we describe the origin of astronomy and the origin of the plate archive in Bonn. We also give an overview of the complete historical collection, to which this archive belongs today. Furthermore, information and ideas of some projects related to astronomy and art are illustrated.

The History of Bonn Astronomy

The first astronomical observatory in Bonn was established in 1845. With financial support from the King of Prussia, Friedrich Wilhelm IV, F.W.A. Argelander (1799–1875) built an observatory with telescopes for visual observations near the centre of the town. The most important project of astronomers in Bonn at that time was the 'Bonner Durchmusterung' (BD), which was a survey of all stars up to a limiting magnitude of $V=10^{m}$.[1] Argelander and his successor E. Schönfeld (1821–1891) published data on about 450,000 stars. The astronomers from Bonn used a simple meridian circle method to determine the positions of the stars. The telescope for this investigation was a refractor (D=7.7cm, f=65cm) from Joseph von Fraunhofer. Maps of the BD were printed by lithography.

Variable stars, distance determinations and reference catalogues represent the other main activities of the astronomers at that time. Due to lucky circumstances, a few of the first observations from 1846 to 1853 are documented. Some years ago, the astronomical diaries of Julius Schmidt (1825–1884), the first collaborator of Argelander in Bonn, were found. These books contain a large number of sketches and drawings of Schmidt's astronomical observations. The books of Schmidt and the print plates of the BD today are good candidates, being not only historical material but also in some respect art objects.

Due to the photographic techniques, and also due to the development of spectroscopic measurements, the orientation of astronomy in Bonn changed after 1900. Stellar classification, the determination of radial velocities, and participation in the

---

[1] F.W.A. Argelander, *Bonner Durchmusterung des Nördlichen Himmels* (Bonn: A. Marcus und E. Weber's Verlag, 1863).

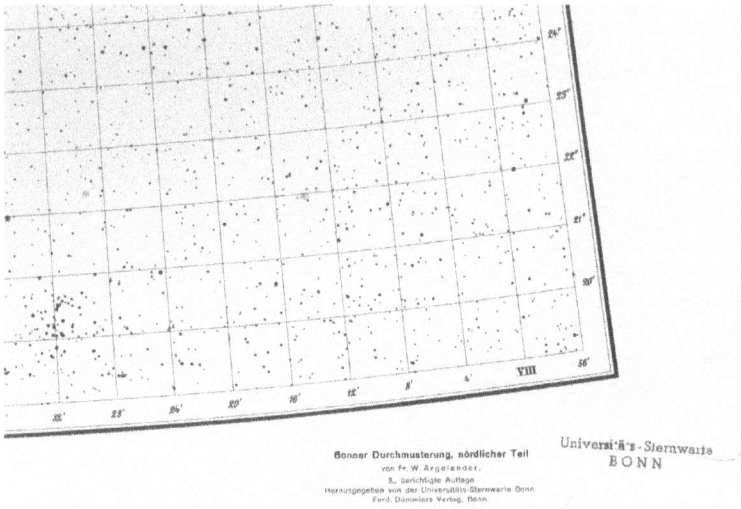

FIGURE 11.1. Map of the Bonner Durchmusterung. Sammlung Historischer Himmelsauf-nahmen der Universität Bonn.

AGK2, a reference catalogue for the determination of positions and proper motions of stars, proved to be the main topics of astronomy in Bonn. After the Second World War, the light pollution in the town hampered further optical observations in Bonn. A new optical observatory (Observatorium Hoher List) was built in the countryside about 80km to the south of Bonn.

At the same time, a new era of astronomical research in Bonn was established with radio astronomy. New telescopes (a 25m telescope on Stockert hill and later a 100m telescope in Effelsberg) started working in 1956 and 1971. With the foundation of the Max-Planck-Institut für Radioastronomy, radio astronomy became one of the main topics of astronomical research in Bonn until today. In addition, theo-

FIGURE 11.2. Observatorium Hoher List in 1966. Sammlung Historischer Himmelsaufnah-men der Universität Bonn.

FIGURE 11.3. The dome of the astrograph (right) at Observatorium Hoher List. Photo: Michael Geffert, Bonn.

retical astrophysics, cosmology and satellite observations on other wavelengths are other important topics of the astronomers of the Argelander-Institut für Astronomie (AIfA) of Bonn University.

Optical observations done in Bonn stopped with the closure of Observatorium Hoher List in 2012. Instruments, clocks, photographic plates and books from the early days of optical astronomy remained and were put – unfortunately not in good order – in some of the rooms of the AIfA. About 15,000 photographic plates with dates ranging from 1900 to 1990 are the main part of this historical material in Bonn.

The Plate Archive of University of Bonn

The first photographic observations were taken in the city of Bonn between 1899 and 1922 with a refractor (D=0.3m; f=5.1m). The enormous efforts of Karl Friedrich Küstner (1856–1936) resulted in 600 photographic plates of excellent quality up to a limiting magnitude of B=16. Already Küstner had the idea, that these plates could be valuable for proper motion work of a future generation of astronomers[2], which was realised after 1980.

However, the major part of the photographic plates of the archive of Bonn was taken between 1952 and 1990 at the Observatorium Hoher List. In this new observatory astronomers used the refractor from Bonn, partly a new 1m Cassegrain telescope and two smaller telescopes for photographic observations. With a new Schmidt telescope (f=1.3m, D=0.36m) and an older astrograph (f=1.5m, D=0.3m)

---

[2] Friedrich Küstner, 'Der kugelförmige Sternhaufen Messier 56', *Veröffentlichungen der Universitätssternwarte zu Bonn* 14 (1920).

FIGURE 11.4. Observation of NGC 7000 with the astrograph at Observatorium Hoher List. The size of the plate is 16cm x 16cm, representing an area of 60 x 60 of the sky. Sammlung Historischer Himmelsaufnahmen der Universität Bonn.

they worked on the astrometry of solar system objects, variable stars, star clusters and the structure of the Milky Way.

The major part of the archive in Bonn contains more than 9000 plates from these two telescopes. The plates, of sizes up to 24cm x 24cm, were thin (1-2mm) glass plates. Since these plates represent negatives, the stars appear as black points on transparent emulsions on the glass plates. The limiting magnitude of some of the plates reaches 18.^{m} One single plate could show more than 50,000 stars. Since measuring engines at that time were not able to digitize and reduce photographic plates within an appropriate time, many of these large field observations (up to 10^0 x 10^0) were analysed only in a very preliminary way.

'Sammlung Historischer Himmelsaufnahmen' and Plate Archive in Bonn

The photographic plates now are part of the 'Sammlung Historischer Himmelsaufnahmen der Universität Bonn'. Meanwhile, the complete historical material of the astronomy in Bonn University belongs to this collection. It is one of twenty collections of the University of Bonn. With these collections, the university is interested

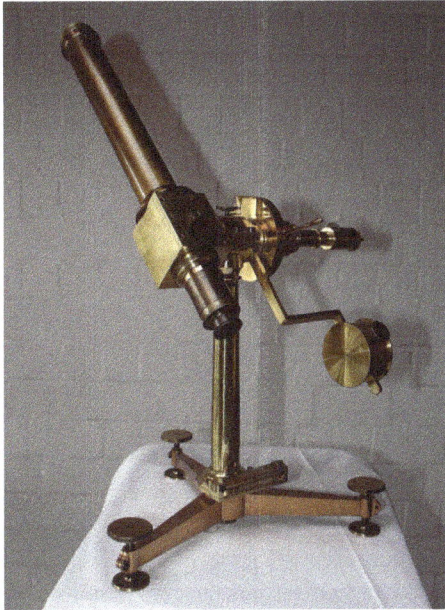

FIGURE 11.5. Zöllners photometer for measuring the brightness of stars. Sammlung Historischer Himmelsaufnahmen der Universität Bonn.

in increasing the value of the historical material. The complete material of our collection is based on the following elements:

* The photographic plates taken in Bonn, Göttingen and at the Observatorium Hoher List.
* A few historical instruments and clocks ranging from small transit instruments up to a refractor of D = 0.16 m, f = 1.95m.
* The astronomical diaries of Julius Schmidt.
* Further historical material like the stones of the BD, and maps from the Carte du Ciel project.

Unfortunately, at this time there is no inventory of all these objects. Plates and instruments are distributed over several rooms in the AIfA. Moreover, there exists no place to install an exhibition of at least some of these objects. Only recently have we been able to separate an area of the library of the AIfA, where some of the telescopes were installed, and where objects may be catalogued and presented to the public. At the moment the future of the collection is safe only for the next few years. From 2019 a solution has to be found, so that the objects of the 'Sammlung Historischer

Himmelsaufnahmen' may find a permanent place, perhaps in a museum or another public institution. This unfavourable situation had some consequences for the users of the plate archive.

Normally, owners of a plate archive are mostly interested in a digitization of their plates. Although we also are in favour of a complete digitization of the books and the plates, we think that our procedure would be more appropriate. Our first goal is to bring the complete historical material to the attention of the University and the public in order to find a permanent solution for preserving the material. Then, digitization would be the next important step. Moreover, due to other activities in public outreach (e.g., public talks, LAB courses for students from schools, teacher training in astronomy) there is not enough time left for a large digitization project.

For this reason, we have chosen the following activities for our material, especially our plate archive:

* **Scientific use of the plates.** The plates from Bonn taken by Küstner are primarily most suited for proper motion work. In addition, series of about 100 plates exist for several fields of OB associations taken with the astrograph at Observatorium Hoher List, which are interesting for the investigation of the long-term behaviour of variable stars.
* **Didactics.** Scans of photographic plates may be used for the work with students from schools. Students from schools determine light curves of variable stars as well as colour magnitude diagrams and high proper motion stars. In the last three years about 50 students performed a LAB course in the AIfA using scans of the photographic plates. Results (e.g. light curves) were published in astronomical journals.
* **Public outreach.** Astronomical images very often are interesting for people, but they also require explanation. In the last three years, in about sixty public talks, some of plates and their scientific 'history' were taken as a topic of a public lecture.

Artists have another view on the historical astronomical material. For them it is not a problem that these objects scientifically have no meaning anymore. They are more interested in the appearance, the shape or material of such objects. Moreover, the historical meaning is also an interesting aspect. Realising the interest of artists in astronomical history and observations, we started a series of exhibitions based on astronomy, the history of astronomy and modern art. Objects of our collection were presented together with paintings, art books and other art-work in exhibitions. For these events, we chose exhibition rooms, especially for art objects. The general idea of such an exhibition is to start a dialogue between three disciplines which normally are separated. With these events we follow also the idea to bring astronomy in contact with people who normally are not in contact with astronomy.

FIGURE 11.6. Küstner 10. Installation art with plates of the moon at the Observatorium Hoher List, 2010. Photo Michael Geffert.

Moon and Meridian
First Exhibitions at Observatorium Hoher List

First attempts of interdisciplinary projects began at the Observatorium Hoher List in the year 2010. During a first review of all the plates taken with the refractor from Bonn, plates of the moon taken by Küstner from 1900 to 1904 were found in the archive. Observations of the moon at that time required very calm air due to the brief observation times of a few seconds. Since observations of the moon for a longer time have not been of scientific interest, they were nearly forgotten. These plates therefore are ideal objects for a transformation of scientific material to exhibits. The presentation of these plates together with paintings of the moon by the artist Helmut Tholen from Cologne was the beginning of a confrontation between photographic plates (history of astronomy) and modern art. Two videos of the plates of the moon completed the exhibition.[34]

Another part of the exhibition was related to the 'meridian'. In former times astronomers used meridian circles – together with accurate celestial coordinates – for a determination of either their exact positions on earth or the accurate determination of time. At that time in Bonn, the clocks of the railway station were adjusted by astronomers. A meridian circle connects the coordinate systems of earth and sky. From the last meridian circle of Bonn only parts are preserved, which were found

---

[3] M. Geffert, ' Der Mond – Küstners vergessene Aufnahmen', 2011, available at https://www.youtube.com/watch?v=x_muV6rXkBQ [accessed 8 February 2016].

[4] M. Geffert, 'Der Mond am 26. März in Bonn', 2013, available at
https://www.youtube.com/watch?v=vQ-i9vMpMaw [accessed 8 February 2016].

FIGURE 11.7. Meridian. Installation art at Observatorium Hoher List, 2010. Photo Michael Geffert.

in the cellar of the institute. We were not able to install completely the instrument, but presented the important parts of the telescope. The aim of the arrangement of tube, micrometer and lens is to provide the visitors with an impression of a sculpture of modern art. Three woodcuts of a portrait of Argelander were printed on a publication of meridian circle measurements. Argelander himself was an enthusiastic meridian circle observer.

The Exhibition 'Dialog der Sterne'

As part of the outreach activities of the Sonderforschungsbereich 956 (Star formation), in 2013 an interdisciplinary project between artists and scientists was launched. A first aim of this collaboration was a common exhibition of astronomers and artists. The project started with discussions where astronomers explained astronomical research. The artists answered using their 'language'. Astronomical objects, e.g. a model of the Herschel satellite, along with paintings and other works of art were presented in a common exhibition.

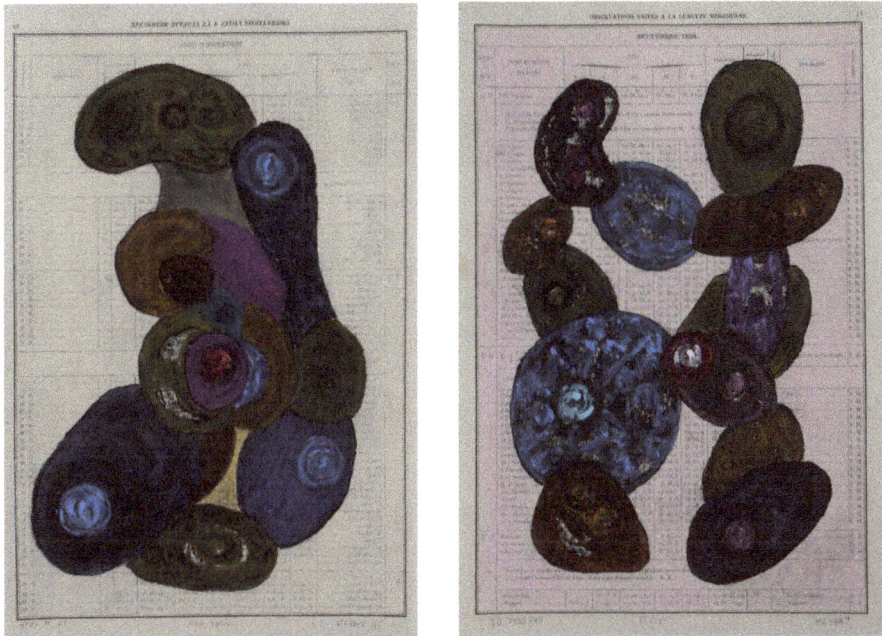

FIGURE 11.8. Gliese 687(left) and 61 Cyg (right), art works in the 'Dialog der Sterne', 2015. Photo Michael Geffert.

The 'Sammlung Historischer Himmelsaufnahmen' also participated in this project. The books of Julius Schmidt with the drawings from 1846 and the photographic

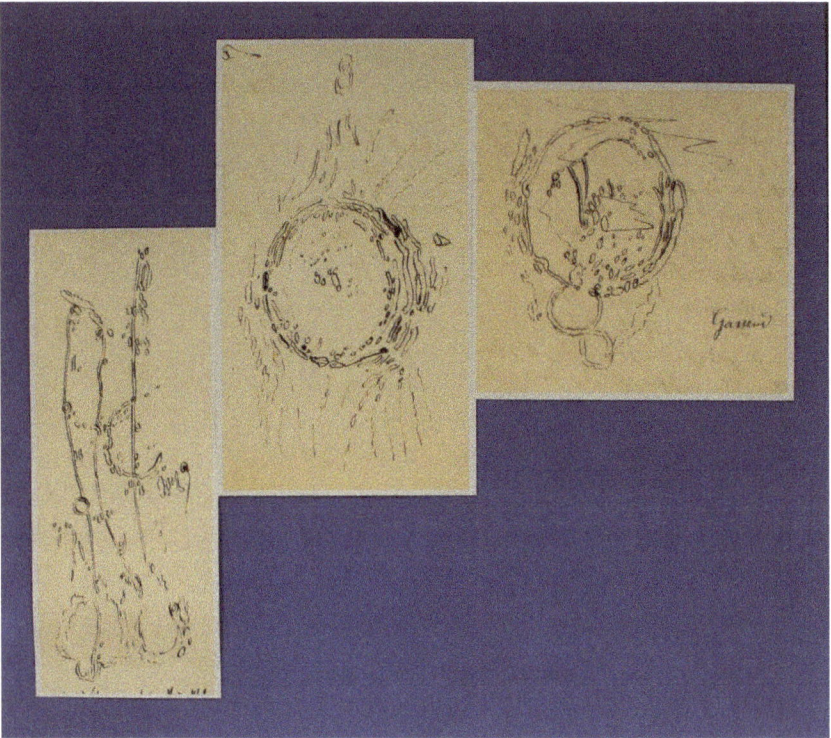

FIGURE 11.9. Julius Schmidt's drawings of details of the surface of the moon from 1846 in Bonn. Photo: Michael Geffert.

material led to new art-objects, which were part of the exhibition. The common exhibition was presented in March 2015 in the rolling mill in Pulheim near Cologne. In this place, normally exhibitions of modern art (e.g. 'ART-Pulheim') were shown. In addition to the exhibition, a large program of public talks, one concert with music of the astronomer and musician W. Herschel (1738-1822), astronomy lessons for school classes took place. Visitors of the exhibition were confronted with art and astronomy. The aim was, to include them in the discussion between the two disciplines.

The Astronomical Diaries of Julius Schmidt

In the nineteenth century, scientists had to use drawings for their research, since photography was not yet developed. Especially in astronomy, the form and shape of extended objects like comets was published using artistic techniques. Julius Schmidt, born 1825 in Eutin, came to Bonn in 1846. In 1853 he moved to Olmütz and later became the director of the Athen observatory. Some years ago his astronomical diaries

Figure 11.10: Constellation invented during an astronomical project with children. Photo Michael Geffert.

from 1843 to 1853 were found in the AIfA. Schmidt created a lot of small drawings and sketches from his observations of the planets, comets, the moon and other objects, especially in his early books.

Since his drawings also represent the first astronomical observations in the observatory in Bonn, we presented the books and the drawings in an exhibition at the Argelander Institut. Although Schmidt's sketches very often look like simple exercises, the large number and the historical meaning of these drawings allow them appear as works of art.

Constellations

Stellar constellations are ideal objects to motivate naked eye observation and orientation of the sky. With the help of constellations, one may find interesting astronomical objects for observation with binoculars or a small telescope.

Constellations are also beautiful artworks of human fantasy. During this INSAP

IX meeting we saw nice examples of constellations from different countries and cultures. It is interesting to see how people invented figures in the sky based on their own surroundings. Such experiments can be easily done with children of all ages. During our astronomy projects, the participants sometimes have to invent their own constellation from paper sheets with black points, without knowing the classical meaning. It is fascinating how the problem of finding an image based on a few black dots can even be solved by very young children.

Conclusion

Although photographic plates mainly do not have a scientific meaning today, they are valuable objects from earlier epochs in astronomy. It is important to find the right context, where photographic plates and other historical material are treated like exhibits. Art projects are one possibility to consider the historical material in another way than scientifically. Well presented, the plates could fascinate people by their appearance alone. It is our experience that artists have a great interest in such plates and also in common projects with astronomers.

'DANCING WITH THE STARS': ASTRONOMY AND MUSIC IN THE TORRES STRAIT

Duane W. Hamacher, Alo Tapim,
Segar Passi and John Barsa

ABSTRACT: Song and dance are a traditional means of strengthening culture and passing knowledge to successive generations in the Torres Strait of northeastern Australia. Dances incorporate a range of apparatuses to enhance the performance, such as dance machines (*Zamiyakal*) and headdresses (*Dhari*). The dances, songs, headdresses and dance machines work together to transfer important knowledge about subsistence survival, social structure, and cultural continuity. This paper explores how celestial phenomena inspire and inform music and dance.

Introduction

The Torres Strait is an archipelago of approximately 275 islands stretching between Cape York Peninsula on mainland Australia and Papua New Guinea, of which seventeen are inhabited by communities. The islands are grouped into five major regions: eastern, central, northern, western, and southwestern (FIGURE 12.1).

Torres Strait Islanders are a Melanesian people indigenous to Australia.[1] As a seafaring people, their traditions contain a significant astronomical component. This is reflected in their traditional laws, creation stories, ceremonies, and daily activities. It is also reflected in the Islander flag (FIGURE 12.1 inset) as a five-pointed star. The five points represent the Strait's five major island groups, and the star itself represents an unspecified navigational star. Two major indigenous languages are spoken: the Papuan *Meriam Mir* language in the eastern islands, and the Austronesian *Kala Lagaw Ya language* (and various dialects) in the central, northern, western, and southwestern islands. Each of the five major island groups have distinct customs, but share commonalities in culture and tradition.

Much of the early ethnographic literature on Islander cultures comes from the London Missionary Society, which established missions in the Torres Strait in the 1840s, and two major ethnographic expeditions from Cambridge University, led by Alfred Cort Haddon in 1888 and 1898.[2] The Haddon expeditions focused on Mer and Mabuyag islands.

---

[1] Nonie Sharp, *Stars of Tagai* (Canberra: Aboriginal Studies Press, 1993).

[2] Alfred Cort Haddon, *Cambridge Anthropological Expedition to the Torres Straits*, 6 vols. (Cambridge: Cambridge University Press, 1901–1935), Vol. 3.

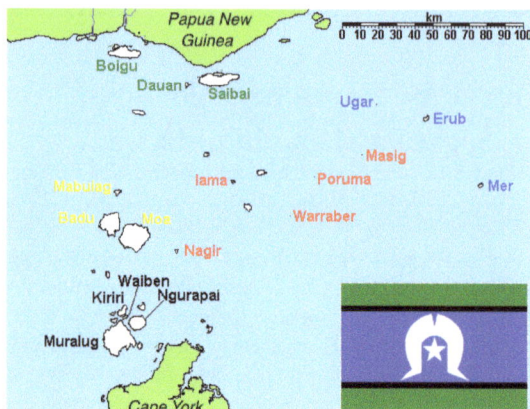

Figure 12.1. Map of the Torres Strait with labels on inhabited islands. Eastern (blue), central (red), northern (green), western (yellow), southwestern (black). Image: Wikipedia. Inset: The Torres Strait Islander flag, developed by Bernard Namok of Waiben in 1992. Green represents the landmasses of Australia and PNG, blue represents the ocean, black represents the people, the headdress is a dhari, and the star represents a navigational star, with each point representing the five major island groups. The colour white represents peace and the Islander's conversion to Christianity.

Astronomy in Torres Strait Traditions

In Islander cultures, ways of being and ways of knowing are closely linked to the stars.[3] This relationship encompasses a sense of identity and belonging to their environment, to the peoples' understanding of themselves, and to their culture and traditions.[4] This sense of belonging links the past, present, and future into a holistic system of knowledge that developed over thousands of years. This knowledge informs Islander traditions: the laws, customs, and practices that are recorded and handed down in the form of story, song, dance, ceremony, art forms, and material culture.[5]

Islander astronomical knowledge also contains practical information about the natural world and encompasses how Islanders perceive and understand the environment in which they live.[6] This is essential for survival and cultural continuity, and is integral to the everyday lives of Islander people. Islander understandings of identity, for instance, are linked to Tagai – the creation deity that is represented by a

---

[3] Sharp, *Stars of Tagai*.

[4] Martin Nakata, 'The Cultural Interface of Islander and Scientific Knowledge', *Australian Journal of Indigenous Education* 39 (Supplement) (2010): pp. 53–57.

[5] Jeremy Beckett, *Torres Strait Islanders: Custom and Colonialism* (Cambridge: Cambridge University Press, 1987).

[6] Nakata, 'The Cultural Interface of Islander and Scientific Knowledge', pp. 53–57.

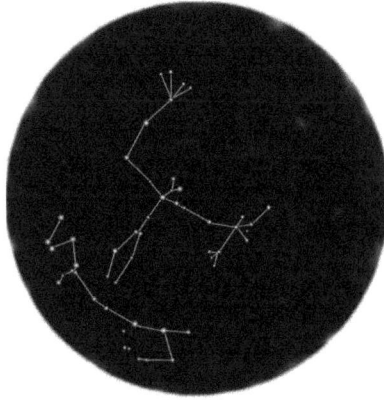

FIGURE 12.2. The stars of Tagai, spanning across the night sky. This image is used as the logo for the Tagai State College primary and secondary schools in the Torres Strait. Image: Wikipedia.

constellation of stars that spans across the sky.[7]

The story of Tagai (aka *Thoegay*) varies across the Strait, but the general theme is that Tagai was a great warrior and fisherman. He had a crew of twelve men called *Zugubals* and his confidant and 'first mate', *Kareg*. The group went fishing on a reef but were unsuccessful in catching any fish. Tagai left to search the reef for a more suitable spot. While he was gone, the Zugubals grew hungry, tired, and frustrated in the heat. They ate all of their food and drank all of their water. They foolishly decided to consume Tagai's food and water. When Tagai returned, he was furious. He tied the men into two groups of six and cast them into the sea, where they drowned. They went into the sky as two groups of stars: the *Usiam* (the Pleiades) and *Utimal/Seg* (belt and scabbard stars of Orion).

Tagai and Kareg went to the other side of the sky to keep away from the Zugubals (this is notably similar to the Greek traditions about Orion and Scorpius being placed on opposite sides of the sky by the gods to keep them away from each other after a battle).

Tagai is a very large constellation spanning across the sky (FIGURE 12.2).[8] His left hand is the Southern Cross (Crux) holding a spear. His right hand is the Western constellation Corvus holding a species of Eugenia fruit. His head and body consist of stars in Centaurus and Lupus. He is standing on the bow of his canoe, traced out by the body of Scorpius. His first-mate, Kareg, is the star Antares at the stern of the

---

[7] Sharp, *Stars of Tagai*.

[8] Peter Eseli, *Eseli's Notebook*, ed. Anna Shnukal and Rod Mitchell, (University of Queensland: Aboriginal and Torres Strait Islander Studies Unit Research Report Series, Vol. 3, 1998), p. 66.

canoe. Their appearance in the morning and evening throughout the year inform seasonal change, food economics, and social structure.

Tagai's influence on Islander social structure encompasses four themes for governing the Islander way of life.[9] The first links the stars of Tagai as custodians of knowledge and spirituality for future generations of Islanders. The second is that Islanders are a sea-faring people who share a common way of life. The third relates to the order of the world (the laws and customs) that are instructed by Tagai. The fourth discusses the cycle of life as a period of time and renewal based on the Sun, the Moon, and the rising and setting of particular stars. Thus, the night sky was recruited, structured, and woven into oral traditions (including song and dance) informing elements of the peoples' morals and values.[10] This is reflected in Malo's Laws, as discussed by the Islander elders. These are the sacred laws followed by all Meriam Islanders. One of Malo's laws is that all stars follow their own path, and so must each person. Do not follow the paths of others, but only of your own.

Music and Dance

Songs and dances are a way of recording knowledge and transmitting it to successive generations. The dance movements are one of the ways in which information is transmitted, and masks, headdresses, dance machines, and other props are an important element to these performances. The instruments used include the *warup* (traditional drum), bamboo drums, or contemporary instruments such as guitars. A warup is an hourglass-shaped drum carved from hollowed driftwood, using the skin of a *goanna* as the drum head.

Mua artist David Bosun created a linocut artwork in 2007 describing the important relationship between Torres Strait Islander astronomers (*Zugubau Mabaig* in western dialects) and the warup (FIGURE 12.3).[11] It is through the warup that these sacred teachings are echoed. The linocut features a Zugubau Mabaig holding two stars in front of a warup, echoing the sacred knowledge. Zugubau Mabaig are taught to read the stars in a *kwod* (a sacred meeting and learning house).

During a visit to Mer in July 2015, Meriam elders and dancers, including co-authors Alo Tapim and John Barsa, performed a dance called *Gedge Togia*. The music was played on two warups, with the lyrics 'Gedge Togia Milpanuka' sung by elders.

---

[9] Sharp, *Stars of Tagai*.

[10] John Whop, 'Stories Under Tagai', (video) (Brisbane: State Library of Queensland, 2012), at https://www.youtube.com/watch?v=5kU4EvV9yI8 [Accessed 4 February 2016].

[11] David Bosun, *Zugubau Mabaig*, Australian Art Network (2007), at http://australianartnetwork.com.au/shop/artwork/zugubau-mabaig/ [Accessed 4 Feb 2016].

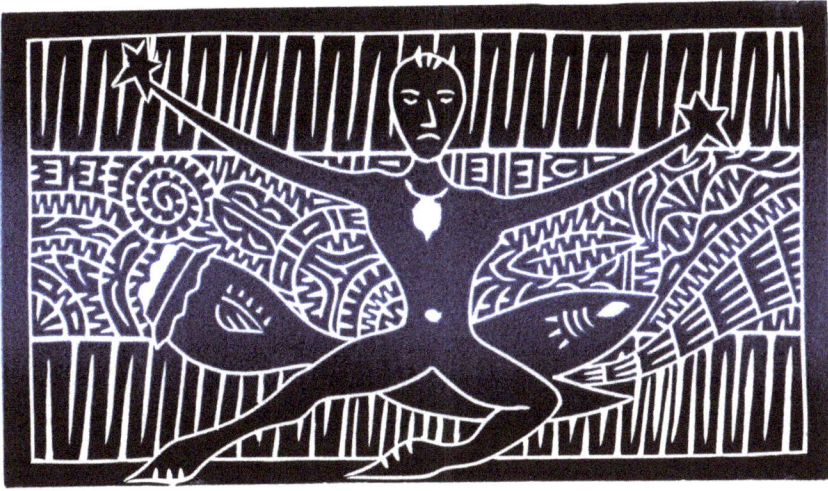

Figure 12.3. A linocut entitled 'Zugubau Mabaig (Astronomer)' by Mua artist David Bosun.

Alo Tapim explained the meaning behind the song. 'Gedge togia' means 'rising over home' (in this case Mer) in the Meriam Mir language. *Milpanuka* refers to the moon in the Mabuyag dialect of the Kala Lagaw Ya language. During the Haddon expedition in the late nineteenth century, vocabulary from Mer and Mabuyag were recorded by linguist Sidney Ray.[12] He recorded the Mabuyag name for the moon as *mulpal*, which referred to the full moon (or near full moon) as opposed to the crescent moon (*inur-dan*). Mer lies 200 km across the sea, at nearly due east, from Mabuyag.

In the Meriam Mir language, for comparison, the moon is called *meb* (which is also the term for a month).[13] The new moon (thin crescent) is called *aketi meb*, the first quarter moon is *meb degemli*, a nearly full moon (waxing or waning gibbous) is *eip meb*, a full moon is *giz meb*, a third quarter moon is *meb zizimi,* and a lunar eclipse is *meb dimdi*. Meriam elder and co-author Segar Passi gave the name *kerkar meb*, which literally translates to 'new moon' – when it first appears as a waxing crescent.

The planting of gardens is associated with lunar phases and the moon's effect on ocean tides.[14] Vegetables are planted during the rising tide; when the tide drops, the gardener stops. Vegetables that bear edible fruit are planted during the new moon, while tubers are planted during the full moon. For example, gardeners plant yams facing the rising full moon.

---

[12] Haddon, *Cambridge Anthropological Expedition*, Vol. 3, p. 152.

[13] Haddon, *Cambridge Anthropological Expedition*, Vol. 3, p. 152.

[14] Sharp, *Stars of Tagai*, p. 60.

Alo Tapim explained that quarter moons are more ideal for fishing than the new or full moon. This is due to the neap tides being low in amplitude, thus not churning up sand and soil that clouds the water (which was confirmed by other community members). This was confirmed by several Islander people. However, hunting crayfish and turtle are more appropriate during the full moon, when the strong moonlight (*meb gerip*) illuminates the water, making the animals easier to see, as noted by co-author John Barsa.

The *Gedge Togia* dancers hold a rod in each hand with a moon at the end: one is of the full moon and the other is of the new moon (the 'new moon' in this context colloquially describes the waxing or waning crescent when the moon is very thinly lit rather than the 'invisible' duration of the new moon familiar to Western astronomers). In addition to passing on important knowledge, Tapim explained how this song is important for demonstrating connections between different islands in the Torres Strait. During a legal battle for sea rights the Australian government in 2005, the opposition lawyer claimed that the islands were all separate enclaves with little or no contact with each other. Tapim sang this traditional song, comprising lyrics in two languages, to the judge as evidence this was incorrect. The Islanders won their case.

Headdresses

Headdresses (*dhari* in Meriam Mir and *dhoeri* in Kala Lagaw Ya) are prominent in Islander dance and ceremony, particularly the *kab kar* (sacred traditional dance). Traditional masks are made from a variety of components, including feathers, wood, and shells. Some headdresses are large and elaborate, representing various things such as animals, stars, and vehicles. Although worn by warriors in pre-colonial times, they are almost exclusively worn during dances and ceremonies today. Dhari are a common representation of Torres Strait culture, which is why they are featured on the Islander flag.

A *Madthubau Dhibal* headdress (FIGURE 12.4, right), created by Jeff and Sedrick Waia from Saibai, is associated with a rain dance and the constellations that comprise Thoegay (Tagai) and *baidam* (the shark). A man is chosen from each clan to watch the night skies for the movements of these constellations. The constellations inform seasonal change, animal behaviour, and gardening. The dance welcomes the coming monsoon season (*Kuki*), which lasts from January to April. Knowledge of Thoegay and other constellations by the Islander astronomer is essential for food production, especially when preparing for the Kuki. Thoegay/Tagai and the Zugubals he cast into the sky are important for denoting seasonal change. Mabuyag elder

FIGURE 12.4. (Left) An eclipse mask by Sipau Gibuma from Boigu ca. 1990. Photo by D. W. Hamacher at the National Museum of Australia, Canberra. (Right) Madthubau Dhibal headdress by Jeff Waia from Saibai ca. 2008. National Gallery of Australia, Accession No: NGA 2009.821.

Peter Eseli explains that because Thoegay placed the Zugubals on the opposite side of the sky from himself and Kang (Kareg), the Zugubals must announce their appearance in the dusk sky with thunder.[15] The dusk rising of Usiam (Pleiades) occurs in mid-November.[16] By early December, both Usiam and Utimal/Seg (Orion) rise at dusk. This coincides with the coming monsoon rains, accompanied by the sound of thunder. At this time, Thoegay and Kang dive into the sea as they set below the horizon, splashing water into the sky. This water causes the first rains of the Kuki.[17]

The *Mayngu dhoeri* (FIGURE 12.4, left) was created by Sipau Audi Gibuma from Boigu. It is used on very special occasions for a performance called *meripal kulkan patan*, meaning 'blood covering the moon'. This performance is conducted during a total lunar eclipse. In Islander traditions, a total lunar eclipse was an omen of war and bloodshed from another island or the Papuan mainland.[18] During the dance, the names of all the islands in the Torres Strait are chanted in a cycle. The island named

---

[15] Eseli, *Eseli's Notebook,* p. 17.

[16] When the stars have an altitude of 5° and the Sun −16°. See Trevor M. Leaman, Duane W. Hamacher, and Mark Carter. 'Aboriginal Astronomical Traditions from Ooldea, South Australia, Part 2: Animals in the Ooldean Sky', *Journal of Astronomical History and Heritage* 19, no. 1 (2016): pp. 61–78.

[17] Sharp, *Stars of Tagai,* pp. 60–61.

[18] Lindsay Wilson, *Kerkar Lu: Contemporary Artefacts of the Torres Strait Islanders* (Brisbane: Queensland Department of Education, 1993), p. 104.

when the moon emerges from totality is the island from which the threat of war is imminent.

Mua artist David Bosun refers to a lunar eclipse as *merlpal maru pathanu*. According to Bosun, this means 'the ghost has taken the spirit of the moon' (from mari/maru = ghost/spirit, and mulpal/merlpal = moon).[19] During a lunar eclipse, warriors prepare for battle while women and children hide. Bosun says the term can be used for both lunar and solar eclipses. A lunar eclipse in the Meriam Mir language is called *meb dimdi*, meaning 'covered moon'.[20]

Lunar eclipses occur about once every 2.5 years from any given location on Earth. The duration of totality can last up to 107 minutes.[21] The relationship between blood and the reddening of the moon during a total lunar eclipse is found in many indigenous traditions. On mainland Australia, a lunar eclipse was seen by many Aboriginal groups as the moon-man covered in blood and was linked to death and omens.[22]

Dance Machines

Dance machines (*zamiyakal*) are dynamic mechanical devices used by dancers to enhance performance and are prominent in Islander dances of the western, northern, and central islands (they are not common in the eastern islands). Zamiyakal serve as a mnemonic for recording and sustaining knowledge and cultural traditions.[23] The machines can mimic the motions of celestial bodies, such as the rising of *Ilwel* (Venus), the movement of a bright meteor across the sky, or the setting of the Southern Cross.

Star designs utilising various mechanical motions are incorporated into dance machines. Some are folded in half, which the dancer expands to reveal, then closes to hide. Some are contracted at the end of bamboo poles. An example is the *Titui* (star) machine from Mua (FIGURE 12.5A). The underside is pushed forward by the dancer, extending the rays of the star outward like an inverted umbrella. Other star designs are an ornamental component of a zamiyakal.

---

[19] David Bosun, *Merlpal Mari Pathanu.* Australian Art Network (2007), available at http://australianartnetwork.com.au/shop/artwork/merlpal-mari-pathanu/ [accessed 4 February 2015]. Used with permission of artist.

[20] Haddon, *Cambridge Anthropological Expedition,* Vol. 3, p. 152.

[21] Karttunen, Hannu, Pekka Kröger, Heikki Oja, Markku Poutanen, and Karl J. Donner. *Fundamental Astronomy* (Berlin, Heidelberg, New York: Springer, 2007), p. 139.

[22] Duane W. Hamacher and Ray P. Norris, 'Eclipses in Australian Aboriginal Astronomy', *Journal of Astronomical History and Heritage* 14, no. 2 (2011): pp. 103–14.

[23] John T. Kris in Robyn Fernandez and Emma Loban, *Zamiyakal: Torres Strait dance machines* (Thursday Island: Gab Titui Cultural Centre, 2009), p. 4.

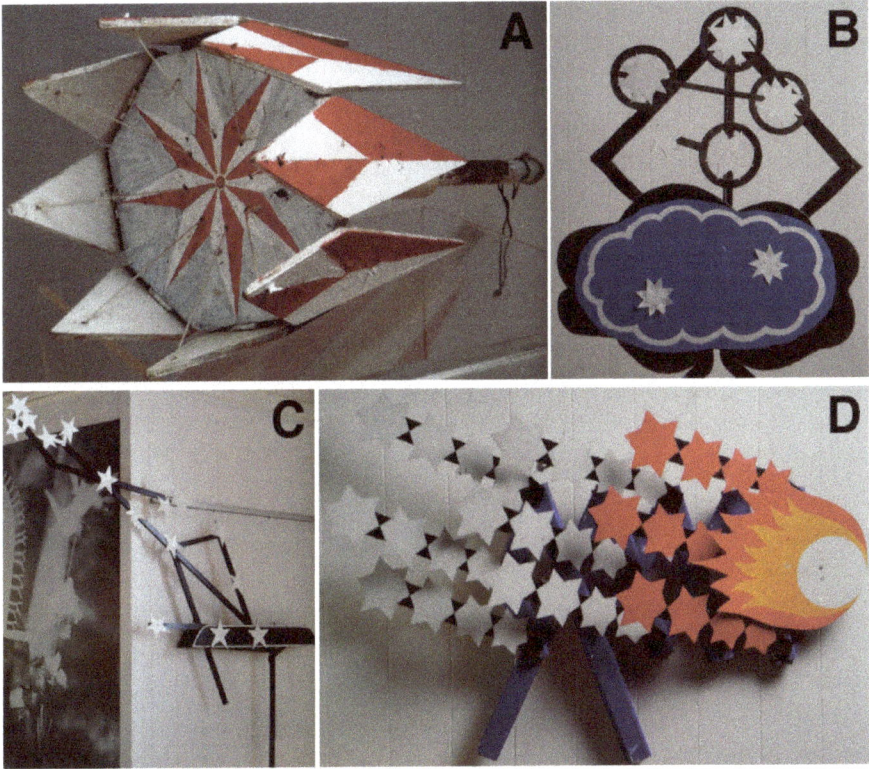

Figure 12.5. Astronomical dance machines (A) Titui, (B) Southern Cross, (C) The Tagai, (D) Comet.[24]

The *Koey Thitui Migi Thithui* dance from St Paul's community on Mua signifies the importance of the Southern Cross (left hand of Thoegay) and its use in navigation.[25] John Barsa notes that Crux is used as a compass, as it points southward. Similarly, a *Southern Cross* machine, developed by Ian Larry of Warraber[26] (FIGURE 12.5B), demonstrates the constellation's importance for denoting seasonal change. The song describes the relationship between the setting constellation and the north winds of the *Nay Gay* season from October to December. The Southern Cross sets at dusk in October, during which time gardeners begin preparing for the coming Kuki.

---

[24] Robyn Fernandez, and Emma Loban. *Zamiyakal: Torres Strait Dance Machines* (Thursday Island: Gab Titui Cultural Centre, 2009), Titui (p. 45), Southern Cross (p. 42), The Tagai (p. 51), and Comet (p. 48).

[25] Will Kepa, Nigel Pegrum, and Karl Nuenfeldt. *Lagau Kompass: Music & Dance From St. Paul's Community (Mua Island) Torres Strait* (CD/DVD) (Cairns: Pegasus Studios, 2013).

[26] Fernandez and Loban, *Zamiyakal*, p. 42

Grass is burned then dug into the soil for planting yams, sugar cane, and banana.[27] This also relates to *The Tagai* machine, made by Tabipa Harry of Dauan (FIGURE 12.5C). It mimics the movement of Tagai is he casts his spear into the sea (held in his left hand, the Southern Cross).

Cultural traditions are also encoded in dance. The *Comet* machine from Iama (FIGURE 12.5D) utilises a mechanical movement that makes it appear to 'shoot' across the sky. This describes a bright meteor rather than a comet (the two are often conflated in language translations). Alo Tapim explains that in Meriam traditions, bright meteors (fireballs) are called *Maier*. They signify that an important person has passed away (a common association across the Straits). The dying person's spirit climbs to the tallest tree and is launched across the sky. If the fireball fragments, the falling 'sparks' (*uir uir*, pronounced wier-wier) tell the observer that the person died leaving a large family behind. The 'booming' sound the fireball makes when it explodes is called *dum* (doom), which is associated with a large, eight-person drum.

Timing of Dances

Another important astronomical element of some Islander dances is that they are held at astronomically significant times. Each year a ceremonial dance is performed on the sacred islet of Pulu off the western coast of Mabuyag. The dance is performed to remember the dead and to celebrate a time of abundance. The timing of the ceremony signifies the yams are ready to be harvested, the coming of the dry season, and the arrival of the cool southeast trade winds. It notifies the people that tubers will increase in size, the rainbow bird (*Merops ornatus*) will migrate north, and the turtles will begin mating. This is ceremony is timed by the rising of the yam-star, *Kek* (Arcturus), which rises at dusk in late April.[28]

Concluding Remarks

Astronomical knowledge and symbolism is prominent in Torres Strait Islander dances, songs, and musical material culture. This knowledge is encoded in the song, the dance performance, and the timing of dances, serving as a mnemonic for retaining culture and passing knowledge to successive generations. Astronomically inspired dances, headdresses, and dance machines remain a vital and important component of Islander musical traditions.

---

[27] Eseli, *Eseli's Notebook.*
[28] Eseli, *Eseli's Notebook*, pp. 20–21.

Acknowledgements: I acknowledge Torres Strait Islander elders and communities, for this is their knowledge and their intellectual property. I thank to Martin Nakata, Doug Passi, Leah Lui-Chivizhe, Michael Passi, Alo Tapim, Segar Passi, Elsa Day, Aaron Bon, Jim Nai, Tommy Pau, and anonymous participants. This project was funded by the Australian Research Council (DE140101600) and the ethnographic fieldwork was approved by the Monash Human Research Ethics Committee (HC15035).

EAST MEETS WEST: SHI ZHIYING'S PICTURING OF ITALO CALVINO'S MR PALOMAR

John Hatch

ABSTRACT: My paper explores an intriguing series of thirty black and white watercolour and ink works by the Shanghai-born artist Shi Zhiying that give visual form to the unusual musings of Mr. Palomar, the title character of Italo Calvino's 1983 book of the same name. In this novel by the Italian writer, we are witnesses to a wonderful and often haphazard exploration of the world from its smallest creatures to the majestic universe it inhabits. Mr. Palomar's attempts to grasp these, to make sense of them, constantly elude and frustrate him, and he yet he never ceases trying despite the growing realization that his futility is the product of a greater sense of things that promises a far richer appreciation of life. It is not surprising to find that Calvino's writing has been long admired in China, as it embraces a contemplation of the world that is more familiar in the Far East than in the West. Shi masterfully demonstrates her grasp of the world of Mr. Palomar with a visual panorama that complements perfectly Calvino's text.

The Cuban born, Italian writer Italo Calvino's interest in astronomy, and science generally, dates back to the 1960s, if not a little earlier. His father was an agronomist and his mother a botanist, and Calvino started his university studies in agronomy before the Second World War interrupted them. Once the war was over, he changed subjects to literature, turning his full attention to writing.

In 1965 he published a collection of short stories, the *Cosmicomics*, in which his fascination for science emerged fully. This occurred during a period of self-imposed exile in Paris in 1964 where he became a frequent visitor to the Parisian Botanical Gardens, the Bibliothèque Nationale and the Astronomical Observatory. He read assiduously *Scientific American*, obtaining ideas and suggestions for his works and fodder for numerous discussions with scientist friends.[1]

Calvino's renewed interest in science is likely the reason why he joined the Oulipo group in 1964.[2] This group was founded four years earlier and was composed largely of French writers and a few scientists. Its name is an acronym for 'Ouvroir de littérature potentielle', which translates roughly as 'workspace for potential literature'. Oulipo held that contemporary fiction needed to address the sciences to be relevant to its time; however, most of the members only did so tangentially, albeit creatively. The most famous example was Georges Perec's *La disparition*, published in 1969 and inspired by mathematical restrictions; it is a three hundred page novel that doesn't

---

[1] Albert Sbragia, 'Italo Calvino's Ordering of Chaos', *MFS Modern Fiction Studies* 39, no. 2 (1993): p. 286.

[2] Sbragia, 'Italo Calvino's Ordering of Chaos'.

use the letter 'e' and whose plot is about the absence of that letter. Perec also used an old chess problem, the Knight's Tour, to describe a Paris apartment building in his 1975 book *Life: A User's Manual*.[3]

It was around the time he joined Oulipo that Calvino wrote *Cosmicomics*. However, another inspiration for *Cosmicomics* needs to be acknowledged, namely Galileo, whom Calvino frequently extolled as Italy's greatest prose writer, in large part because of Galileo's ability to take the complex and simplify it, while also being able to communicate it – simplification and communication being two different things.[4] According to Calvino, these qualities are represented, for example, in Galileo's *Dialogue Concerning the Two Chief World Systems* by the characters of Salivati and Sagredo, the former embodying the power of reasoning, the latter of the imagination.[5]

The stories in *Cosmicomics* use as their starting point various scientific theories, current and past, which are cited at the opening of each story, upon which a fable is woven, narrated by a character whose name is impossible to pronounce, Qfwfq. For example, the story 'All at One Point' starts out with the following scientific statement:

> Through the calculations begun by Edwin P. Hubble on the galaxies' velocity of recession, we can establish the moment when all the universe's matter was concentrated in a single point, before it began to expand in space.

Then the tale begins:

> Naturally, we were all there, old Qfwfq said, where else could we have been? Nobody knew then that there could be space. Or time either: what use did we have for time, packed in there like sardines?
>
> I say 'packed like sardines', using a literary image: in reality there wasn't even space to pack us into. Every point of each of us coincided with every point of each of the others in a single point, which was where we all were. In fact, we didn't even bother one another, except for personality differences, because when space doesn't

[3] On the Oulipo, see Peter Consenstein, *Literary Memory, Consciousness, and the Group Oulipo* (Amsterdam and New York: Faux Titre, 2002) and Alison James, *Constraining Chance: Georges Perec and the Oulipo* (Evanston, IL: Northwestern University Press, 2009).
[4] Italo Calvino, 'Two Interviews on Science and Literature', in *The Uses of Literature: Essays*, trans. Patrick Creagh (San Diego, CA: Harcourt Brace Jovanovich, 1986), pp. 28–38.
[5] Pierpaolo Antonello, 'The Myth of Science or the Science of Myth? Italo Calvino and the "Hard Core of Being"', *Italian Culture* 22, no. 1 (2004): pp. 81–82. Franco Gallippi, 'Calvino and the Nature of Style: an Interaction Between Literature and Science', *Rivista Di Studi Italiani* 21, no. 2 (2003): pp. 134–50; and, forthcoming, Franco Gallippi, 'Italo Calvino and Galileo: the Book of Nature', unpublished MS, available at https://www.academia.edu/8411417/Italo_Calvino_and_Galileo_the_Book_of_Nature [accessed 11 August 2015].

exist, having somebody unpleasant like Mr. Pbert Pberd underfoot all the time is the most irritating thing.

How many of us were there? Oh, I was never able to figure that out, not even approximately. To make a count, we would have had to move apart, at least a little, and instead we all occupied that same point. Contrary to what you might think, it wasn't the sort of situation that encourages sociability... [6]

In a charming way, Calvino's *Cosmicomics* connect science with mythology as fable, reigniting a past when science and the arts were an inseparable part of culture. This is a theme that will persist throughout Calvino's writings, culminating with Mr. Palomar in Calvino's last novel who sees the universe as providing a type of ethical model for human behaviour.

Calvino's *Invisible Cities*, published in 1972, relates Marco Polo's accounts of the cities of Kublai Khan's eroding empire. Khan is seeking some kind of structure, order and meaning to his crumbling empire through these accounts. Polo describes 55 cities in all, none that he has actually seen, and all based on Polo's hometown of Venice. That said, Polo's refashionings do have some links to far eastern practices. A number of his cities incorporate in their description the tradition of paralleling the celestial with the terrestrial.

The city of Isidore is dominated by spirals and is a place where one can find the finest violins and telescopes made. Thekla is described as under construction; when a visitor peers through a fence to glimpse the city's construction, the question is asked:

'Where is the plan you are following, the blueprint?'
'We will show it to you as soon as the working day is over; we cannot interrupt our work now', is the answer.
'Work stops at sunset. Darkness falls over the building site. The sky is filled with stars. "There is the blueprint", they say'.[7]

The description of Andria reads as follows:

Andria was built so artfully that its every street follows a planet's orbit, and the buildings and the places of community life repeat the order of the constellations and the position of the most luminous stars... thus the days on earth and the nights in the sky reflect each other.

The astronomers, after each change takes place in Andria, peer into their telescopes and report a nova's explosion, or a remote point in the firmament's change of colour from orange to yellow, the expansion of a nebula, the bending of a spiral

---

[6] Italo Calvino, *Cosmicomics* (Orlando: Houghton Mifflin Harcourt, 1968), p. 43.
[7] Italo Calvino, *Invisible Cities* (London: Pan Books, 1979), p. 101.

of the Milky Way. Each change implies a sequence of other changes, in Andria as among the stars: the city and the sky never remain the same...

Convinced that every innovation in the city influences the sky's pattern, before taking any decision they calculate the risks and advantages for themselves and for the city and for all worlds.[8]

And sometimes things don't go so well, as we find with the city of Perinthia:

Summoned to lay down the rules for the foundation of Perinthia, the astronomers established the place and the day according to the position of the stars... Perinthia – they guaranteed – would reflect the harmony of the firmament, nature's reason and the gods' benevolence would shape the inhabitants' destinies.

...In Perinthia's streets and square today you encounter cripples, dwarfs, hunchbacks, obese men, bearded women. But the worse cannot be seen; guttural howls are heard from cellars and lofts...

Perinthia's astronomers are faced with a difficult choice. Either they must admit that all their calculations were wrong and their figures are unable to describe the heavens, or else they must reveal that the order of the gods is reflected exactly in the city of monsters.[9]

Invisible Cities became a popular book in China and it attracted the attention of the Shanghai-born artist Shi Zhiying to Calvino's work.[10] Shi was born in 1979; she graduated with a Masters degree in oil painting from Shanghai University Fine Arts College in 2005.[11] Initially overwhelmed by various artistic influences after graduating, and quite anxious about the direction her painting should go, Shi found her voice during a trip to California when she looked over the ocean from a lighthouse in San Francisco and was washed over by a feeling of immense calmness.[12] The initial result was a series of works titled *High Seas* begun in 2008, which led her to focus on and study the work of likeminded and like-eyed artists ranging from the Sung Dynasty artist Ma Yuan to the contemporary Japanese photographer Hiroshi Sugimoto;[13] at the same time, Shi began to immerse herself in Buddhism.[14]

[8] Calvino, *Invisible Cities.*, pp. 116–17.
[9] Calvino, *Invisible Cities.*, pp. 113–14.
[10] Christopher Moore, 'Shi Zhiying's "Infinite Lawn"', *Randian*, 1 June 2012, available at http://images.jamescohan.com/www_jamescohan_com/Randian_Moore_June_1_2012.pdf [accessed 14 Oct.2016].
[11] Press release, 'Shi Zhiying', *Undo.net*, 13 June 2015, available at http://www.undo.net/it/mostra/191598 [accessed 6 July 6 2015].
[12] Moore, 'Shi Zhiying's 'Infinite Lawn'.
[13] Luise Guest, 'The Art of Now: Teaching Chinese Art', available at http://teachingchineseart.blogspot.ca/p/shi-zhiying.html [accessed 6 July 2015].
[14] Luise Guest, 'The Paintings of Shi Zhiying: Searching for the Sublime', *The Culture Trip*, available

Figure 13.1. Shi Zhiying, *Palomar – Forward I*, 2011–2012, watercolour and ink on paper, 31×41cm. © Shi Zhiying. Courtesy James Cohan, New York.

Her interest in meditations on the world from its smallest parts to its largest, her fascination with science and astronomy, and the character of Mr. Palomar himself combined to inspire Shi to produce a series of 30 works, all roughly 41 x 31cm in size: one for each of the chapters of *Mr. Palomar*, a cover and two prefatory images that are the bookends of Mr. Palomar's meditations, namely the celestial, with an image of the famous observatory on Palomar mountain in San Diego County, California, and the terrestrial (Fig. 13.1).[15]

Published in 1983, two years before Calvino's death, *Mr. Palomar* is a collection of 27 musings on diverse aspects of the world and our relation to it. The main character believes in an inherent harmony, which he tries to reveal through a dogmatic analytical approach, but his efforts are consistently and often comically thwarted by reality.

Mr. Palomar is an everyman in the medieval sense of the term: an explorer, a journeyer, whose encounters we might all experience in some fashion or other, not the same but similar. In a way he is the common person's Marco Polo. This is another aspect that attracted Shi, as she relates:

> Mr. Palomar can be any one of us; there is nothing special about him. He lives a normal life like us. He goes on vacation, is close to nature, shops in the city, travels around the world. In facing the world, he comes up with many thoughts and ideas, so here the emphasis is not on Mr. Palomar but on 'the experience of Mr. Palomar' – each of our experiences in everyday life.[16]

at http://theculturetrip.com/asia/china/articles/the-paintings-of-shi-zhiying-searching-for-the-sublime/ [accessed 6 July 2015].

[15] Sue Wang, 'Shi Zhiying: The Infinite Lawn Featuring Her Recent Watercolor and Ink Works', *Cafe Art Info*, 30 May 2012, available at http://en.cafa.com.cn/shi-zhiying-the-infinite-lawn-featuring-her-recent-water-color-and-ink-works.html [accessed 13 February 2015].

[16] Karolle Rabarison, 'The Infinite Lawn, Shi Zhiying: Interview', *The Morning News*, 19 June 2012,

In order to maybe give us some hope that Palomar's observations are not entirely fruitless, there is an index at the end of the book that supplies us a key to the underlying structure of Calvino's work. The preface to the key explains how it works, with the number '1' referring to visual experiences, '2' the cultural, and '3' the more conceptual, addressing larger questions of the cosmos, time, infinity, etc. Curiously enough, given that there have been a number of comparisons between the two, I have yet to find referenced in the Calvino scholarship the fact that the index pattern in *Mr. Palomar* echoes what one finds in the structure of Dante's *Divine Comedy*, which is composed of 3 books, 33 cantos per book and 3 lines per canto (plus the addition of 1 extra canto to represent the divine). *Mr. Palomar*, though not as substantial, nevertheless has three parts with each part subdivided into 3 sections that are each subdivided into 3 chapters. As Dante's book is a representation of the medieval world in the fourteenth century, Calvino's is a representation of our contemporary world; as Dante's book is an allegory of the soul's journey toward God, Calvino's is a journey of the self and our relation to the world. As Dante's text was a poetic rephrasing of St. Thomas of Aquinas' *Summa Theologica*, for example, Calvino's is a poetic representation of some of the philosophical and scientific concerns of his own day.[17]

Mr. Palomar, the book, begins with the famous chapter 'Reading a Wave' (FIG. 13.2) that spells out the objective of the title character's meditations and their inevitable pitfalls. He tries to isolate and observe a single wave, with the goal of mastering 'the world's complexity by reducing it to its simplest mechanism'.[18] Palomar runs into the inevitable problems of where does the wave begin and end, how does one account for the variations of each individual wave, etc. Most of us would simply relax to the sound and images of the waves, the experience, whereas Palomar does the opposite in taking a surgical, analytical approach to the wave, and his relaxation will come 'Only if he manages to bear all the aspects in mind at once... ', since only then 'can he begin the second phase of the operation: extending this knowledge to the entire universe'.[19] An operation which does not happen in the way Mr. Palomar anticipates or hopes, since singling out something like a wave is not the best place to start if one is intending to objectively dissect the world and derive its fundamental principles. However, this is not so surprising in light of Calvino's interest in Chaos theory during the 1970s and early 80s.[20]

available at http://www.themorningnews.org/gallery/the-infinite-lawn [accessed 6 July 2015].
[17] Guy P. Raffa, 'Eco and Calvino Reading Dante', *Italica* 73, no. 3 (1996): pp. 396–403.
[18] Italo Calvino, *Mr. Palomar* (Orlando: Harcourt, 1985), p. 6.
[19] Calvino, *Mr. Palomar*, p. 8.
[20] Sbragia, 'Italo Calvino's Ordering of Chaos', pp. 286–89.

FIGURE 13.2. Shi Zhiying, *Palomar – Reading a Wave*, 2011–2012, watercolour and ink on paper, 31×41cm. © Shi Zhiying. Courtesy James Cohan, New York.

If there is a problem with what Palomar selects to be observed, the next chapter, 'The Naked Bosom', spells out the difficulties with the observer. Walking on the beach, Mr. Palomar approaches a young woman sunbathing topless. He is not sure how to react. Mr. Palomar walks by once, focusing exclusively on the seascape and ignoring the young woman, then he goes by another time feeling that his first response wasn't the appropriate one; after a few more variations of how he thinks he should act passing the topless sunbather, she finally covers herself up and leaves, frustrated by this old man who keeps walking past her. The mental debate Palomar undertakes illustrates beautifully contemporary gender issues, and highlights the fact that a so-called objective observer can never truly be objective. 'The Blackbird's Whistle' highlights the difficulty of human communication that ends with Palomar and his wife failing to understand each other in a common domestic conversation despite having the advantage of a more complex language than that of blackbirds. Yet, Calvino does remind us that the inherent hurdles of human communication can be overcome, but this can only happen through a careful weaving of the scientific precision and poetic qualities of that language, as Calvino believed Galileo had succeeded in doing, a point we are reminded of with Calvino's reference to Jupiter and the Medicean stars or Galilean moons in the chapter 'The Eye and the Planets',[21] an image Shi reproduces in her image for the chapter 'The Contemplation of the Stars' (FIG. 13.3).

In an interesting way, Calvino demonstrates this merging of the analytical and the poetic as a teaser in the chapter 'Moon in the Afternoon' (FIG. 13.4) with its wonderful opening paragraph:

Nobody looks at the moon in the afternoon, and this is the moment when it would most require our attention, since its existence is still in doubt. It is a whitish shadow

---

[21] Calvino, *Mr. Palomar*, p. 41.

FIGURE 13.3.(LEFT) Shi Zhiying, *Palomar – The Contemplation of the Stars*, 2011–2012, watercolour and ink on paper, 31×41cm. © Shi Zhiying. Courtesy James Cohan, New York.

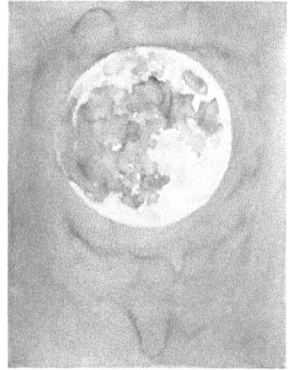

FIGURE 13.4. (RIGHT) Shi Zhiying, *Palomar – Moon in the Afternoon*, 2011–2012, watercolour and ink on paper, 41×31cm. © Shi Zhiying. Courtesy James Cohan, New York.

that surfaces from the intense blue of the sky, charged with solar light; who can assure us that, once again, it will succeed in assuming a form and glow? It is so fragile and pale and slender; only on one side does it begin to assume a distinct outline, like the arc of a sickle, while the rest is all steeped in azure. It is like a transparent wafer, or a half-dissolved pastille; only here the white circle is not dissolving but condensing, collecting itself at the price of gray-bluish patches and shadows that might belong to the moon's geography or might be spilling of the sky that still soak the satellite, porous as a sponge.[22]

The chapter 'The Infinite Lawn', returns us to the problem encountered in 'Reading a Wave' while introducing us to the dilemma of detail versus totality. Mr. Palomar is pulling the weeds from his patch of lawn and he wonders, can he see the whole of his lawn as a collection of discreet parts, in other words can he think the lawn rather than see it as we normally tend to do. He inevitably gets bogged down in the detail and before he can answer the question, Mr. Palomar extends his musings about his lawn to the universe. 'Is the universe regular and ordered or a chaotic proliferation; perhaps it is finite but countless, unstable within its borders, disclosing other universes within itself. The universe as a collection of celestial bodies, nebulas, fine dust, force fields, intersection of fields, collections of collections.... '[23] But then his mind wonders off again, of course.

For Shi Zhiying, the *Infinite Lawn* (FIGURE 13.5) is a meditative ideal, not a site of

[22] Calvino, *Mr. Palomar*, p. 34.
[23] Calvino, *Mr. Palomar*, p. 33.

Figure 13.5. Shi Zhiying, *Palomar – The Infinite Lawn*, 2011–2012, watercolour and ink on paper 31×41cm. © Shi Zhiying. Courtesy James Cohan, New York.

anxiety. The same is true in the chapter 'The Sand Garden' where Mr. Palomar visits a Zen sand garden, which Shi portrays as a sight of tranquillity, as one imagines it to be (FIGURE 13.6).

Figure 13.6. Shi Zhiying, *Palomar – The Sand Garden*, 2011–2012, watercolour and ink on paper, 31×41cm. © Shi Zhiying. Courtesy James Cohan, New York.

In Calvino's hands, the tranquillity is broken up by a massive crowd of tourists jostling Mr. Palomar to catch a glimpse of the Japanese sacred garden. In a way, what I like about Shi's approach is that she draws out, in a sense, the Eastern interpretation of the Palomar scenes, leaving it to the novel to describe the Western anxiety. In an ingenious fashion, Mr. Palomar's Western objectivity and attention to detail in the

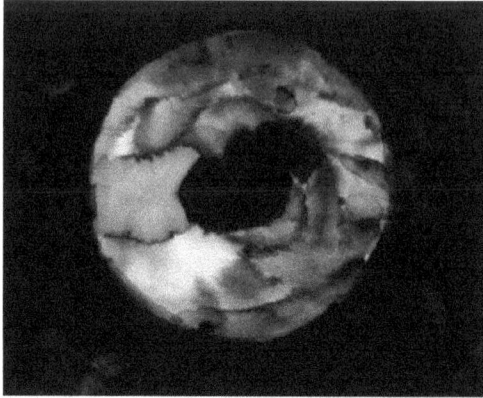

FIGURE 13.7. Shi Zhiying, *Palomar – The Eye and the Planets*, 2011–2012, watercolour and ink on paper, 31×41cm. © Shi Zhiying. Courtesy James Cohan, New York.

hopes of finding meaning are contrasted by Shi to an Eastern concept of self-awareness and attentiveness that are the meaning. Her work essentially takes the subjects of Calvino's novel but removes Mr. Palomar from the images, in a way removing his anxieties, frustrations, etc., but replacing him with us, the viewer; we are Mr. Palomar, just as Shi herself noted in the quote earlier referring to Mr. Palomar as an every person. In fact, the choice of working in black and white was done in part, as Shi notes in an interview, to avoid prescribing how one should read the images, thus leaving room for the viewer's imagination to fill in substantial parts of the picture.[24]

As mentioned earlier, the infinite lawn poses the problem of detail versus the whole. To see the lawn in its totality you must ignore the individual parts; to see the individual parts, you must ignore the whole. It is a problem Calvino reiterates in 'The Eye and the Planets' (FIG. 13.7): when Mr. Palomar turns his telescope to Mars, he sees the planet but the details are fuzzy, he can focus on some details but loses others – a common issue using your average telescope in observing Mars. But this is yet another version of that problem of the impossibility of seeing both at the same time, which he feels plagues our understanding of the world and ourselves. It may even plague our search for a unified theory, of bringing together the infinitely small and the infinitely large.

'The Cheese Museum' illustrates wonderfully our classification impulse and its failures when applied to the real world. Mr. Palomar visits a cheese shop where each cheese is carefully labelled; mentally he tries to catalogue what he sees. In essence, he now gives in to enumeration as his best hope to reveal order. Yet Mr. Palomar's

---

[24] Rabarison, 'The Infinite Lawn, Shi Zhiying: Interview'.

FIGURE 13.8. Shi Zhiying, *Palomar – The Universe as Mirror*, 2011–2012, watercolour and ink on paper, 31×41cm. © Shi Zhiying. Courtesy James Cohan, New York.

classification is unable to comfortably include the tastes, the smells, or how these cheeses may be consumed, all eroding his tidy image of a cheese universe. Again, he is caught between two modes of seeing the world, two ways of comprehending it.

Although Calvino's book is quite short, each chapter unpacks a myriad of questions and ideas related to some of our greatest challenges in understanding the world around us, phrased in a wide assortment of imagery that include an albino gorilla playing with a car tire, unmatched slippers, mating turtles and so on. In turning to the cosmos in the last part of the book, Mr. Palomar embraces ultimately the middle road; there just can't be a single model of the universe – its complexities and our own simply undermine arriving at such a model. However, he wonders why our images of the harmony of universe can't be turned back on ourselves (FIG. 13.8); what we have arrived at so far in terms of our knowledge of reality and the beauty of it all is breathtaking.

Whether we believe our understanding of the structure of the universe is our creation or revealed, it is still something we have accomplished. Here is a moment when Mr. Palomar adopts a view that has dominated much of our history on this planet, of looking to the stars for guidance; Mr. Palomar feels we can continue to do so but in ethical rather than religious terms. It is soon after this realization that Mr. Palomar sees the light, a light that was always there for Shi Zhiying; it is the moment when he tries to imagine what the universe would be like were he dead – at that instant he dies.

'LIFE IS ASTRONOMICAL':
CONNECTING ART, ASTRONOMY & PHOTOGRAPHY AT ROYAL MUSEUMS GREENWICH

Melanie Vandenbrouck and Marek Kukula

ABSTRACT: The Royal Observatory Greenwich is one of London's leading tourist attractions and an internationally recognised centre of excellence for public engagement with astronomy. As a museum of the history of astronomy, navigation and timekeeping, as well as a modern science centre and planetarium, it is also an integral part of Royal Museums Greenwich, alongside the National Maritime Museum, the Queen's House art gallery and the clipper ship *Cutty Sark*. The Observatory has always used astronomical images as powerful tools for explaining scientific ideas but in recent years it has increasingly been working with art historians and contemporary artists to explore the aesthetic appeal of space imagery as part of its public programme. Two prominent examples of this approach are the Insight Astronomy Photographer of the Year competition and the exhibition *Visions of the Universe*. These projects give equal weight to the artistic and scientific aspects of astronomical images and have proved to be invaluable new catalysts for public engagement, both by giving a fresh perspective to traditional astronomy audiences and by reaching out to new audiences who would not normally attend a science-related event. They have also had a profound effect on the Royal Observatory and the wider Museum, prompting a new willingness among the astronomy team to explore and discuss the aesthetic content of astronomical photographs and their relationship to the arts. In this they have been met enthusiastically by the Museum's art curators, who recognise a deep potential in the collision of art and science, and in contemporary art as a means to engage audiences and strengthen the institution's relevance to the present. Collaborations with living artists on interventions, commissions and other projects are creating new avenues of investigation for the Museum, while the formulation of a (thus far overlooked) contemporary art collecting policy opens up new perspectives on the Museum's historical collections and core themes. The Museum is developing a collection, both historical and contemporary, which reflects the relationship between representation and understanding. Objects relating to the cultural impact and dissemination of astronomy show the penetration of astronomical ideas into the broader history and culture of the period in which they were made. In the past, the fine arts have been used as a way to record scientific information and the work of astronomers. Now contemporary artists' fascination with the cosmos and our place within it enables diverse audiences to engage with science in a variety of ways, while also encouraging scientists to look at their own subject with fresh eyes. This chapter will describe our experiences and explore how they have had a powerful impact on the public engagement strategy of the Royal Observatory and on the collecting strategy of the wider Royal Museums Greenwich.[1]

Introduction:
The Royal Observatory Greenwich and Royal Museums Greenwich

Founded in 1675 by King Charles II, the Royal Observatory Greenwich (ROG)

---

[1] Dr Marek Kukula is the Public Astronomer at the Royal Observatory Greenwich; Dr Melanie Vandenbrouck is Curator of Art, post-1800 at Royal Museums Greenwich. Both are judges on the Insight Astronomy Photographer of the Year competition.

served as a working observatory for almost three hundred years. Its original mission was to improve the state of positional astronomy and thus solve the problem of finding longitude at sea, a quest whose legacy remains in the form of the Prime Meridian and Greenwich Mean Time.[2] In the nineteenth and early twentieth centuries the work of the Observatory expanded to encompass the new discipline of astrophysics, embracing techniques such as spectroscopy and astrophotography and establishing itself in fields such as solar physics and the study of double stars. Smoke and light pollution from the expanding city of London eventually led, in the mid twentieth century, to the research functions of the Observatory being relocated elsewhere while the historic site in Greenwich was handed over to the neighbouring National Maritime Museum and opened to the public in the 1960s as a museum of the history of astronomy, navigation and timekeeping. Increasingly this focus on history was supplemented by content reflecting contemporary astronomy and astrophysics and, after significant investment in 2007, the Royal Observatory site acquired new state-of-the-art facilities for presenting the latest astronomical discoveries to the public, including galleries, learning spaces and the 132-seat Peter Harrison Planetarium.

Combining the functions of both a museum and a science centre the Royal Observatory now forms an integral part of Royal Museums Greenwich (RMG), an organisation that also includes the National Maritime Museum, the *Cutty Sark* (a nineteenth century tea clipper preserved in dry dock), and the Queen's House (Britain's earliest example of Palladian architecture, and now a world class art gallery).[3] This unusual and eclectic museum therefore covers topics in history, art and science, all linked by themes of travel, exploration and discovery and embedded within the wider cultural context of the Maritime Greenwich UNESCO World Heritage Site.

The Observatory's Science Learning & Public Engagement department is recognised as a leader in the dissemination of astronomy and space science to schools, the public and the media, and is active in developing and implementing best practice in the fields of science communication and public engagement. However, the unusual context of a science engagement team working alongside specialists in the arts and humanities also presents unique opportunities for cross-disciplinary research and public engagement. In particular, the Museum's collection contains examples

---

[2] This was the object of an exhibition at Royal Museums Greenwich in 2014, *Ships, Clocks and Stars: The Quest for Longitude*, with a catalogue published by Collins, edited by Richard Dunn and Rebekah Higgitt.

[3] While Royal Museums Greenwich is the brand name assigned to the group of sites in 2012, the National Maritime Museum remains the legal entity under which the four sites, including their collections, are administered.

of astronomically-themed fine art, including paintings, drawings and sculpture, as well as globes, star charts, scientific diagrams, lantern slides and photographic plates, many of which have a strong aesthetic appeal in addition to their historic and scientific significance. Increasingly RMG is recognising the value of considering this material from a combination of artistic, historical and scientific perspectives and of exploring the links between these historic collections and contemporary astronomical photographs and data. This enables the Museum to broaden its audiences and challenge their preconceptions about both science and art, and the on-going interaction between RMG's astronomy and arts specialists is the main theme of the current chapter.

The Insight Astronomy Photography of the Year Competition and Exhibition

Astronomy and photography have been closely intertwined since the earliest days of the photographic medium, with pioneering figures in photography such as John Herschel, Charles Piazzi Smyth, Annie Maunder and Warren De la Rue coming from astronomical backgrounds, and modern advances in photographic technology, including digital camera chips and image processing software, owing much to developments made for the purposes of advancing astronomical research. Equally, the impact of photography as a tool for astronomers, particularly from the latter quarter of the nineteenth century onwards, is difficult to overstate.[4] Indeed, one of the main buildings on the Royal Observatory site in Greenwich was constructed during the 1890s largely for the purpose of facilitating the ROG's astrophotographic research.

Astronomers were quick to recognise the potential of astrophotography as a tool for engaging and educating the public and the RMG collection includes several sets of lantern slides featuring early photographs of the Sun, Moon, eclipses, comets, stars and nebulae as well as telescopes, observatories and expeditions.[5] These slides were used in the late nineteenth and early twentieth centuries to illustrate public lectures by observatory staff such as Edward Walter Maunder and other members of the British Astronomical Association. Since then, astrophotography has continued to exert a defining influence on the public and media perception of astronomy, from

---

[4] On the relationship between modern science and photography, see the exhibitions (and their catalogues), *Brought to Light: Photography and the Invisible, 1840-1900*, SF MOMA, San Francisco, 2008 and *Revelations: Experiments in Photography*, Science Museum, London, 2015. For a survey of the relationship between science and photography, see Kelley Wilder, *Photography and Science*, Reaktion Books, 2009.

[5] The Royal Museums Greenwich collection can be accessed and searched online at http://collections.rmg.co.uk [accessed 1 February 2016].

David Malin's influential full-colour plates of galaxies and nebulae in the 1970s and 80s to the foundation of the Hubble Heritage Project in 1998 and the high-definition digital images beamed back by the current generation of space probes and planetary rovers.[6]

As a museum and science centre with a remit to disseminate and explain astronomical research to the public, the Royal Observatory has always made extensive use of these professional astronomical images but, from the 1990s onwards, the availability of affordable digital camera technology, the huge increase in the processing power and storage capacity of home computers and the arrival of the internet as a tool for disseminating images and sharing expertise have led to an explosion in the field of amateur astrophotography. The Observatory was eager to engage with these developments for several reasons: to harness an abundant new source of beautiful and appealing images for the purposes of communicating astronomy to the public, to forge closer links with the vibrant community of amateur astrophotographers and showcase their work to a wider audience, and to explore the cultural and aesthetic impact of this 'democratisation' of astronomy. As a significant new chapter in the story of astrophotography it was felt that this would offer a fresh, contemporary window on the long association between astronomy and photography, both at the ROG and beyond.

In 2009, as part of its celebrations for the UNESCO International Year of Astronomy, the Royal Observatory Greenwich launched the first Astronomy Photographer of the Year competition[7], attracting some 450 entries from around the world, with categories and prizes covering the Solar System and deep space objects as well as upper atmospheric phenomena such as aurorae and meteor showers and landscape shots in which the night sky is a prominent feature. A separate category was open to photographers under the age of sixteen.

Community engagement was a primary goal of the competition from the outset: entries were uploaded and submitted via the photo-sharing website Flickr, and online message boards allowed photographers to converse with each other as well as with the competition organisers and other experts in astrophotography. A media partnership with *BBC Sky At Night Magazine* helped the competition to engage with the amateur astronomy community in the UK as well as other Anglophone territories. Once the 24 prizewinning images had been announced they were displayed in a

---

[6] See *Coloring the Universe: An Insider's Look at Making Spectacular Images of Space*, by Travis Rector, Kimberly Arcand and Megan Watzke (Fairbanks: University of Alaska Press, 2015).

[7] Insight Astronomy Photographer of the Year page on the RMG website, available at http://www.rmg.co.uk/astrophoto [accessed 1 February 2016].

free exhibition held in the Royal Observatory's temporary exhibition gallery.

The public and media response was extremely positive, with the images creating a social media 'buzz' and garnering extensive press coverage in the UK and beyond, encouraging the Museum to continue to support the competition as a regular part of the Royal Observatory's programme. Since 2009 the competition has grown year-on-year, with over 4500 entries being received from 80 countries in 2016, and the project has expanded to include calendars and other merchandise, an annual book featuring the year's winning and shortlisted photographs and a planetarium show, produced in-house by the Royal Observatory team and distributed for free to planetaria around the world.

For the 2015 competition sponsorship was secured from Insight Investment, a London-based asset management company that is also a long-term sponsor of the Royal Academy's Summer Show and other 'blockbuster' art exhibitions. This partnership is recognition of both the powerful inspirational appeal of astronomical imagery and the strong connections between the mathematical, problem-solving mind-set of astrophysics and the similar requirements of the financial industry. As a result, the competition has been renamed as Insight Astronomy Photographer of the Year, with an increase in the number of categories and prizes to reflect the (already significant) developments in amateur astrophotography since 2009. Sponsorship also prompted the commencement of a more ambitious exhibition programme, involving an upgrade of the gallery at the ROG and moves towards implementing an international touring version of the exhibition from 2017 onwards.

Since the competition's inception, the judging panel has always reflected a broad range of expertise, ranging from experienced astrophotographers and professional astrophysicists to science communicators, historians of science, magazine picture editors, curators and artists. The diversity of opinions and expertise embodied in the panel has helped to keep the competition open to change, and over time the range and scope of both entrants and photographs has expanded and evolved. While the amateur astronomy community still constitutes the core demographic of entrants, increasingly the competition is also engaging amateur and professional photographers with little or no prior involvement in either amateur astronomy or astrophotography, including wildlife photographers, photojournalists and fine art photographers.

This has broadened the style and approach of the photographs in interesting and unexpected ways, including more ambitious and imaginative uses of landscape and skyscape photography, experimentation with colour palettes other than the now traditional 'Hubble palette', and the pushing of the boundaries of astrophotographic convention, for example by choosing to show the moment of third contact of a total solar eclipse as a dazzling blaze of light rather than a carefully exposed 'diamond ring' effect, or by using camera motions to draw out the refractive colour variations of the star Sirius (See Fig. 14.1). In turn, these developments have served to widen the popular appeal of the competition, provoking (largely positive) debate and engaging the attention of the photography and fine art communities. Press coverage has also expanded to include over 400 media outlets, including reviews by art critics in newspapers such as *The Times* and periodicals such as the art magazine *Aesthetica*.

In 2016 further sponsorship from Insight Investment increased the competition's prize fund, with the value of the first prize rising to £10,000 – comparable to global photography competitions such as the Sony World Photography Awards and Wildlife Photographer of the Year. The composition of the judging panel also continues to diversify, with the addition in 2016 of a representative of the European Southern Observatory, as well as Turner Prize-winning artist and Royal Academician Wolfgang Tillmans.

It is hoped that these developments will enable the competition to continue expanding both the pool of entrants and its audience over the coming years, but already Astronomy Photographer of the Year has had a profound effect on public engagement at the Royal Observatory. The public and media response has demonstrated once more the appeal of beautiful images and their ability to make complex scientific concepts accessible. But our experience has also shown that foregrounding this aesthetic appeal and openly discussing the artistic aspects of the images is a powerful way to attract audiences that would not normally engage with scientific topics. This in turn has prompted us to think more carefully about connections between art and science and how we can meaningfully explore the two in our permanent galleries, public programmes and exhibitions across Royal Museums Greenwich.

Visions of the Universe

In 2011 the construction of the Sammy Ofer Wing at the National Maritime Museum site provided Royal Museums Greenwich with an 800m² state-of-the-art gallery in which to stage large scale temporary exhibitions relating to all of the topics covered by RMG. Early uses of this space included a contemporary art show examining the effects of climate change in the arctic, an exhibition on the cultural and social history of the River Thames and a survey of the role of water in the landscape photography of American artist Ansel Adams – all speaking to the Museum's maritime, historical and artistic themes. However, for the summer of 2013 the gallery was made available to the team at the Royal Observatory for an exhibition on astronomy, entitled *Visions of the Universe* (7 June–15 September 2013).

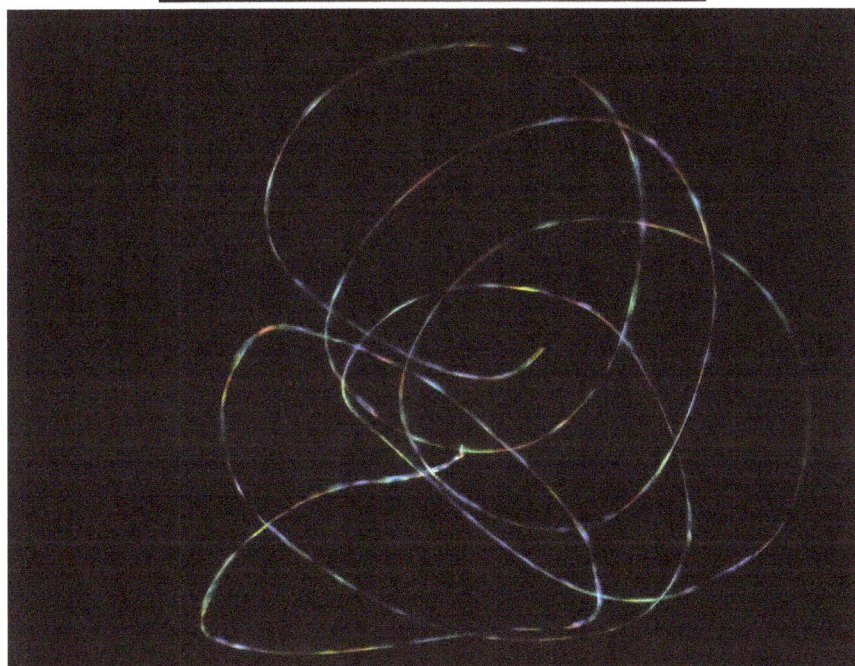

FIGURE 14.1. Top, *Totality Ends*, by David Wrangborg (2015), runner-up in the category 'Our Sun'. Bottom, *Sirius 9798* by David Pye, 2015, joint-runner up in the category 'Stars and Nebulae'. Insight Astronomy Photographer of the Year Competition, 2015. National Maritime Museum and the photographers.

Drawing on the lessons learned from the Astronomy Photographer of the Year competition, the brief for *Visions of the Universe* was to display amateur and professional astronomical images in a way that was both informative and aesthetically appealing and to present them as much as works of art as scientific documents and data. One important narrative thread was the story of how the development of the camera and of photography have been intimately connected to advances in astronomy over the past two centuries, while another narrative – that of a journey leading further into the depths of space – was conveyed by structuring the exhibition in sections covering the Moon, the Sun, the planets, stars and nebulae, and extragalactic objects. Two final sections explored the relationship between humans and the night sky and the prospects for new discoveries that may be opened up by future developments in spacecraft and telescope technology.

Just over 100 photographs were selected for inclusion in *Visions of the Universe*, including a range of recent images from major professional telescopes and space missions such as the Hubble Space Telescope, Cassini/Huygens and the European Southern Observatory, as well several previous prize winners from the Astronomy Photographer of the Year competition, to highlight the contribution of amateur astrophotographers. The exhibition also contained reproductions of historically important photographs such as William Draper's early daguerreotype of the Moon, Arthur Eddington's photograph of the 1919 solar eclipse, with which he attempted to test Einstein's General Theory of Relativity, and Edwin Hubble's 1921 photograph of a variable star in the Andromeda Nebula, which demonstrated the existence of galaxies outside the Milky Way.

As well as informing and educating the visitor, the project team wanted to convey the experience of being in an art gallery, to showcase the aesthetic qualities of the images and to stimulate a sense of awe and wonder, and it was essential that the design and layout of the exhibition was sympathetic to these aims (FIG. 14.2). Two London-based design consultancies were engaged to realise this aspiration, both of which were known for their work on a range of cutting edge cultural and commercial projects as well as having previous experience in designing science exhibitions. The 2D and 3D gallery designs, as well as graphics and layouts, were carried out by Bibliothèque, whose portfolio includes the *Treasures* gallery at London's Natural History Museum, the Victoria & Albert Museum's *Cold War Modern* exhibition (2008) and work for iconic design and lifestyle magazine *Wallpaper**.[8] Digital design and interactive elements were provided by Kin Design, who had previously worked

---

[8]   http://www.bibliothequedesign.com/projects/exhibition-and-environment/visions-of-the-universe/ [accessed 1 February 2016].

FIGURE 14.2. Installation shots of Royal Museums Greenwich's *Visions of the Universe* exhibition (2013) with gallery design by London-based design consultancy Bibliothèque. National Maritime Museum.

with the Manchester Museum of Science & Industry as well as designing the hi-tech interactive 'Denim Studio' for Selfridge's department store and window installations for the Ted Baker fashion brand.

In choosing the photographs and deciding how to display and interpret them, the *Visions of the Universe* project team were influenced by a number of key sources.

FIGURE 14.3. '*Mars Window*' digital installation in the *Visions of the Universe* exhibition showing a projected NASA Mars Science Laboratory rover panorama of Gale Crater and Mount Sharp. National Maritime Museum.

Artist Michael Light's *Full Moon*, a groundbreaking exhibition of photographs from the NASA Apollo missions, first displayed concurrently in 1999 at London's Hayward Gallery and the San Francisco Museum of Modern Art and also available as a book, was an important touchstone.[9] The work of art historian Elizabeth A. Kessler was also influential, particularly in its focus on the aesthetic dimensions of contemporary astronomical images and the debt that they owe to nineteenth century landscape painting and artistic traditions of the Sublime.[10] Design cues were also taken from popular science fiction movies and TV shows such as *2001: A Space Odyssey* (1968), *Moon* (2009) and *Doctor Who* (1963–present) to provide a stylish, contemporary feel. A consistent design motif applied to the walls throughout the gallery was the cross-shaped 'reseau mark' familiar from Apollo photography of the Moon.

---

[9] Michael's Light's website, available at http://www.michaellight.net/fm-intro/ [accessed 1 February 2016].

[10] Elizabeth A. Kessler, *Picturing the Cosmos: Hubble Space Telescope Images and the Astronomical Sublime* (Minneapolis, MN: University of Minnesota Press, 2012).

Most images were displayed either backlit on lightboxes or printed on 'Dibond' aluminium composite display boards. However, from the beginning it was intended that the exhibition should be punctuated by large-scale projections at key points to enhance and focus the narrative and provide moments of awe and wonder. Here a key inspiration was artist Olafur Eliasson's installation *The Weather Project* in the Turbine Hall at Tate Modern, London, in 2003.[11] This vast indoor evocation of the Sun shining through mist beneath a mirrored ceiling was immensely popular, creating an environment in which visitors gathered to sit, gaze, and contemplate. For *Visions of the Universe* a similar effect was accomplished with several animated sequences of astronomical data showing phenomena such as the phases of the Moon, solar activity and galaxy formation. But perhaps the exhibition's most arresting moment was provided by the 'Mars Window', a 13m by 3m curving wall onto which were projected panoramic images of the Martian landscape, as photographed by NASA's Spirit, Opportunity and Curiosity rovers. Here we worked with Professor Sanjeev Gupta of Imperial College London, a member of the Mars Science Laboratory mission team who acted as our liaison with the science and imaging teams from the various rovers.

Audience response to the exhibition was extremely positive: more than 70,000 visitors attended during the three-month run of the show in summer 2013 (almost double the projected attendance of 40,000) with a broad demographic which included young adults and children with families but also older, more traditional museum audiences. The media response was also extremely favourable, with positive reviews from both science and arts journalists – none more so than *The Guardian* newspaper's art critic Jonathan Jones, who awarded the show a rare five-star review in a piece which ran across the broadsheet's centre pages, saying 'to put it bluntly, this is the most beautiful and significant exhibition I have seen in ages'.[12] Press reviews also focused on the power of astronomical images to inspire as well as explain, with a leader column in *The Independent* suggesting that 'If there were any question… as to the wisdom of spending public money on such optional extras as sending Major Tim Peake to the International Space Station, so inspirational a display must surely dispel them'.[13] A supporting programme of events and activities allowed the Muse-

---

[11] Tate website, available at http://www.tate.org.uk/whats-on/exhibition/unilever-series-olafur-eliasson-weather-project/olafur-eliasson-weather-project [accessed 1 February 2016].

[12] Jonathan Jones, 'Visions of the Universe exhibition reveals full wonder of space images', *The Guardian*, 11 June 2013, available at http://www.theguardian.com/artanddesign/2013/jun/11/visions-of-the-universe-greenwich-space-photography [accessed 1 February 2016].

[13] Editorial, 'Every reason to aim for the stars', *The Independent*, 23 May 2013, available at http://www.independent.co.uk/voices/editorials/editorial-every-reason-to-aim-for-the-stars-8629696.html

um to explore some themes in more depth and in more daring or unusual ways: a highlight was a one-day symposium on astronomy and the Sublime, with artists, art historians, writers, amateur and professional astronomers all attending.

Visions of the Universe has now toured to several other UK venues but perhaps its most influential legacy has been in the way it has helped to forge a new spirit of exploration and collaboration between the astronomers of the Royal Observatory and the art curators of the wider Royal Museums Greenwich. One marker of the profound shift in thinking that *Visions of the Universe* entailed for the team at Greenwich was the inclusion in the exhibition of seven photographic artworks by Wolfgang Tillmans: six C-type prints from his *Venus transit, 2004* series and the inkjet print *In flight astro (ii)* (2010). With hindsight, by placing these works alongside scientific images made by professional astronomers the Museum was making a statement of intent and signalling a new commitment to contemporary art that explores scientific themes.

Collaborations with artists: showcasing and commissioning

In the words of *The Art Newspaper*'s editor, Anna Somers Cocks, 'art, like science, is our tool for understanding the world'.[14] The link between art/aesthetics and science is not only a powerful one in the dissemination of astronomical knowledge, it is also of significant import in the way artists are themselves inspired by space and the related progress in science and exploration. Indeed, the ROG is regularly approached by artists interested in its site, history, collections and staff expertise, while the historical collections themselves reflect this fruitful collision. The challenge remains to devise a sound strategy for collecting for the future, and to illustrate the contemporary relevance of astronomy through artistic engagement. Crucially, it is these collaborations and conversations with artists that are helping us articulate our ambitions and shape the direction of the Museum's contemporary art strategy.

Visions of the Universe galvanised the RMG's approach to both astronomy and contemporary art but for some time the ROG had already been playing a prominent role in supporting creativity by showcasing and commissioning artworks as well as facilitating access to the collections and expertise of its staff, who function as an advisory group or sounding board. In fact, since 2012, we have been working to develop these connections, sometimes taking their projects outside of the Museum

---

[accessed 1 February 2016].

[14] Anna Somers Cocks, *The Art Newspaper* 275, 1 January 2016, p. 12, available at http://theart-newspaper.com/news/news/art-provides-the-answers-to-questions-no-one-has-asked/ [accessed 24 September 2016].

to test out how artists respond to astronomical themes, and to build up relationships with contemporary practitioners, curators, collectors and gallerists.[15]

Within its buildings and permanent galleries, the ROG showcases art – both historical and contemporary – alongside objects illustrating the history of astronomical knowledge. On long term loan to the ROG since 2012, sculptor Matthew Luck Galpin's *Anvilled Stars* are displayed within the 'Astronomy Questions' gallery, alongside objects and instruments which have played a key role in our search for knowledge about the Universe, such as the first edition of Sir Isaac Newton's *Principia* (1687) and the flight-spare mirror of the ESA's Hipparcos astrometric satellite. Craftsmanship and conceptual art meet in *Anvilled Stars*: Galpin used his blacksmithing skills to rework iron meteorites by heating, hammering, grinding and polishing them into mirrors. This process echoes the formation of the planets, pulled together and shaped by gravity, heat and rotation, and thus re-forming these fragments of space debris becomes akin to recapitulating the origins of the Solar System. Galpin's meteorites fell to Earth about 6,000 years ago, having formed when the Solar System was young. They are leftover remnants of the material from which the planets were made, the building blocks of the Earth and everything on it. The haptic quality of the objects – between mirror and small scale, handheld sculpture – makes them beguilingly intimate pieces. The imperfection of their mirroring surface offers a blurred, moving vision of ourselves, while their making process also echoes the grinding of telescope mirrors and lenses for telescopes, vital instruments for astronomers' exploration of the Universe.[16]

The ROG also supports artists (and other creative practitioners, such as filmmakers, authors and poets) by launching their work. In 1716, Edmond Halley had suggested that the observation of the transit of Venus would enable the distance between the Earth and the Sun to be calculated, a proposition that could not be verified

---

[15] Such projects include the exhibition *dark frame / deep field* at London's Breese Little gallery, available at http://www.breeselittle.com/past-dark-frame-deep-field/4589314270 [accessed 1 February 2016]; also see M. Kukula and M. Vandenbrouck, Essay in *dark frame/deep field* e-catalogue, pp. 134–58, available at https://issuu.com/breeselittle/docs/dark_frame_deep_field_-_e. catalogue_021e95975378ba?e=8048082/13972233 [accessed 24 September 2016]; and public speaking, for instance for Wolfgang Tillmans's Hasselblad Award symposium in Gothenburg, 1 December 2015, see http://www.hasselbladfoundation.org/wp/wolfgang-tillmans-3/ [accessed 24 September 2016].

[16] These works have also been shown at the Science Museum, London, and at the Museum of the History of Science, Oxford. See Matthew Luck Galpin's website and an essay by Brian Catlin, 'Anvilled Stars', 2010, available at http://www.matthewluckgalpin.com/anvilled-stars/ [accessed 1 February 2016].

until the next occurrence of this natural phenomenon, in 1769. In 2012, the ROG programmed a series of events to celebrate the second (and last) transit of Venus of the twenty-first century. This programme reflected the ROG's long involvement in the observations of these events, with Charles Green, assistant to Astronomer Royal, Nevil Maskelyne, embarking on a voyage to the Pacific in 1769, and with photographic expeditions organised by the Observatory's solar department in 1874 and 1882. On the day after the 2012 transit, the UK premiere of Lynette Wallworth's *Coral, Rekindling Venus* took place in the Observatory's Peter Harrison Planetarium.[17] The making of this video and sound piece was inspired by an encounter between Wallworth and astrophotographer David Malin, during which they compared an image of the birth of a star with that of a micro coral spawn taken on the Great Barrier Reef.[18] As an immersive video installation, *Coral, Rekindling Venus*, connects the depths of the oceans with the wider Universe, the underwater world shown above the viewers' heads to striking effect on the dome of the planetarium. The mass coral spawning of the Great Barrier Reef recalls stellar nurseries and distant galaxies, their luscious colours conjuring up the most extravagant hues from the Hubble palette.

But this piece has deeper, philosophical links to astronomy. 'Coral is the canary in the coalmine of the ocean,' Wallworth said. 'They can handle very little temperature change. It is impossible for us to imagine a sky without stars but we have to be able to contemplate an ocean without coral.'[19] Wallworth was struck by the way that Halley's call to action led to a concerted response by the European scientific community – arguably the first example of international scientific collaboration for the common good – a situation analogous to what is now required to tackle the significant threat of climate change, for which the plight of the coral reef is emblematic. Her piece thus evokes the fragility and timeliness of coral against the astronomical timescales of space. The eighteenth century was also the point of origin of the Sublime – the idea that an awesome spectacle may provoke feelings of fear and wonder in equal measure, and evince the transience of humankind in comparison to the grander scale and natural splendour of the Universe. In the immersive space of the planetarium, Wallworth's haunting video piece provokes similar feelings as those described as 'astronomical sublime' by art historian Elizabeth Kessler. The majestic corals hovering

[17] The film's website, available at http://coralrekindlingvenus.com/ [accessed 1 February 2016]. The film was later presented to great acclaim at the Sundance Festival, in 2013.
[18] Lynette Wallworth's talk, 'Coral rekindling Venus', TedxSydney, 21 June 2012, available at https://www.youtube.com/watch?v=Ll0iPE2uXwU [accessed 1 February 2016].
[19] Mark Brown, 'Lynette Wallworth: a timely journey into the alien world of coral reefs', *The Guardian*, 28 May 2012, available at http://www.theguardian.com/artanddesign/2012/may/28/lynette-wallworth-alien-world-coral-reefs [accessed 1 February 2016].

above the audience, slowly unfolding in unfathomable blackness, have an alien familiarity, showing how the oceans and space both remain at the frontier of knowledge. The marine focus of this piece resonates further within the wider RMG, as Captain James Cook's voyages to the Pacific (the first of which was mounted to facilitate the 1769 observations and took him to the Great Barrier Reef) are a key chapter in the histories told at the ROG's sister site, the National Maritime Museum. In fact, this interconnectedness between the sites' histories is often privileged in RMG's public programme, as can be seen with the following three commissions/exhibitions.

Indeed, beyond simply showcasing or launching work, the institution has also taken a more proactive involvement in artistic practice, by actively commissioning new work from contemporary artists. In 2006, British photographer Dan Holdsworth (born 1974), was invited to show at the NMM as part of the contemporary arts programme *New Visions* (1999–2009). His solo exhibition, *At the Edge of Space part 1–3* (8 June 2006–7 January 2007), combined a selection of works drawn from two existing series alongside new work specifically commissioned by the NMM. The first series, *At the Edge of Space*, was shot at the European Space Agency's spaceport in Guiana in 1999. Depicting Ariane rockets on a background of verdant jungle landscape alongside pristine interiors, reminiscent of the film *2001: A Space Odyssey*, the series showed the natural world and advanced technology sitting side by side. The second series, *Gregorian*, took Holdsworth to the National Astronomy and Ionosphere Centre in Puerto Rico in 2005, where he was given access to the Arecibo Telescope, with its polyhedral Gregorian sub-reflector. The world's largest single-aperture telescope, Arecibo picks up the faintest radio signals from deep space, like whispers uttered from the edges of the known Universe, the telescope peering back in time to reach them. Working with an analogue camera, Holdsworth used very long exposures to capture what the human eye cannot see: the surreal quality of the site and the movement of the Earth evoked in delicate star-trails. The new commission, *Hyperborea*, was shot in Iceland and Norway in 2006. The pastel veils of glowing aurorae, sometimes punctuated by the trails of satellites, contrast with the frozen land below, where feeble traces of human presence (vehicle tracks, artificial light or diminutive human habitat) bring a touch of reality to a fantasy scene.

With these three series, Holdsworth revealed different levels of human interaction with the cosmos: aurorae shimmering above ice-topped dwellings, rockets piercing the Earth's atmosphere, and gigantic instruments scrutinising the farthest reaches of the Universe. By colliding nature with man-made edifices and astronomical phenomena with humankind's attempts to understand them, Holdsworth explored 'the possibilities of human knowledge', the inconceivable expanses of time and space. The collaboration was fruitful to both the artist and the Museum. The new commission enabled Holdsworth to speak to local researchers, engineers and scientists to develop his project. But it affected him on a profound level. In the artist's words,

> … the experience of photographing the Northern Lights felt like I was entering a different time space. Whilst being alone in the arctic wilderness, I became aware of the cycle of the Earth. The lights are a visual representation of everything that we cannot see but which goes on around us all the time. It's like being given a glimpse of the rhythm of the universe.[20]

In fact, Holdsworth continued engaging with the eerie, supernatural landscapes of Iceland in his *Blackout* series, published in 2010.[21] He also served on the judging panel of Astronomy Photographer of the Year from the beginning of the competition until 2012, and in 2016, he donated a work from the *Hyperborea* series to the Museum.

New Visions endeavoured to link RMG's sites and themes, and this 'creative alignment' is an ethos that remains key to current projects, including the more recent Contemporary Arts Programme (2012–present). Following on from the acquisition of Yinka Shonibare MBE's *Nelson's Ship in a Bottle* in 2012, a series of this artist's work was shown across the Museum's sites in the exhibition, *Yinka Shonibare MBE at Greenwich* (18 September 2013–23 February 2014). Shonibare's work resonates with the stories told at RMG in the way in which it interrogates origins, cultural exchange and authenticity through an exploration of Britishness, imperialism and historical narratives. The exhibition encouraged visitors and staff alike to look at the collections across RMG with fresh eyes, and ask different questions about Britain's maritime and stargazing past. For display at the ROG, the Museum commissioned a new piece from the artist's *Planets in my Head* series, in which celestial sphere-headed children partake in learned activities typical of the Enlightenment period: music, writing or scientific observation. Sited within the context of the Observatory's Flamsteed House, his resulting *Cheeky Little Astronomer* (FIG. 14.4) highlighted the dual nature of the ROG site: its crucial relationship with the stars at a research level, and the Astronomer Royal's home from 1675 to the early twentieth century. Enlivening the domestic apartments with a playful presence, the sculpture reminded the visitors that this was once a family home, with children growing up cheek-by-jowl with working astronomers and their instruments. Dressed in Shonibare's signature Dutch wax fabric, this figure also suggested a darker undertone: that by helping to improve

[20] 'At the Edge of Space', National Maritime Museum, 08/06/2006 – 07/01/2007, http://www.danholdsworth.com/exhibitions/attheedgeofspace/ [accessed 1 February 2016].
[21] Dan Holdsworth, *Blackout*, with an essay by Oliver Morton (Göttingen: Steidl, 2012).

FIGURE 14.4. *Cheeky Little Astronomer* by Yinka Shonibare MBE (2013, mixed media), in situ in Flamsteed House during *Yinka Shonibare MBE at Greenwich*; © the artist and National Maritime Museum, Greenwich, London.

maritime navigation, astronomers also contributed to Britain's imperial project. Inspired by Indonesian batik and first produced by the Dutch for a European market, this brightly coloured printed cotton has now become a powerful symbol of West African identity, but also remains a marker of a complex trade and colonial history. Upon his election as Royal Academician in 2014, Shonibare chose *Cheeky Little Astronomer* as his reception piece, thus establishing the significance of this commission, and the considerations that came into making it, in his overall practice.

What is crucial in such collaborations is to lead both the artist and the institution in new directions. When successful these dialogues offer fruitful, sometimes necessary, grounds for investigation (and indeed self-reflection) for both parties. By helping artists to develop new work, RMG is also part of the intellectual process. The ROG's expertise, collections and advisory role was instrumental for the production of artist Dryden Goodwin's first feature-length film, *Unseen: The Lives of Looking* (2015), commissioned by RMG with support from the Wellcome Trust, and showcased in the Queen's House (5 March–26 July 2016) as part of the Contemporary Arts Programme.[22] Goodwin is known for his exploration of portraiture in intensely

[22] 'Unseen: The Lives of Looking (2015)', http://www.drydengoodwin.com/unseen_installation.

sketched drawings. *Unseen* furthers his practice of close observation through drawings and film, focusing on three individuals who each have a compelling relationship
to looking: an eye surgeon, Sir Peng Tee Khaw, a human rights lawyer, Rosa Curling,
and a planetary scientist, Professor Sanjeev Gupta. This work proved a new departure for Goodwin, not only as he engaged with a new medium, but also because of
the close collaboration he established with scientists – both Gupta and the astronomy team at the ROG.

Gupta works with NASA and the European Space Agency, using his geological
knowledge to study the surface of Mars via images gathered by the Mars Science
Laboratory 'Curiosity' rover. Like Goodwin, Gupta uses observational drawing as
part of his practice, and the artist followed the scientist on a field trip to the Jurassic
Coast in Dorset, where the latter produced analytical drawings of rock formations.
The exhibition in the Queen's House brought together Goodwin's drawings and the
objects that formed part of the filmmaking process. A showcase presented the tools
and papers of Gupta's trade, such as measuring instruments (from a simple tape
measure to a compass clinometer for measuring the tilt of rock beds), observing
and recording instruments (from notebook and pencil to digital camera and 3D
glasses), as well as rock samples and a NASA Curiosity viewing device. Gupta has
a long-standing relationship with the ROG (having worked on the 'Mars Window'
for *Visions of the Universe* in 2013), and Goodwin filmed and drew Gupta using the
ROG's 1894 Great Equatorial Telescope to observe Mars in April 2014.

Crucial to this first unveiling of Goodwin's *Unseen* was the context of Greenwich
itself. The exhibition showed how the three contemporary 'lookers' related to historical Greenwich 'lookers'. The latter included the first Astronomer Royal, John
Flamsteed (1646–1719), who created his catalogue and charts of the fixed stars from
nightly observations, initially on the roof of the Queen's House and then from the
Royal Observatory. Also showcased were objects relating to the Astronomer's Assistant, Edward Walter Maunder (1851–1928), whose appointment as Photography
and Spectroscopy Assistant in 1872 brought the Observatory into the age of astrophysics. Analogous to Gupta's contemporary investigations, in the late nineteenth
and early twentieth centuries Maunder became involved in debates about the nature of Mars, responding to some astronomers' claims that they could see 'canals'
or channels on the surface. With the help of the Royal Hospital School (which was
based in the buildings that became the NMM in 1937), Maunder carried out a vision
experiment with some of its boys; the results made astronomers think more carefully
about the limits of telescopic observation.

htm [accessed 1 February 2016].

Beyond illustrating the place of the ROG in the development of astronomical knowledge, *Unseen* illuminated this discipline in original ways. The film and related objects revealed how by combining intense scrutiny over interplanetary distances with his tactile and forensic understanding of Earth's sedimentary rock formations, Gupta's work reminds us of how we look for the familiar in the alien. As for Goodwin, this project allowed him to explore the act of looking from his own as well as his sitters' multi-facetted perspectives. After a successful launch at Greenwich, the film toured the US in 2016, thus expanding further the reach of the Royal Observatory.

Collecting

While temporary exhibitions and commissions are inspiring for both visitors and artists alike (and certainly for the Museum staff), as a national museum RMG has an important remit as custodian, whose aims include collecting for the future. RMG holds a collection of over two and a half million objects relating to its mission of illustrating the importance of sea, ships, stars and time, and their relationship to people. As such, part of the historical collection relates to astronomical knowledge and the role of the ROG astronomers in its development, through scientific instruments, records of observations, scientific papers and also fine art. Indeed, for centuries, art had a crucial role in visualising astronomical knowledge and recording scientific information and the work of astronomers for dissemination to their peers and the wider public.

A Royal Academician and Crayon Painter to King George III and to George, Prince of Wales, John Russell (1745–1806) was one of the most renowned portraitists of the eighteenth century. He had a large and fashionable clientele, which included astronomers, and sensitive portraits of his friends William Herschel and Sophia and Nevil Maskelyne are now in the Museum's collection.[23] Russell was also a keen amateur astronomer. His obsession with the Moon led him to spend thirty years observing its features through his telescope, producing detailed pencil and pastel studies.[24] Russell's pastel drawing of the Moon (Fig. 14.5) in the collection at Greenwich was completed with a micrometer, and his remained the most accurate delineations of the Moon until the introduction of photography. These were produced to

---

[23] The former oil on panel, the latter two pastels on paper, and respectively museum nos. BHC2764, http://collections.rmg.co.uk/collections/objects/14237.html, ZBA5101, http://collections.rmg. co.uk/collections/objects/562247.html and ZBA5100, http://collections.rmg.co.uk/collections/objects/562246.html [accessed 1 February 2016].

[24] The graphite sketches and one large pastel are now in the collection of the Museum of the History of Science in Oxford. Other Moon pastels are held in the collections of the Royal Astronomical Society, London, Science Museum, London, and Birmingham Museum and Art Gallery.

FIGURE 14.5. The artist as amateur astronomer and the astronomer as amateur artist. *The Moon* by John Russell (ca .1787, pastel on paper, museum no. PAJ3148) and *The Great Comet of 1843* by Charles Piazzi Smyth (1843, oil on canvas, museum no. BHC4148). National Maritime Museum.

perfect his map of the Moon, and conceive, in 1797, an extraordinarily sophisticated instrument, the *Selenographia,* a scientific lunar globe constructed to reproduce the libration of the Moon with respect to the Earth.[25]

If the artist Russell was an amateur astronomer, until the invention of photography astronomers were often amateur artists, as they needed to record their observations both in writing and visually. Among them, the astronomer and self-taught artist Charles Piazzi Smyth (1819–1900) produced landscapes (or skyscapes) that were as scientifically useful as they are beautiful. Before becoming Astronomer Royal for Scotland (1846–1888), from the age of sixteen Smyth was assistant to Sir Thomas Maclear, Her Majesty's Astronomer at the Cape of Good Hope, where he observed Halley's Comet in 1835–36 and the Great Comet of 1843, C/1843 D1. He depicted the latter in two paintings showing the comet over Table Bay, by day and by night. Roberta Olson and Jay Pasachoff argue that, while he strove for 'accurate positional observation, he was also eager to note the physical characteristics'.[26] His drawings were primarily produced for scientific aims, but, as Olson and Pasachoff suggest, his paintings '... were clearly intended to be works of art, not merely scientific

---

[25] Museum no. GLB0140, http://collections.rmg.co.uk/collections/objects/19827.html [accessed 1 February 2016].

[26] Robert Olson and Jay Pasachoff, *Fire In The Sky* (Cambridge: Cambridge University Press, 1998), pp. 202–3.

FIGURE 14.6. *The Sun*, issued by the Working Men's Education Union (about 1850–60, cotton, Museum no. ZBA4551); *Searching for Halley's Comet at Greenwich Observatory*, by William Heath Robinson (pen, ink and watercolour, 1909, Museum no. ZBA5194). National Maritime Museum.

recordings'.[27] In a suitably Sublime approach, in the night scene (FIG. 14.5), Smyth evokes the awe at observing astronomical phenomena: the diminutive figure in the foreground provides a sense of wonderment and suggests the small place of humankind in the Universe. However, the painting also accurately records the relative positions of the stars, the overall brightness of the comet in relation to them, and the scale of its bifurcated tail.

The Museum's collection also reflects the use of astronomical drawings for educational purposes, with objects such as a group of nine wall hangings produced for the Working Men's Educational Union in the 1850s (FIG. 14.6).[28] This philanthropic society was founded in 1853, for 'the elevation of the working classes, as regards their physical, intellectual, moral, and religious condition'. Their chief means of achieving this was to encourage popular literary and scientific lectures 'by preparing suitable Diagrams and other aids to lecturers' and to promote lending libraries and 'mutual instruction classes'. Equally attractive for their design sense and visual simplicity, these visuals are borrowed from instructional prints, such as those published by James Reynolds in the mid nineteenth century. It is likely that they were reproduced on cotton to avoid paper duty, but also to be easily transportable. Their early educational aim reflects the ROG's current role as learning and public engagement centre.

---

[27] Olson and Pasachoff, *Fire In The Sky*, p. 206.

[28] Museum nos. ZBA4550 to ZBA4558, http://collections.rmg.co.uk/collections.html#!csearch;authority=agent-173313;browseBy=maker;makerFacetLetter=W [accessed 1 February 2016].

Not all depiction of knowledge is serious, however. RMG holds an outstanding collection of caricatures which, while mostly of naval subjects, also includes such staple satires as aged astronomers distractedly looking through their telescope while their young wives are preoccupied with their lovers.[29] Some caricatures relate to the ROG specifically. The recently acquired sketch by William Heath Robinson (FIG. 14.6) shows a group of cultural figures, astronomers and Greenwich regulars (two of these may be identifiable as George Bernard Shaw and 'The Greenwich Time Lady' Ruth Belville) observing Halley's Comet on the roof of Flamsteed House. The cartoon was reproduced in *The Sketch* on 17 November 1909, anticipating the 1910 return of the comet on 20 April 1910. In true 'Heath Robinson' fashion (the caricaturist was best known for his drawings of absurd mechanical inventions), three astronomers are seen climbing a large telescope, additionally holding a magnifying glass, binoculars, a spyglass and spectacles to obtain maximum magnification, while the comet smiles mischievously at them. This light-hearted take on astronomy's endless quest for bigger and better telescopes illustrates the growing popular interest in astronomy, but also demonstrates that the ROG has long been intimately connected in the popular imagination with momentous cosmic events.

Despite the positive engagement with living artists described above, and the richness of the astronomy-themed art collection, contemporary art remains conspicuous by its absence in the ROG collection, a lack we are currently endeavouring to remedy. One direct consequence of *Visions of the Universe* was a sense among both the astronomy and the art teams at the Museum that we should acquire examples of Wolfgang Tillmans' works. Tillmans' practice is distinguished by an observation of his surroundings and an on-going investigation of the foundations of the photographic medium. By his own account, astronomy was his 'childhood obsession', when he first grappled with the use of a camera, and was initiated with a discipline that required intense scrutiny – a useful learning experience for someone who would become one of the world's leading art photographers.[30] His work frequently reflects this early interest with the inclusion of various astronomical subjects and themes, and in 2014, the Museum acquired three photographic prints from the *ESO* series (FIG. 14.7).

---

[29] Such examples include Jean-Baptiste Le Prince's 1776 *Le marchand de lunettes*, Museum no. ZBA4477, http://collections.rmg.co.uk/collections/objects/386459.html; and Thomas Rowlandson's 1811, *Looking at the comet till you get a criek in the neck*, Museum no. ZBA4569 http://collections.rmg.co.uk/collections/objects/460864.html [accessed 1 February 2016].
[30] Wolfgang Tillmans in conversation with Dimitar Sasselow, 'The Edge of Visibility: A Night of Astronomy', Serpentine Gallery, 20 August 2010. (Transcript of the event provided by the artist).

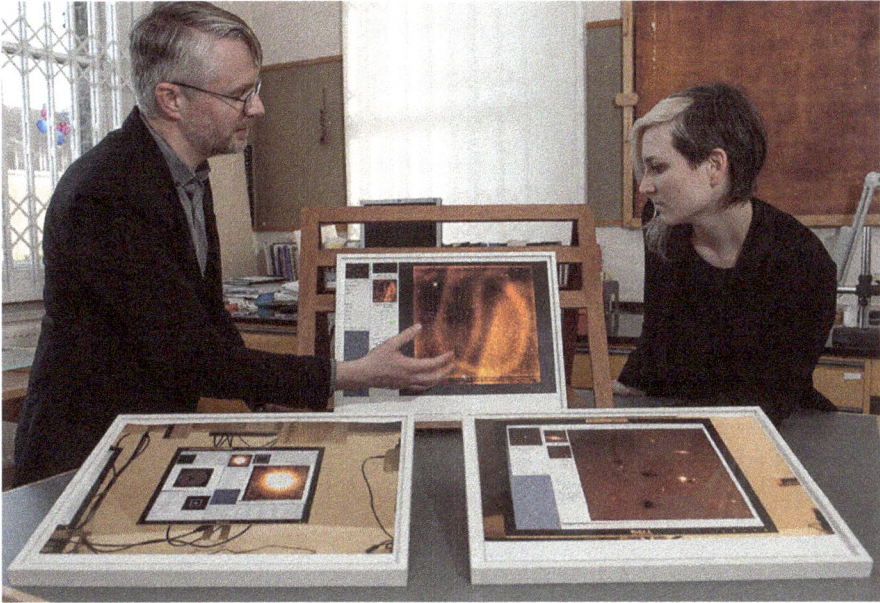

FIGURE 14.7. The authors examine Wolfgang Tillmans' *ESO, 2012* series, acquired by Royal Museums Greenwich in 2014. National Maritime Museum.

Tillmans' artistic approach to astronomy, foregrounding the abstract and aesthetic appeal of the images themselves while exploring and questioning the photographic techniques by which they are produced, speaks directly to the long tradition of astrophotography as practiced in Greenwich. The *ESO* series explores the work of the European Southern Observatory's telescopes at Cerro Parañal – a twenty-first century equivalent of Greenwich in the nineteenth and early twentieth centuries. These are photographs of photographs which are in the process of being created from raw data – astronomical knowledge in the process of construction. What scientists may consider to be mere intermediate stages in the data reduction process – rather than a finished product – Tillmans presents as images in their own right: digital data given materiality by their physical manifestation on a computer display screen. These works also make explicit the limitations and distortions inherent in photography – factors which need to be understood by astronomers in order to interpret their data. Just as the ROG astronomers Walter and Annie Maunder commented publicly on the problems and pitfalls of photographing the heavens, Tillmans' astronomical artworks show how the debate about the representation of the heavens continues to be relevant in the twenty-first century.

Acquiring the *ESO* triptych effected a turning point on the part of the Museum.

189

It was a statement of intent that we would from now on embrace the collision of science and art within the collection itself. It also prompted an invitation to Tillmans to join the judging panel of Insight Astronomy Photographer of the Year in 2016, a move which, it is hoped, will raise the profile of the competition in the fine art and photography communities.

The next 'ROG' art acquisition, in 2015, was *Drawing on Space*, by American artist Michelle Stuart (born 1933). Stuart is best known as a member of the Land Art movement, with Robert Smithson and Nancy Holt, in the 1960s–70s. Her recent practice has involved the use of photography (shot by herself or collected from archival sources and the internet) to create large conceptual grid arrangements. She describes the grids as being like 'silent movies' which have a narrative structure. These photographic grids include 'real history, imaginative history and natural history.'[31] Visually similar to the grids used on early modern star charts, her photographic installations connect to Stuart's career-long fascination with exploration. Mixing images of stars, galaxies and nebulae with those of fireworks, *Drawing on Space* takes celestial and man-made lightshows as its focus, suggesting both a contemplative and euphoric aspect to stargazing. As Stuart questions, 'Do we put a lens on space because we feel alone? Or is it curiosity about the incomprehensible vastness of the universe? Does that cosmic energy, in fact, make us feel, not alone, but in some way complete?'.[32] Some of the components of *Drawing on Space* actually show antique lenses, prompting us to consider the importance of both the telescope and the camera in the advancement and recording of astronomical knowledge. The ways in which Tillmans' work reveals the process behind the imaging of distant galaxies correspond to the ways in which *Drawing on Space* connects the methods of historical and contemporary imaging. While Stuart commands a large audience and is represented in international collections, this is the first time her work enters a public collection in the UK, a very exciting 'first' for RMG.

The latest acquisition, in 2016, *Starfield Shoreline* (2010) by Susan Derges, (born 1955) merges two core RMG subjects: astronomy and maritime history. Derges's work is distinguished by the use of the camera-less technique of the photogram, with which she started working in the 1980s; she is best-known for her work on water, central to her practice since 1990. The photographs from the *Shoreline* series (1997–2010) are taken at night, with the photosensitive paper directly resting onto the shore, combining a microsecond of flash exposure with a longer exposure under

---

[31] Michelle Stuart, with an introduction by Ben Tufnell, *Michelle Stuart: Trace Memory* (London: Parafin, 2015), p. 3.
[32] Stuart, *Michelle Stuart*, p. 10.

ambient moonlight. Taking this experimentation with the medium (reminiscent of the ROG's own historical involvement in experimental photography) further, *Starfield Shoreline* is a composite of two photographic techniques, in which the artist unites a view of the heavens taken in a backyard observatory in South Taunton with a 5/4 Linhof field camera and that of a wave sweeping the shoreline. By overlaying the seashore with the image of a star field, she connects the ocean and the cosmos, life on Earth to the wider Universe. The acquisition was all the more timely for the Museum, as the dye destruction photographic process used, Cibachrome, is becoming increasingly rare (the materials required were discontinued by manufacturers Ilford, with the last production batch delivered in 2012).

These three recent acquisitions were first showcased in the Queen's House, when it reopened for its 400th anniversary renovation and redisplay in 2016. Stuart's and Tillmans' works joined Piazzi Smyth's *Comet of 1843* and astronomical hangings from the Working Men Educational Union, in a room entitled 'Cosmic Sublime', themed around the collision of science and art in the visualisation of astronomical knowledge. Derges' piece formed part of a display focussed on life on the British coast and the evolving practices of photography over a century and a half, showing the documentary capacity of the camera as well as the artistic appeal of the medium. Together, these displays are the first public manifestation of the Museum's new commitment to acquiring contemporary art that engages with astronomy and, we intend, the beginning of an ongoing conversation between astronomers, curators, artists and the public.

Conclusions

As a public institution encompassing history, art and astronomy, Royal Museums Greenwich has a remit to create public programmes and to curate collections that address the cultural as well as the purely scientific significance of astronomy. Recent initiatives such as Insight Astronomy Photographer of the Year and *Visions of the Universe* have challenged the Museum's traditional policy of treating the sciences and humanities separately, while demonstrating the popular appeal of a cross disciplinary approach. Now these insights are also reshaping the ways in which the Museum envisions its collecting policy and its relationship with the sphere of contemporary art. The Museum is opening itself to engagement with contemporary artists, who are inspired by its collections and expertise, and in return it is learning to see its own role and potential in a more nuanced and creative way.

As of 2016 the Museum is formally acquiring the winning photographs from Insight Astronomy Photographer of the Year into its collection, where they will reside alongside historic examples of scientific astrophotography as well as an expanding body of contemporary art. In the future there are opportunities to expand the range of media that we acquire beyond that of photography and to continue to work with artists who have a collaborative approach with scientists.

New discoveries in astronomy continue to have profound implications for the way in which human beings conceive of themselves and their place in the Universe, a concern which unites astronomy with the arts. Indeed, as the artist Wolfgang Tillmans says, 'life is astronomical'.[33] The opportunity to build a national collection of art that engages with astronomy is both exciting and challenging, taking the Museum forward into uncharted territory. For three centuries the Royal Observatory Greenwich was at the forefront of research in astronomy and the development of astronomical instrumentation. Now, in the twenty-first century, it is once more breaking new ground.

---

[33] 'Wolfgang Tillmans in conversation with Beatrix Ruf', preface to Wolfgang Tillmans, *Neue Welt* (Cologne: Taschen, 2012), unpaginated.

THE ZODIACAL LIGHT AND ITS USE IN CULTIC PRACTICE

George Latura

ABSTRACT: In Europe, the zodiacal light was first investigated scientifically toward the end of the seventeenth century. The ethereal glow that appears at a specific time of the night has been known in Islam from the beginning of that religion, around 600 CE. How did the Muslim world know about the zodiacal light, while its scientific discovery in the West had to wait until a thousand years later? Its close link to the equinox, which has been connected to cultic practice over millennia, suggests that the zodiacal light had the potential to have been likewise employed in ancient cultic practice.

Rarely seen today due to light pollution, the zodiacal light is rendered even more elusive because, in temperate climes, it is best seen at opposite seasons of the year, at disparate hours of the night. The rarity of its appearance suggests an opportunity for the zodiacal light to be part of cultic activity. David Kelley and Eugene Milone explain this heavenly phenomenon:

> Dust is a common feature of the solar system… The dust is located predominantly in the ecliptic plane. Sunlight reflected from this dust is visible just after sunset or before sunrise along a tapered cone pointing upward from the horizon, along the ecliptic. It is called zodiacal light. It is best seen at the times of the year when the ecliptic rises most steeply from the horizon. This means early evening in spring, and before dawn in the fall.[1]

The appearance of the zodiacal light at different periods of the night, at different times of the year, makes it difficult to keep track of this ethereal glow. Westerners especially were unfamiliar with this celestial sight.

After he sailed with Commodore Perry to the Orient in the 1850s as a chaplain, chronicler, naturalist and astronomer, George Jones' observations of the zodiacal light were published by the US Senate, and he too reports on its seasonal appearance and disappearance.

> It appears to best advantage when the ecliptic makes its highest angle with the spectator's horizon; at which times, in moderate latitudes, it reaches to his zenith or beyond it, having near the horizon a striking brilliance, and thence fading upward… As those seasons advance when the ecliptic is declining gradually towards

---

[1] David H. Kelley and Eugene F. Milone, *Exploring Ancient Skies: An Encyclopedic Survey of Archaeoastronomy* (New York: Springer, 2005), 138.

the horizon, the Zodiacal Light fades away…[2]

Jones assigns the first Western investigation of the zodiacal light to Cassini in the 1680s.[3] From before Cassini, no scientific investigation of this ethereal sight appears to have survived in the West.

Because this celestial apparition was unfamiliar to Westerners, it would at times be confused with dawn, with daybreak, as related by Jones.

> In the lower latitudes, however, and especially in those near the equator, it has often an exceeding brilliancy at the horizon, ascending from this, a striking object, far into the sky; and I have in several instances known the *reveille* to be beaten in our ships evidently from mistaking the Zodiacal Light for the dawn.[4]

Europeans were not the only ones apt to be misled by the zodiacal light before dawn. After a distinguished career in Britain's Foreign Office as interpreter and translator of Oriental languages, James Redhouse became the Secretary of the Royal Asiatic Society from 1861 to 1864. Writing to that august body in 1880, Redhouse documents the various responses to his query as to whether the zodiacal light of Europeans could be the same phenomenon as the 'false dawn' of the Muslims. One reply claimed this was doubtful because Arab astronomers had not written about this phenomenon.

> I have recently had doubts expressed to me as to the possibility of the false dawn's having been known to the legists and poets of Islam, because no mention has been found, by my correspondent, of this phenomenon in the works of the great Arabian astronomers.[5]

Redhouse's informant seems to have been unaware of al-Biruni's work (c. 1029 CE) that discussed the zodiacal light, perhaps because the title indicated it to be a work on astrology.[6] A somewhat similar disconnect appears to persist to the present day.

---

[2] George Jones, *United States Japan Expedition: Observations on The Zodiacal Light from April 2, 1853 to April 22, 1855, made chiefly on board The United States Steam-Frigate Mississippi. Vol. III* (Washington: Beverley Tucker, Senate Printer, 1856), III.

[3] Jones, *United States Japan Expedition*, IV; See also Eberhard Grun, Bo Gustafson, Stan Dermott, Hugo Fechtig, eds., *Interplanetary Dust* (Berlin: Springer, 2001), p. 6.

[4] Jones, *United States Japan Expedition*, III.

[5] J.W. Redhouse, 'Identification of the "False Dawn" of the Muslims with the 'Zodiacal Light' of Europeans.' In *Journal of the Royal Asiatic Society of Great Britain and Ireland*, Vol. XII (London: Trubner, 1880), p. 334.

[6] Abu al-Raihan al-Biruni, *The Book of Instruction in the Elements of the Art of Astrology*, trans. Ramsay Wright, (London: Luzac, 1934), p. 51–52.

FIGURE 15.1. Zodiacal light rises in the east along the ecliptic, and envelops Saturn, Venus, and waning Moon, shortly before dawn, October 7, 2007 (Skinakas Observatory, Crete). Photo courtesy Dr. Stefan Binnewies.

Relatively recent tomes on Arabic astronomy do not list the zodiacal light in their index, nor seem to discuss it in the text.

A more productive direction for Redhouse's inquiry would be indicated by a Muslim official in Zanzibar, and the final proof was provided by a Muslim cleric in Istanbul.

Dr. Kirk, Her Majesty's Agent and Consul-General at Zanzibar, also sent me an answer, as follows:

'I have made all endeavours to settle the question raised as to the 'false dawn.' I found that Seyyid Bargash, the Sultan, knew more of the thing than any one else here, and got him to tell me exactly where and when to notice it. In these climates it is a thing known at all seasons… Seyyid Bargash wished me to read up on some theological books…'[7]

The only direct and decisive proof I have received… is contained in the following letter… from the commanding officer of Her Majesty's man-of-war steamer *Fawn* (Tuzla Bay, *September* 26, 1879):

'Dear Capt. Evans, – For the information of Mr. Redstone [*sic*], I have to tell you that I can satisfactorily answer his question as to the false dawn of the Turks. On

---

[7] Redhouse, 'Identification of the "False Dawn"…', p. 331.

the morning of 20th inst., at 3:30 AM, I went to a mosque at Biyukdera and inter-viewed the Imam, who, on being asked for the 'fejri kyazib (false dawn),' at once pointed out the zodiacal light, then brightly shining in the east... There can be no doubt as to the coincidence of the two. Yours sincerely, W.J.L. Wharton.'[8]

An interesting conundrum: the zodiacal light seems to be found not in Arabic as-tronomical texts but, besides al-Biruni's 'astrological' work, in Islamic 'theological books' and in a religious context.

The zodiacal light appears to have been known in Islam since the time of Mu-hammad, who gives the timing of fasting during Ramadan at dawn in the Koran (Surah 2, *Al-Baqara*). In order to warn the faithful against mistaking the false dawn (zodiacal light) for the true dawn – the proper time for prayer and fasting – Muslim tradition (*hadith*) explains that there are two kinds of dawn (*fajr*) – the false dawn that is vertical (zodiacal light) and the true dawn that is horizontal.

'Knowing the prayer-times is a prescribed duty for discerning Muslims. This is summarized in the *Qu'ran*, my friend, and was explained by Ahmad [*i.e.*, the Prophet Muhammad, referred to as Ahmad in the *Qu'ran*], the most outstanding of men... Bear in mind that there are two stages of daybreak according to our doctrine: distinguish between them carefully – you are the one who decides this. The first daybreak looks like a wolf's tail rising in the sky: this is the false dawn. The later one is the true dawn: you see it illuminate the sky like a fire.'[9]

Samurah bin Jundab narrated that the Messenger of Allah said: 'The Adhan of Bilal should not stop you from eating, nor the whiteness in the horizon. You should stop eating when whiteness spreads like this (horizontally).' Hammad [Muhammad] motioned his hand to show the horizontal position of the streaks of light.[10]

This important information about the false dawn – the whiteness in the horizon that is vertical – was transmitted, from the earliest days of Islam (c. 600 CE) to the pres-ent day, because in the sacred month of Ramadan devout Muslims must fast during daylight hours, but may eat at night. A careful watch must be kept on the horizon. One should not be fooled by the vertical 'tail of the wolf' or false dawn – the zodia-cal light. You can keep eating until the appearance of the true dawn, the horizontal

---

[8] Redhouse, 'Identification of the "False Dawn"...', p. 333.

[9] David A. King, *In Synchrony With the Heavens: Studies in Astronomical Timekeeping and Instru-mentation in Medieval Islamic Civilization, Vol. 1: The Call of the Muezzin* (Leiden: Brill, 2004), p. 215.

[10] Zakiuddin Abdul-Azim Al-Mundhiri, *The Translation of the Meanings of Summarized Sahih Mus-lim* (Ryadh, Saudi Arabia: Darussalam, 2000), p. 320.

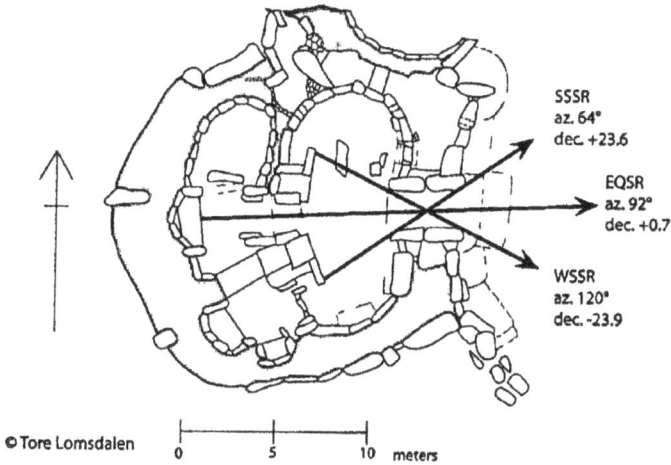

FIGURE 15.2. South Temple at Mnajdra (c. 2500 BCE) aimed eastward to the equinox sunrise. Plan courtesy Tore Lomsdalen, from his *Sky and Purpose in Prehistoric Malta: Sun, Moon, and Stars at the Temples of Mnajdra* (Ceredigion, Wales: Sophia Center Press, 2015), p. 132.

dawn indicated by Muhammad himself with a hand gesture. Then eating stops and prayer begins. Awareness of the zodiacal light in Islam seems to have been viewed not as astronomical knowledge but as part of cultic practice.

Islam's lunar calendar does not match the solar cycle of the seasons, and the month of Ramadan gradually drifts through the different seasons of the year. The zodiacal light, on the other hand, is closely tied in temperate zones to the equinoxes: 'early evening in spring, and before dawn in the fall,' as Kelley and Milone explained.[11]

In Redhouse's account, the Imam in Istanbul pointed to the zodiacal light on the morning of September 20, and the connection between equinox and zodiacal light proves most useful. The equinox does appear in cultic context over millennia and across continents, which suggests that the zodiacal light that is closely linked to the equinox has the opportunity for similar usage in cultic practice.

In the past some doubt has been expressed concerning the use of the equinoxes in megalithic architecture.[12] However recent work is revising this opinion. For example, Lomsdalen documents the equinoctial alignment of one of the oldest cultic sites in Europe, at Mnajdra in Malta.[13] The main axis of the South Temple (c. 2500 BCE) is aligned to the east, to the equinoctial sunrise (FIG. 15.2). In line with

---

[11] Kelley and Milone, *Exploring Ancient Skies*, p. 138.

[12] Clive Ruggles, *Ancient Astronomy: An Encyclopaedia of Cosmologies and Myth* (Santa Barbara: ABC-CLIO, 2005), pp. 151, 213.

[13] Tore Lomsdalen, 'Mnajdra Was Not Built in A Day,' in *Proceedings* of SEAC 2012 Conference (Ljubljana: Slovene Anthropological Society, 2013).

this thinking the organizers of the SEAC Conference in Malta offered an 'organized tour to observe the spectacular Autumn Equinox Sunrise at the Mnajdra Temple' on Wednesday, September 24, 2014.[14]

Through aligned portals pointing east, the equinoctial rising of the sun is revealed for several days, a remarkably accurate planetary clock that has registered the equinox for about 5000 years. If the bus carrying the SEAC delegates to the Mnajdra equinox sunrise had arrived ninety minutes earlier, they might have witnessed (depending on light pollution) another amazing sight: the zodiacal light that foretells the fall equinox sunrise – illustrating the close relationship between these two celestial apparitions. Seen from the same location, looking in the same direction (east) about an hour apart.

Krupp saw equinoctial alignments in the geometrical stone patterns found in the British Isles.

> Equinoctial alignments can also be found at Long Meg and Her Daughters, where two conspicuously larger boulders, on opposite sides of the ring, establish the east-west line; in a line of 18 stones at Eleven Shearers, in southeast Scotland; and among the parallel lines of small stones on Learable Hill, near the border between the Highland counties of Sutherland and Caithness. One of the five lines extending from the main ring of Callanish, the 'Scottish Stonehenge,' is oriented equinoctially.[15]

Mesoamerica seemed to have likewise recognized the cultic power of an equinox astronomical alignment. Krupp points to the eastward orientation of the Maya pyramid and temple group at Uaxactun, and Aveni gives a vivid reconstruction of the equinox sunrise in Aztec cultic practice at Templo Mayor in Mexico.

> From the top of Pyramid E VII sub, or better still, from a lower spot on its east stairway, solstice sunrises appear at the outer corners of the temples on the north and south. The line of sight over the center temple agrees very nicely with equinox sunrise.[16]
>
> Evidently the priest and worshippers faced to the east to view the sunrise in the gap between the twin temples atop the great pyramid… The extensive plaza fronting the twin temples atop the Templo Mayor platform would have provided an excellent vantage point for witnessing the main event that usually took place there. High on the terrace above, a victim, stretched out on his back over a carved stone and held fast by four assistants, could be seen awaiting sacrifice to the Aztec

---

[14] http://www.seac2014.com/tours.html

[15] E. C. Krupp, *Echoes of the Ancient Skies: The Astronomy of Lost Civilizations* (New York: Harper & Row, 1983), p. 167.

[16] Krupp, *Echoes of the Ancient Skies*, p. 249.

Figure 15.3. Equinoxes in Mithras cult: Attendants with torches and intersecting legs purportedly signify the equinoxes. Photo courtesy Reza Assasi (Royal Ontario Museum).

sun god. This took place when the High Priest plunged a flint knife into the victim's chest and tore out the beating heart, offering it to the sun, thus ensuring the continuation of the movement of the solar deity on his course.[17]

The use of the equinox in cultic practice has also been proposed in Ulansey's identification of the two attendants at the Mithraic tauroctony with the equinoxes.[18] The intersecting legs of Cautes and Cautopates suggest to these authors the intersections of the ecliptic and the celestial equator that indicate the equinoxes (Fig. 15.3). Beck likewise interprets the layout of a mithraeum in equinoctial terms.

The cult-niche is thus the spring equinox, the entrance the autul equinox, and the aisle is the diameter of the universe.[19]

---

[17] Anthony F. Aveni, *Skywatchers: A Revised and Updated Version of Skywatchers of Ancient Mexico* (Austin: University of Texas Press, 2001), p. 238.

[18] David Ulansey, *The Origins of the Mithraic Mysteries: Cosmology and Salvation in the Ancient World* (New York: Oxford University Press, 1989), 64; See also Elizabeth Wayland Barber and Paul T. Barber, *When They Severed Earth From Sky: How The Human Mind Shapes Myth* (Princeton: Princeton University Press, 2004), pp. 197f, 206f.

[19] Roger Beck, *The Religion of the Mithras Cult in the Roman Empire: Mysteries of the Unconquered Sun* (Oxford: Oxford University Press, 2006), p. 108.

With the cult of Mithras carried by Roman legions from Syrian Dura-Europos to the British Isles, the equinox was apparently used in cultic practice across the Roman Empire.

The equinox therefore appears to have been used in cultic practice over thousands of years and thousands of miles. The close connection between the equinoxes and the zodiacal light suggests that potential exists for cultic usage of the celestial light that precedes the autumnal equinox sunrise, and follows the vernal equinox sunset, by about an hour in temperate zones.

Various authors have also proposed links between the zodiacal light and cultic practice in the ancient world. Gandz pointed to cultic names that suggest the zodiacal light among ancient Egyptians, Arabians, Syrians and Phoenicians.[20] He also proposed the cultic use of this celestial phenomenon in Hebrew temple sacrifice.

> We know that even in the temple of Jerusalem the Hebrew priests were anxious to slaughter the daily morning sacrifice just at the time of the appearance of the pillar of the dawn, which marked the beginning of the ritual and sacrificial day.[21]

Gandz references Habakkuk 3: 3–4, verses 'which', he says, 'have hitherto defied all attempts at a rational interpretation,' verses that he recasts as a description of the zodiacal light.

> 3. 'God cometh from Teman, and the Holy One from Mount Paran, Selah. His majesty covers the heavens and the earth is full of His glory.
> 4. And a brightness appeareth like that of the light-pyramid of the dawn...'[22]

In an earlier paper Gandz describes the discovery of a hieroglyph shaped like an upward-pointing triangle that purportedly represented the Egyptian god Sopt, the embodiment of the zodiacal light.

> From various hieroglyphic texts and inscriptions Brugsch was able to establish that the Egyptians not only knew, but also deified and worshipped the pyramid of the zodiacal light as the messenger, the herald and harbinger of the rising sun.[23]

---

[20] Solomon Gandz, 'The Zodiacal Light in Semitic Mythology', in *Proceedings of the American Academy for Jewish Research*, Vol. 13 (New York: 1943), pp. 1–39.
[21] Gandz, 'The Zodiacal Light in Semitic Mythology', p. 34.
[22] Gandz, 'The Zodiacal Light in Semitic Mythology', p. 16.
[23] Solomon Gandz, 'The Zodiacal Light in Ancient Hebrew Literature' in *Proceedings of the American Academy for Jewish Research*, Vol. 9, 1939. Reprinted in *Studies in Hebrew Astronomy and Mathematics* (New York: Ktav, 1970), p. 219.

Along similar lines, Gary and Talcott advanced the hypothesis that since the cult of the Sun god in Egypt was a continuum – with the journey of the Sun below the earth at night depicted in Pharaonic tombs – one of the special moments along this circular path (ecliptic) is the foretelling of the Sun's imminent appearance by the zodiacal light that manifests itself shortly before sunrise. [24]

> The first signs of the appearance of the sun god peaking over the akhet and becoming visible to the world… This we interpret to be the zodiacal light, sometimes referred to as 'false dawn'.[25]

The Sun cult was of paramount importance in Egypt over thousands of years, with the Pharaoh closely connected to Ra. Each night, the Sun traveled through the Underworld, the Duat, facing daunting foes and numerous obstacles. Then the unvanquished Sun rises at dawn to recreate the world.. The appearance of the zodiacal light could therefore have heralded this moment.

Another hypothesis that connects the zodiacal light to cultic practice proposed that the Mysteries of Eleusis held near Athens for a thousand years had an astronomical link to that celestial sight.[26] The Lesser Mysteries of Eleusis were held in spring, in preparation for the Greater Mysteries that were celebrated in the fall, the two opposite equinoctial seasons when the zodiacal light is best witnessed.

Boutsikas and Ruggles suggested that the Mysteries had an astronomical component.[27] The ten-day festival included a *pannychis* (all-night ritual) that is described by Euripides:

> …the all-night torch of the twentieth day
> When the star-gleaming heaven of Zeus
> Strikes up the dance
> And the Moon dances…[28]

That the zodiacal light is visible when the waning Moon is in a dark sky (FIG.15.1)

---

[24] Erik Hornung, *The Ancient Egyptian Books of the Afterlife* (Ithaca: Cornell University Press, 1999).

[25] Patricia Blackwell Gary and Richard Talcott, 'Illuminated in Lightland: The Archaeoastronomical Origins of the Seat of the First Occurrence in the Egyptian Solar Cult Religion', *American Research Center in Egypt Bulletin* (Spring 2006): p. 28.

[26] George Latura, 'Plato's X & Hekate's Crossroads: Astronomical Links to the Mysteries of Eleusis', *Proceedings* of SEAC 2013 Conference, *Journal of Mediterranean Archaeology & Archaeometry* 14, no. 3/4 (2014): pp. 37–44.

[27] Efrosyni Boutsikas and Clive Ruggles, 'Temples, Stars, and Ritual Landscapes: The Potential for Archaeoastronomy in Ancient Greece', *American Journal of Archaeology* 115, no. 1 (2011): p. 56.

[28] Euripides, *Ion*. Trans. David Kovacs, (Cambridge, MA: Harvard University Press, 1999), p. 449.

was confirmed by Stefan Binnewies, who took the photograph shortly before sunrise, a few weeks after the fall equinox. Since the ethereal glow of the zodiacal light along the ecliptic is best seen in temperate zones at a specific time of the night – at a particular time of the year – it would seem well suited to religious ritual and recurrent cultic festivals.

To conclude, the zodiacal light was first investigated scientifically in Europe by Cassini in the 1600s. Although most Arabic astronomical works do not mention the zodiacal light, it has been known in Islamic texts since the time of Muhammad as the 'false dawn,' as opposed to the true dawn, when eating stops and the morning prayer begins in the month of Ramadan. In temperate zones, the zodiacal light is best seen at the equinoxes. Likely links to the cultic aspect of the equinoxes have been located in ancient Malta, in the British Isles, in the cult of Mithras across the Roman Empire, and in Mesoamerican Maya and Aztec rituals, a seasonal solar event that seems to have been connected to cultic practice over millennia. The close temporal spacing – about an hour – between the best sighting of the zodiacal light and the autumnal equinox sunrise (or the vernal equinox sunset) suggests a potential for use of the zodiacal light in cultic practice inspired by astronomical phenomena.

THE COSMOS AS VIEWED THROUGH THE LENS OF A NATIVE-AMERICAN ASTRONOMER-ARTIST

Annette S. Lee

ABSTRACT: In Ojibwe the Morning Star is called *I'kwe Anung*, (Women's Star). In D(L)akota the same planet, Venus, is called *Aŋpetu Luta*, (Red Star). Both cultures have rich and interesting understandings of Venus that relate to other indigenous cultures throughout the world. Yet tragically the Native American star knowledge is disappearing. My research and programming project, *Native Skywatchers*, focuses on revitalizing the Ojibwe and D(L)akota star knowledge and bringing it back to our communities. Ojibwe elder, Paul Schultz, shared this vision of the native star knowledge returning, especially through the youth. He called them 'star readers'. As an astronomer, I am deeply inspired by and fascinated by the cosmos. As an artist, I am passionately engaged in creating art that aids in our perception of the cosmos, to see things that can't or aren't seen by the naked eye by most people today. This might be native constellations like *Maang* (Loon) or *Wakiŋyaŋ* (Thunderbird) or patterns in the motion of celestial objects, like the Dance of Venus....or it might be the teaching of *Kapemni* (As it is above; it is below.) Looking at the cultural and historical knowledge, specifically the Ojibwe and D(L)akota star knowledge, remembering it and honoring it, I use it to inspire the present and guide the future. I will present here selected paintings that speak to the idea of transforming the human experience through a relationship with the cosmos as viewed through the lens of a Native American astronomer-artist.

Native Skywatchers

For nearly a decade the *Native Skywatchers* initiative has addressed the crisis of the loss of the Ojibwe and D(L)akota star knowledge, the indigenous peoples of Minnesota.[1] There is urgency to this projectas Elders are passing. No one person

---

[1] For background to this paper see Frederic Baraga, *A Dictionary of the Ojibway Language* (St. Paul: Minnesota Historical Society Press, 1992). (First published as *A Dictionary of the Otchipwe Language* (Montreal: Beauchemin and Valois, 1878, 1880)); Eugene Buechel and Paul Manhart, *Lakota Dictionary* (Lincoln: University of Nebraska Press, 2002); Gregory Cajete, *Native Science: Natural Laws of Interdependence* (Santa Fe: Clear Light Publishers, 2000); Linda Miller Cleary and Thomas Peacock, *Collected Wisdom: American Indian Education* (Boston: Allyn and Bacon Publishing, 1997); Michael Johnson, *The Tribes of the Sioux Nation* (Oxford: Osprey Publishing, 2000); George E. Lankford, *Reachable Stars: Patterns in the Ethnoastronomy of Eastern North America*. Tuscaloosa: University of Alabama Press, 2007; Nancy Maryboy. and David Begay, *Sharing the Skies, Navajo Astronomy* (Tucson: Rio Nuevo Publishers, 2007, 2010); Ron Morton Carl Gawboy, *Talking Rocks: Geology and 10,000 Years of Native American Tradition in the Lake Superior Region* (Minneapolis: University of Minnesota Press, 2000, 2003); John Nichols Earl Nyholm, *A Concise Dictionary of Minnesota Ojibwe* (Minneapolis: University of Minnesota Press, 1995); Stephen Riggs, *A Dakota-English Dictionary* (St. Paul: Minnesota Historical Society Press, 1992, (1890)); Albert White Hat Sr., *Reading and Writing the Lakota Language* (Salt Lake City: University of Utah Press, 1999).

holds all of the details of the native Ojibwe and D(L)akota star knowledge, and much has been lost. The work is therefore necessarily collaborative and many voices are needed. At the same time due to the history of colonization, there are many layers of social upheaval on some reservations. This research and programming is dedicated to reclaiming the native star knowledge, documenting it, disseminating it. The ideal outcome is that more native people have a meaningful connection to the stars. In this connection to the stars a sense of cultural pride, a sense of connectedness and purpose is nurtured. Inherently interdisciplinary, this work includes astronomy, culture, language, art, science education, history, and community wellness. The implications of this foundational work are many, including encouraging more native young people to graduate from high school and possibly choose a career in STEM (Science, Technology, Engineering and Math). Currently Minnesota has some of the lowest graduation rates and highest achievement gaps for native students in the United States.[2]

I founded the *Native Skywatchers* research and programming initiative in 2007; I am a mixed-race D(L)akota Sioux astronomer and artist. This is a native-led initiative and the team is composed of persons of different expertise, scientists, artists, educators, writers, and historians. The other team-members include Carl Gawboy (Ojibwe), Jeff Tibbetts (Ojibwe), William Wilson (Ojibwe), Jim Rock (Dakota), and Charlene O'Rourke (Lakota). Acknowledgement goes out to two elders that were also part of the team and that have since passed away: Paul Schultz (Ojibwe) and Albert White Hat Sr. (Lakota). Also, a note about terminology and the use of Dakota or D(L)akota. Some say that Dakota means all three bands of the 'Sioux'– Dakota, Lakota, and Nakota. We made the map to include Dakota and Lakota. So for consistency we can use D(L)akota for all cases if that makes sense. The exception would be if someone specifically is referring to one of the bands.

Two Native Star Maps

At the foundation of this work are two native star maps that were created in 2012: *Ojibwe Giizhig Anung Masinaaigan* and *D(L)akota Makoċe Wiċaŋhpi Wowapi.*[3] Wilson, who was also the language expert. The map was based on the unpublished work

---

[2] T. Post, *Minnesota Near bottom in on-time graduation for students*, Minnesota Public Radio (MPR) News, Feb. 19, 2015; Alejandra Matos, *At Bemidji Charter, American Indian Students Excel*, Star Tribune, Minneapolis, MN, Nov. 7, 2015.
[3] See Annette Lee, William Wilson, Jeffrey Tibbetts and Carl Gawboy, *Ojibwe Sky Star Map Constellation Guide. An Introduction to Ojibwe Star Knowledge* (Cloquet: Avenue F Productions, 2014); Annette Lee, Jim Rock and Charlene O'Rourke, *D(L)akota Star Map Constellation Guide. An Introduction to D(L)akota Star Knowledge* (Cloquet: Avenue F Productions, 2014).

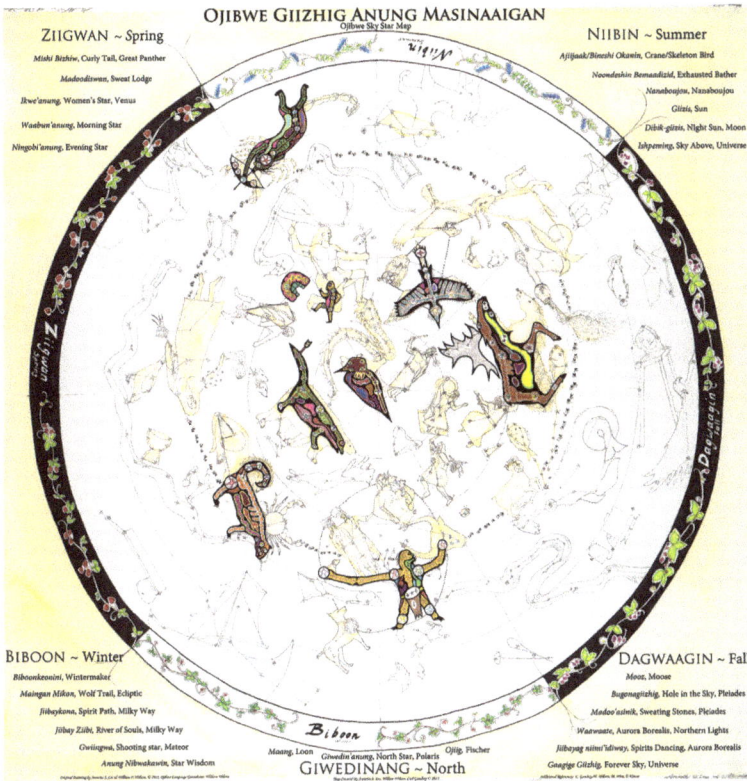

FIGURE 16.1. *Ojibwe Giizhig Anung Masinaaigan*, A. Lee, W. Wilson, C. Gawboy, 2012.

of Carl Gawboy. Since the 1960's Carl had been interviewing elders and researching the Ojibwe star knowledge. Carl was the first person to identify the pictographs at Lake Hegman, Boundary Waters as Ojibwe constellations.[4] The D(L)akota Star Map – *D(L)akota Makoċe Wiċaŋḣpi Wowapi,* was painted by Annette S. Lee. The language expert was Jim Rock. The map was based on the work published by Ron Goodman, who interviewed many Lakota elders from the South Dakota area in the 1980's, and published his results in 1992.[5] Several of the elders quoted in Goodman's book are a part of the *Native Skywatchers* project.

I painted the Ojibwe Sky Star Map- *Ojibwe Giizhig Anung Masinaaigan* with William. Both maps are astronomically accurate and visual works of art created in order to communicate an indigenous perspective of the night sky. Located at the center of

---

[4] Carl Gawboy, Personal Communication, Duluth, Minnesota, April 2012.

[5] Ronald Goodman, *Lakota Star Knowledge: Studies in Lakota Stellar Theology* (Mission: Sinte Gleska University, 1992).

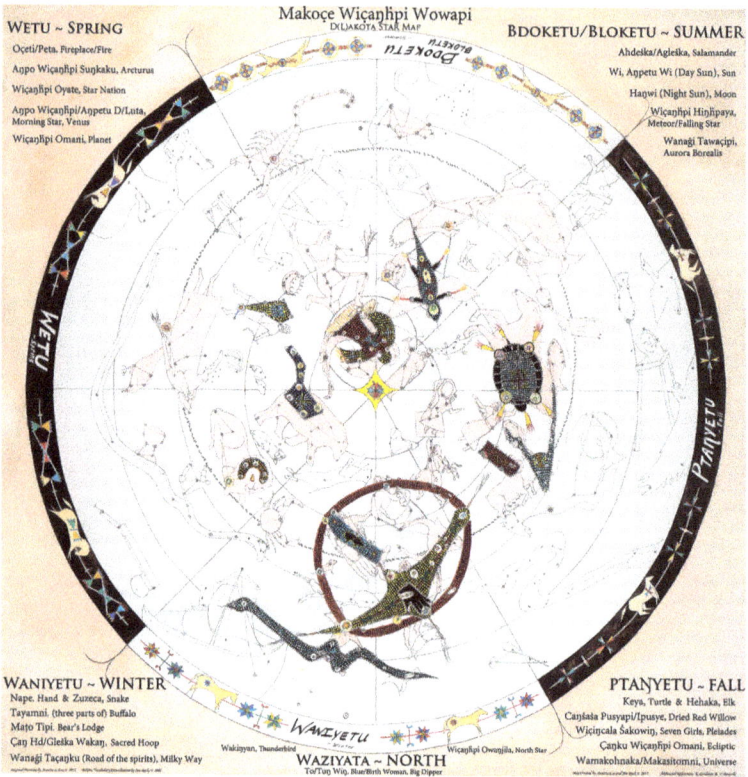

Figure 16.2. *D(L)akota Makoċe Wiċaŋhpi Wowapi*, A. Lee, J. Rock, 2012.

both maps is the north celestial pole (NCP) and the North Star (Polaris), known as *Giiwedin Anung* in Ojibwe, and *Wiċahhpi Owaŋjila* to the D(L)akota. Moving outwards from center are the north circumpolar stars, assuming a viewing location of approximately 40–50° north latitude. Surrounding the central area are the seasonal stars. The four seasons can be seen written on the outer border of the map: Fall, Winter, Spring, Summer. The maps illustrate the native constellations in brightly colored, clearly marked areas correlated with their seasonal association. Wilson painted the Ojibwe constellations in a traditional woodland x-ray style, and I painted the D(L)akota constellations in pointillism style in order to suggest beadwork. In both star maps the 'Greek' constellations are lightly painted in a quiet wash to allow the map-reader some sense of common ground between the Greek and the native Ojibwe and D(L)akota constellations. Sometimes the Ojibwe/D(L)akota constellations and western constellations coincide. One example. Ojibwe moose, Mooz, which falls mostly in Pegasus, using the four bright stars of the square, although the horns of the moose are the same stars as the modern constellation, Lacerta.

206

My Personal Statement

As an astronomer, I am deeply inspired by and fascinated by the cosmos and the ways in which mathematical thinking looks for and sometimes finds patterns in chaos. What I regard as a linear, logical, divide-and-conquer approach to astronomy has allowed us to send men to the Moon, spacecraft to the outer edges of the Sun's heliosphere, and to detect gravity's elusive waves. This is miraculous, and I am thrilled to participate in and practice science.

As an artist, I am passionately engaged in creating art that aids in our perception of the cosmos, to see things that can't or aren't seen by the naked eye by most people today. I am a painter who is passionately connected to the stars. Both science and art have always been at my core. I am an astronomer and an artist who is passionately interested in the intersection between art-science-culture and I believe that giving back to community is essential. I am a mixed-race native person, my tribe is D(L)akota-Sioux and my communities are D(L)akota and Ojibwe. Always included in my art are elements of the reverence for life and for mystery. My paintings are the meanderings of my spirit and heart. As an artist I am also a reflexive researcher, paying attention to my own inner processes, such as dreams.[6] For example, last month I had a dream:

> I was looking at the night sky, but it was all blurry. I had forgotten my glasses. Nearsighted. As I stared at the fuzzy patches thinking disappointedly, 'Where are my glasses? I'm missing a great view of the stars on this perfect night.' Suddenly I noticed that the stars had come into focus. They were more than crystal clear. Sparkling pinpoints of dancing light held my gaze. The longer I looked, more layers of shifting lights and darks became apparent. I could see an infinite, chaotic, and glorious pattern of space, stars, and darkness. Without lenses, and just for a moment, I could see a richness in the cosmos even beyond the 8-meter Hubble Space Telescope. Equally vast and powerful, I witnessed a different kind of beauty and complexity than could ever be seen with science alone.

Kapemeni

One of the most important underlying ideas in both the Ojibwe and D(L)akota star knowledge, is the idea of *kapemni*, the equivalent of the Hermetic maxim 'as above, so below'.[7] The idea can be visualized by two triangles (or tipis) stacked vertically connected at their apexes. The bottom tipi represents the material, or

---

[6] Charlotte Aull Davies, *Reflexive Ethnography: a Guide to Researching Ourselves and Others* (London and New York: Routledge, 1999).

[7] Jabir ibn Hayyan, 'The Emerald Tablet of Hermes Trismegistus' in E.J. Holmyard, *Alchemy*, (Harmondsworth: Penguin Books, 1957), pp. 97–98.

physical world. The top tipi represents the sky, star, and spirit world. The first part of the teaching is that both realms are real, important and meaningful. Furthermore, traditional native people should strive to acknowledge this relationship, either in ceremony or everyday life. When they do, it is like standing at the apex between the two triangles, which functions like a doorway, and there is a flow between the two worlds. In D(L)akota there is a single word, *kapemni* to describe this teaching. As Goodman wrote,

> Traditional Lakota believed that ceremonies done by them on earth were also being performed simultaneously in the spirit world. When what is happening in the stellar world is also being done on earth in the same way at the corresponding place at the same time, a hierophany can occur; sacred power can be drawn down.[8]

In D(L)akota many constellations have *kapemni pairs* in which a geographical site located on Earth that is the counterpart to the constellation in the sky. For example, when the Sun is in the constellation *Maŧo Tipila* (Bear's Lodge corresponding to Gemini), this is the traditional time for people to meet at *Pte he Ġi* (Grey Horn Butte, Devil's Tower) and pray. There is a direct pairing between the Sundance, *Wi'waŋyag wacipi*, in the stars, (the Sun is located in *Maŧo Tipila)*; and the Sundance, *Wi'waŋyag wacipi*, on Earth (the people are located at *Pte he Ġi).* Both above and below are in sync. There is flow.

Selected Work

Fall Stars

This painting tells the story of *kapemni* as experienced in the Fall. The upper half of the painting portrays three cultural views of the four bright stars commonly known as 'the great square of Pegasus'. On the left is the Ojibwe *Mooz*. On the right is the D(L)akota *Keya* (turtle). In the center is the Greek constellation Pegasus (winged horse). Where the above and the below meet the D(L)akota 'star tree' or 'cosmic tree' (cottonwood) is seen holding the doorway open. This deciduous tree stays close to water and is used in D(L)akota ceremonies. Seen here the cottonwood–star tree is losing its leaves with only a whisper of the green light of summer remaining. On the lower half is a close up of a star tree's leaf. White lines etched into the panel hint of a skeletal framework amongst decay. The entire image is done in typical fall colors with the interplay of lights and darks, lines and shades, working to balance

---

[8] Goodman, *Lakota Star Knowledge*, p. 1.

FIGURE 16.3. *Fall Stars,* Annette Lee, 2012

the whole. A closer inspection reveals lightly layered 'pencil' lines underneath the main figures. The layering suggests memory or a whisper of what has past. Pinpoints of light emerge from the very bottom of the page suggesting stars born in the cold darkness.

To/Tuŋ Wiŋ - Blue Spirit Woman

This painting tells the story of the journey of a newborn child from the spirit world into the physical world. In D(L)akota, there is a spirit named 'To/Tuŋ Wiŋ (Blue Spirit/Birth Woman)' that resides at the bowl of the Big Dipper (Plough). It is taught that the spirit comes from the Star Nation. In the center of the image *To/ Tuŋ Wiŋ* - Blue Spirit Woman – can be seen holding the unborn child. When a baby is born, a ceremony is done to honor the umbilical cord, the baby's lifeline while inside their mother's womb. After birth, a person's lifeline is then their connection with the stars. The umbilical cord, along with other medicines, is placed inside a

FIGURE 16.4. *To/Tuŋ Wiŋ Blue Spirit Woman*, Annette Lee, 2013.

special leather pouch made either in the shape of a *Keya* (turtle, a Fall constellation) for the girls, or an *Agleśka* (salamander, a summer constellation) for the boys. The accompanying prayers then secure the connection to the stars. In this painting, the infant is seen preparing to come through the birth doorway in the springtime. Looking southward a few hours after sunset, a person would see the spring stars (Leo, *Mishi Bizhiw* – Curly Tail) overhead and south, the winter stars setting (Orion, *Biboonkeinini* – Wintermaker), and the summer stars rising (Corona, *Madoodiswan* – the Sweat Lodge, Hercules). Upon close inspection there are small footprints (top right) walking along the *Wanaġi Taċaŋku* – Milky Way, marching through the D(L)akota womb – *Ċaŋ Hd/Gleśka Wakaŋ* (bottom center), and then to the *Oċeti* – fireplace for the Sweat Lodge ceremony in Leo. Finally the footsteps arrive at the place of To/Tuŋ Wiŋ (Blue Spirit/Birth Woman) in the scoop of the Big Dipper (Plough) and the child is ready for birth.

Figure 16.5 *Spring Lions in the Stars,* Annette Lee, 2014.

Spring Lions in the Stars

This painting tells the story of springtime unfolding in the night sky. Facing south a few hours after sunset we see the Greek lion constellation Leo and the Ojibwe lion constellation *Mishi bizhiw*-Curly Tail sharing the sky on the top half of this painting. On the bottom half are the small red buds of newly awakened trees in springtime. A figure (bottom center) sitting close to earth prays with a drum and smudgebowl nearby. A clear intention is made by the figure's upraised arm in solidary and strength with the stars above. The figure is simply painted with little detail, except a bright red heart. The human heart connects with the lion's heart. The bright star, Regulus, is directly overhead. The smoky fire colors are reminiscent of warmth coming out of darkness. The centermost object is a group of trees in almost complete darkness except for starlight raining down faintly.

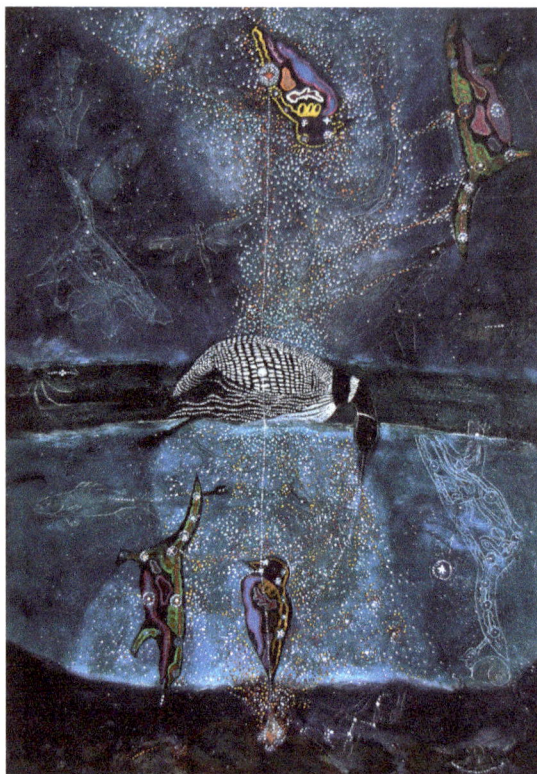

FIGURE 16.6. *Maang,* Annette Lee, 2012

Maang - Doorkeeper of the North

This painting tells the story of the *Maang*-Loon both as a constellation (the Little Dipper) and the Ojibwe teachings of the *Maang*. The *Maang* (Loon) constellation encompasses the same set of stars as the Little Dipper or the brightest seven stars in Ursa Minor. The first, most obvious question is: why did the Ojibwe associate this North American aquatic diving bird, the loon, with the most central place in the night sky? There are several answers to this question. The first reason is that the Loon were one of the two leaders in the Ojibwe clan system (a framework of government and organizing society). As Benton recorded, 'The Crane and the Loon Clans were given the power of chieftainship. They were given the people with natural qualities and abilities for leadership'.[9] Shown here is the mirroring of Earth and Sky. The North Star being within one degree of the north celestial pole (NCP) appears almost motionless as viewed from the ground. All other stars and celestial objects in

---

[9] E. Benton-Banai, *The Mishomis Book: The Voice of the Ojibway.* Hayward: Indian Country Communications, Inc., Minneapolis: University of Minnesota Press, 1988, reprinted 2010, p. 74.

the entire day and night sky appear to be circling around it. The motionless star is a leader. This leadership is reflected in the star knowledge by naming *Maang*, the Ojibwe clan leader, as the constellation containing the seven brightest stars nearest the motionless point (NCP) or the Little Dipper. *Maang* (Loon) is a leader in the sky, and a leader on the Earth. The teachings of the loon are many. One example is that the loon has a very close connection to the water. It avoids going on land, except to nest. It is not uncommon that loons become stranded in parking lot or pasture puddles. They need a quarter mile of water to take off.[10] In addition, loons survive physically by diving and spearing fish. Loons need clear, calm water to survive. The sacredness of water for life and survival is found throughout Ojibwe culture. Lastly, the loon has the stars of the night sky reflected on its back. The pattern of a black background with many small white dots mirrors a starry night sky. When an Ojibwe person hunts the loon, respect is shown by offering tobacco to the loon for giving it's life, and the loon is never to be turned upside-down (the backside of the loon must always be facing the sky).[11] The earth-sky mirroring is respected and maintained even after its death. Ojibwe first language speaker and *Native Skywatchers* team-member, W. Wilson, explains that the word *maang* is closely related to the word, *a maa(ng)* which means 'listen or pay attention'. [12]

Community Art Events

In 2015 the *Native Skywatchers* project was awarded an Arts Learning grant from the Minnesota State Arts Board. This funding supported the design and delivery of twelve participatory art making workshops in northern and central Minnesota and the partners included: K-12 schools, a library, several history museums, a tribal center, and an art institute. Four native artists led four unique art events that wove together art, science and culture. This programming, *Native Skywatchers – Earth Sky Connections*, allowed the native voice to act as the lead voice and then invited both native and non-native participants to explore their human connection to the sky and earth through the practice of art.

The workshop I led was called *As it is Above; it is Below*. The framework of the workshop was threefold: (1.) to observe and respond to a professional painting, (2.) to make art and practice a new approach and (3.) to share work and have the opportunity to exhibit work.

---

[10] David C., Evers, James D. Paruk, Judith W. Mcintyre and Jack F. Barr. Common Loon (Gavia immer), *The Birds of North America Online* (A. Poole, Ed.). Ithaca: Cornell Lab of Ornithology; at http://bna.birds.cornell.edu/bna/species/313, 2010.

[11] William Wilson, Personal Communication. St. Cloud, Minnesota, January 2012.

[12] Wilson, Personal Communication 2012.

FIGURE 16.7. Participants at the *Native Skywatchers – Earth Sky Connections* art workshop, photo by Annette Lee.

FIGURE 16.8. Participant at the *Native Skywatchers – Earth Sky Connections* art workshop, photo by Annette Lee.

214

FIGURE 16.9. Participants at the *Native Skywatchers – Earth Sky Connections* art workshop, photo by Annette Lee.

Both native and non-native participants showed up to make art related to the cosmos. Many people both on and off the reservation were eager to learn another culture's connection to the stars and both native and non-native participants showed up to make art related to the cosmos.. One native participant offered this insight, 'Learning about the Ojibwe constellations was like finding long lost relatives'. The framework of the workshop was threefold: (1) to observe and respond to a professional painting, (2) to make art and practice a new approach and (3.) to share work and have the opportunity to exhibit work.

In this one-year of programming almost 400 participants made earth and sky art and an electronic audience of over 50,000 was reached. Clearly this kind of programming is in high demand. As one participant said, 'The teacher (Annette) presented science as though it is connected to us - to us in our everyday lives. The art we created seemed philosophical - as though the science of the skies had a connection to how we think and feel about the world. And our art reflects that feeling'.

Conclusion

My work has many branches, making interdisciplinary connections in science and culture, formal and informal science education, artwork and art programming, history and heritage, outreach and community wellness. *Native Skywatchers – Revitalization of Ojibwe and D(L)akota Star Knowledge* research and programming initiative has worked with community members to create meaningful recourses and programming that communicate an ancient and living relationship with the cosmos.

It is our goal to build community around the native star knowledge.

All cultures, throughout human history have had a connection to the stars and a global and timeless human connection to the cosmos can be seen in the powerful idea of earth-sky mirroring.[13] The Egyptians, for example, understood the Nile River to be the reflection of the Milky Way.

The D(L)akota people know *Pte he Gi* (Grey Horn Butte – 'Devil's Tower') as the earthly counterpart to *Ma' o Tipila* – the Bear's Lodge constellation (Gemini). The Hawaiians experience Mauna Kea as 'the navel string of *Wakea* (Father Sky)'.[14] The Greek gods resided in Mount Olympus, which was what the Ojibwe and D(L)akota would call the *kapemni* reflection of Hesiod's *Ouranos* – the Starry Heaven (Hesiod). Clearly, there is a strong connection between earth, sky and being human. The focus point between the two worlds is the apex or a doorway. We stand at this doorway. Our humanity implores our participation in the cosmos.

It is intended that the *Native Skywatchers* research and programming initiative will help individuals and communities in rebuilding and remembering the native Ojibwe and D(L)akota connection to the stars. It is our hope that this work will serve as a stepping-stone to honor and remember all indigenous ways of knowing and ultimately our human connection to the cosmos.

---

[13] Giulio Magli, *Architecture, Astronomy and Sacred Landscape in Ancient Egypt* (Cambridge: Cambridge University Press, 2013), p. 19; Campion, *Astrology and Cosmology*, p. 13.

[14] Rubellite Johnson, John Mahelona and Clive Ruggles, *Na Inoa Hoku: Hawaiian and Pacific Star Names* (West Sussex: Ocarina Books, 2015).

Tables 1 & 2 - Star Vocabulary

Table 1.   Ojibwe Celestial Vocabulary

	Ojibwe	Related Greek Constellations
Fall	*Dagwaagin*	
Moose	*Mooz*	Pegasus
Hole in the Sky	*Bugonagiizhig*	Pleiades
Sweating Stones	*Madoo'asinik*	Pleiades
Winter	*Biboon*	
Wintermaker	*Biboonkeonini*	Orion, Canis Minor, Taurus
Spring	*Ziigwan*	
Curly Tail, Great Panther	*Mishi bizhiw*	Leo, Hydra
Sweat Lodge	*Madoodiswan*	Corona
Summer	*Niibin*	
Crane/Skeleton Bird	*Ajiijaak/Bineshi Okanin*	Cygnus
Exhausted Bather (Person)	*Noondeshin Bemaadizid*	Hercules
Nanaboujou	*Nanaboujou*	Scorpio
North	*Giwedinang*	
Loon	*Maang*	Little Dipper
Fisher	*Ojiig*	Big Dipper
North Star	*Giwedin'anung*	Polaris

Table 1cont. Ojibwe Celestial Vocabulary

Celestial Object	Ojibwe
Star	Anung
Star World	Anung aki
Moon	Dibik-giizis (Night Sun)
Sun	Giizis
Sky	Giizhig
Venus	Ikwe'anung (Women's Star)
Venus – Evening Star	Ningobi'anung
Venus – Morning Star	Waabun'anung
Ecliptic	Maingan Mikan (Wolf Trail)
Milky Way	Jiibaykona (Spirit Path)
Milky Way	Jiibay Ziibi (River of Souls)
Meteor/shooting star	Gwiingwa
Universe	Gaagige Giizhig (Forever Sky) and Ishpeming (the Sky Above)
Aurora Borealis (Northern Lights)	Waawaate
Star Knowledge (Wisdom)	Anung Nibwakawin
(Sky) Star map	Giizhig Anung Masinaaigan
East	Waabunong
West	Ningobinong
North	Giwedinong
South	Jawanong

Table 2. D(L)akota Celestial Vocabulary

	D(L)akota	Related Greek Constellations
Winter	*Waniyetu*	
Hand	*Nape*	Orion, Eridanus
Bear's Lodge	*Mato Tipi/Mato Tipila*	Gemini
Racetrack	*Ki Iŋyaŋka Oçaŋku*	Winter Circle
Sacred hoop	Çaŋ Hd/Gleśka Wakaŋ	Winter Circle
Sweat lodge	*Inipi/Initipi*	Winter Circle
Buffalo (three parts of)	*Tayamni*	Orion, Canis Major, Pleiades
Buffalo embryo head	*Tayamni pa*	Pleiades
Buffalo embryo ribs	*Tayamni cutuhu*	Betelgeuse & Rigel
Buffalo embryo backbone	*Tayamni caŋkahu*	Orion's belt
Buffalo embryo tail	*Tayamni siŋte*	Sirius
Snake	*Zuzeca/Zuzuheca*	Columbia, Puppis, Canis Major
Fall	*Ptaŋyetu*	
Turtle	*Keya*	Pegasus
Dried Red Willow	*Caŋśaśa Pusyapi/Ipusye*	Aries, Triangulum
Elk	*Heḣaka/Upaŋ*	Pisces
Seven Girls	Wičiŋyaŋna *Śakowiŋ/Wičiŋcala Śakowiŋ*	Pleiades
Summer	*Bdoketu/Bloketu*	
Salamander	*Ahdeśka/Agleśka*	Cygnus
Spring	*Wetu*	
Fireplace/Fire	*Oćeti/Peta*	Leo
Arcturus	*Itkob u* (going toward), *Ihuku Kigle* (Under went it), *Aŋpo Wičaŋhpi Suŋkaku* (younger brother of Morning star)	Bright star in Bootes
North	*Waziyata*	
Blue Woman/Birth Woman	*To Win/Tuŋ Wiŋ*	Big Dipper - inside Bowl
Stretcher	*Wičakiyuhapi*	Big Dipper - Bowl stars
Mourners	*Waśihdapi/Waśiglapi*	Big Dipper - Handle stars
Skunk	*Maŋka/Maka*	Big Dipper
Dipper/Wooden Spoon	*Wicakiyuhapi/Can cinkska*	Big Dipper
Seven sacred rites/Council fires	*Oćeti Śakowiŋ*	Big Dipper
North Star	*Wičaŋhpi waziyata/Wičaŋhpi Owaŋjila* (Star which stands in one place)	Polaris
Thunderbird	*Wakiŋyaŋ*	Draco, Ursa Minor

Table 2 cont. D(L)akota Celestial Vocabulary

Celestial Object	D(L)akota
Star	*Wičaŋḣpi*
Star Nation	*Wičaŋḣpi Oyate*
Moon	*Haŋhepi Wi /Haŋyetu Wi/Haŋwi (Night Sun)*
Moon	*Anog Ite (Double Faced Woman)*
Sun	*Wi, Aŋpetu Wi (Day Sun)*
Venus - Morning Star	*Aŋpo Wičaŋḣpi /Aŋpetu D/Luta*
Ecliptic	*Çaŋku Wičaŋḣpi Omani/Maḣpiya Maka Iciyagle*
Milky Way	*Wanaǧi Tačaŋku (Road of the spirits/Ghost trail)*
Meteor/Falling star	*Wičaŋḣpi Hiŋḣpaya/Wiahpihinḣpaya/Woḣḣpe Wakaŋ*
Universe	*Wamakohnaka/Wamakhognaka/Makasitomni*
Aurora Borealis (Northern Lights)	*Wanaǧi Tawačipi (Spirit Dancers)/Maḣḣpiyataŋiŋ/Wiyosaya*
Comet	*Wičaŋḣḣpi Siŋtetuŋ/Wičaŋḣḣpi Siŋte Yukan/Wicaŋpisiŋteton*
Star map	*Makoče Wičaŋhḣpi Wowapi*
Planet	*Wičaŋḣpi Omani/Wičaŋḣpi Nuni/Wičanḣpi Sa*
Sundogs	*Wiaceic' iti (Sun making fire)*
Solar Eclipse	*Witha Wit' e (Sun dies)*
Lunar Eclipse	*Haŋwitha*
Constellations	*Wičaŋḣpi Tiospaye (Extended family)*
Galaxies	*Okakše Taŋka Wičaŋḣpi Ota/Wičaŋḣpi Optaye Taŋka*
Groups of galaxies	*Wičaŋḣpi Oyate (Nation)*
Summer solstice	*Bdoke cokaya/Bloke cokaya/Anpawi (Morning Sun)*
Winter solstice	*Waniyetu cokaya/Nahomni (Swing around)*
Spring Equinox	*Wetu Aŋpa Haŋyetu Iyehaŋtu*
Fall Equinox	*Ptaŋyetu Aŋpa Haŋyetu Iyehaŋtu*
Seasons	*Omaka/Makoncage (Earth grows with time change)*
North	*Waziyata*
South	*Itokagata*
East	*Wioḣinyanpata*
West	*Wiyoḣpeyata*
Above	*Wankantu/Waŋkatika*
Below	*Kutakiya/Kutkiya*
Center	*Čokata/Čokaya*
Buffalo embryo head	*Tayamni pa* (Pleiades)
Buffalo embryo ribs	*Tayamni cutuhu* (Betelgeuse and Rigel)
Buffalo embryo backbone	*Tayamni caŋkahu* (Orioṅs belt)
Buffalo embryo tail	*Tayamni siŋte* (Sirius)

CHRIST AND THE CELESTIAL SPHERE: A UNIQUE MOSAIC IN SAINT ISAAC'S CATHEDRAL?

Michael Mendillo and Ethan Pollock

ABSTRACT: While celestial imagery appears in many religious paintings, Christ holding a celestial sphere is not commonly seen in Church decorations. Saint Isaac's cathedral in Saint Petersburg, Russia, was built between 1818–58, with its architecture and decorative motifs given strong oversight by professors in the Imperial Academy of Arts and by the personal interests of three Tsars. The interior wall (called the iconostasis) that separates the sanctuary from the public portion of the building contains a remarkable icon – a huge mosaic of Christ holding a large glass sphere containing stars and constellations. The motif of Jesus holding a sphere with a crucifix atop is called *Salvator Mundi*. In this paper, we explore the possible origins of the unusual *Salvator Mundi* in Saint Isaac's that adds stars and asterisms – most prominently Orion and Ursa Major – to the glass sphere held by Christ.

Introduction

During the 2003 tercentennial celebration of the city of Saint Petersburg, Saint Isaac's Cathedral emerged as a showpiece of the new Russia. It was refurbished and icons were returned to their original locations. Special attention was given to the wall of paintings and mosaics that separates the inner sanctuary from the outer areas of the cathedral. To the right of this iconostasis, with its huge gold doors at the entrance to the sanctuary, sits Christ in the classic pose of *Salvator Mundi* – Jesus seated with his right hand raised in blessing, his left hand holding a sphere representing the world saved by his suffering on Earth. In this extraordinarily large version, done in spectacular mosaic, Jesus is not holding a sphere of gold nor one of solid glass, but rather a large transparent crystalline globe populated with stars and the asterisms they form (FIGURE 17.1). Within Saint Isaac's Cathedral, Christ is Saviour of the Universe.

Beyond its physical size and the mannerist style for Jesus, certainly the most unusual component of the *Salvator Mundi* within Saint Isaac's is the transformation of a glass sphere into a celestial globe with easily recognized constellations. Most obvious is Orion to the left of centre, and then Ursa Major ('the big dipper') on the upper right (FIGURE 17.2). These star patterns have their orientation following the 'external viewpoint' used for celestial globes. Thus, the three belt stars of Orion that we see from Earth going from lower left to upper right are reversed when seen from beyond the celestial sphere – as God would see them.

FIGURE 17.1. A view of the iconostasis within Saint Isaac's Cathedral in Saint Petersburg (left) and its mosaic of *Salvator Mundi* (right), from Glavnyi ikonostas Isaakievskogo sobora (The Main Iconostasis of Saint Isaac's Cathedral) at the Harvard Fine Arts Library.[1]

The handle of the Big Dipper (when the bowl is upright) is on the right, while we on Earth see it on the left. Even a casual inspection of FIGURE 17.2 reveals quite a bit of artistic license in the details of these two constellations. The placements of the component stars are not quite right, and their spatial separation is clearly one of convenience – they are shown much closer together than seen on an accurate celestial globe (as displayed in FIGURE 17.2, right). Not paying attention to equivalent details would have been unthinkable for the religious content of icons (e.g., if Saint Peter held a carpenter's saw rather than the keys to heaven). The intended message appears to have been that St. Isaac's *Salvator Mundi* is the Saviour of the Universe, with rigorous celestial cartography given lower priority. We have not yet identified the person(s) who made the decision to add these asterisms to this *Salvator Mundi*; this brief chapter describes our attempts to do so.

---

[1] Alexander Vitushkin, *Glavnyi iconostas Isaakievskogo sobora* (St. Petersburg: Gos. Muzei-pamiatnik 'Isaakovskii sobor', 2008).

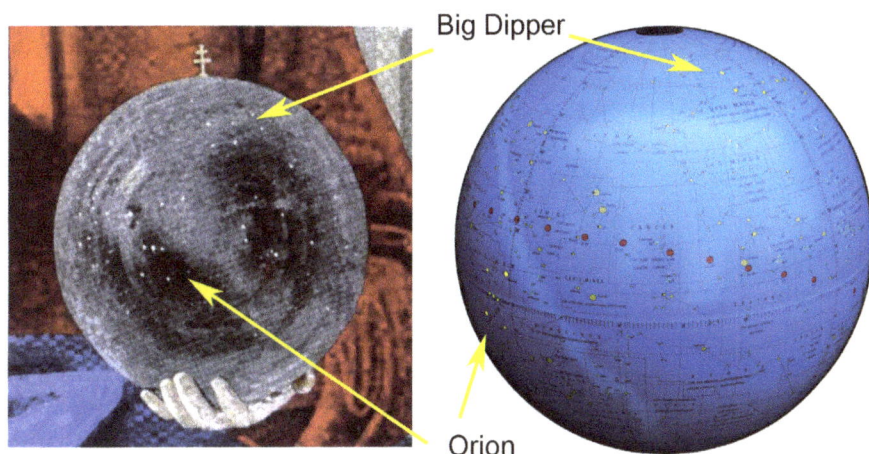

FIGURE 17.2. Comparison of the constellations Orion and Ursa Major as depicted in the *Salvator Mundi* in Saint Isaac's Cathedral (left) with their typical portrayals upon a modern celestial sphere (right).

Building a Cathedral in Tsarist Russia

The story of Saint Isaac's is one of many grand building projects that, quite literally, helped define the Tsarist architecture of the nineteenth century. The cathedral took forty years to build (1818–58), spanning the reign of three Tsars.[2] The design and groundbreaking occurred under Alexander I, construction under Nicolas I, and dedication during Alexander II's reign. The cathedral cost a fortune for the times and served as a showpiece for Russian architecture, engineering and artistic ornamentation. The overall design was by Auguste de Montferrand (1786–1858), a man who had served with distinction in Napoleon's army, but not with the Grande Armée that suffered a humiliating defeat by Russian forces in 1812.[3] Just four years later, in 1816, a high point for Montferrand's career occurred when he was selected as chief architect for a new Saint Isaac's. The scale of his overall plan, his choices of expensive building materials, and the mechanical infrastructure techniques he created to bring the project to fruition amounted to a nineteenth-century Imperial Russian equivalent to Brunelleschi's achievements. Once the dome was completed in 1841, attention could finally be devoted to decorating the enclosed structure. That final phase would take eighteen years.

---

[2] See V. K. Shuiskii, *Ogiust Monferran: istoriia zhizni I tvorchestva* (St. Petersburg-Moscow: MiM-Delta, Tsentrpoligraf, 2005); and Vitushkin, *Glavnyi iconostas Isaakievskogo sobora*.

[3] Shuiskii, *Ogiust Monferran: istoriia zhizni I tvorchestva*.

Montferrand made it clear from the onset that he was not simply issuing sub-contracts for interior furnishings. In fact, he had little choice since the bureaucracies of empire and its various academies of art and architecture demanded review and approval for virtually every decision concerning ornamentation. Russian Orthodoxy called for a church's ceiling, floor and walls to be adorned by religious images. For a building the size of Saint Isaac's, this amounted to an enormous amount of surface area to cover. To both spread the commissions among the favoured artists of the day, as well as to keep to the schedule for completing so many paintings, the decoration of the new cathedral became a focal point of the entire artistic community of St. Petersburg. The project provided commissions spanning two decades. The artist Peter Basin (1793–1877), for example, devoted a considerable portion of his career to provide no fewer than forty paintings for the cathedral.[4]

Montferrand was to enlist the talented Bryullov family to help with the project. Generations of Bryullov artists had achieved fame in France and Russia. In 1841, the most famous artist in Russia (and professor at the Imperial Academy of Arts) was Karl Bryullov (1799–1852). He was attracted to the Saint Isaac's project by the offer to paint the dome and ceiling – the quintessential goal of every artist since Michelangelo.

Karl Bryullov's younger brother Alexander Bryullov (1798–1877) was famous for his portraits as well as architecture – including his design of the Pulkovo Astronomical Observatory.[5] Yet, it does not appear that he assumed any type of leadership role at Saint Isaac's when Karl became ill. There were, nevertheless, others waiting in the wings. Just as Pope Julius II hired both Michelangelo and Raphael at the same time and under the same roof – to keep them working in competitive spirits – so too was the Bryullov's rival Fedor Bruni (1799–1875) offered competing spaces to adorn.[6] His initial contributions were frescoes for the walls. Some he painted himself, others were done by assistants, and still others by professional artists hired under his supervision.

As for the mosaics, it was probably Fedor Bruni and Timothy A. Neff (1804–76), also a professor at the Academy, who played the pivotal roles in the designs used for the iconostasis.[7] Their plan for Christ as Saviour of the World amounted to a bold departure from the traditional depiction scheme of the Salvator Mundi motif. Where might Bruni or Neff have found the image to give to the mosaicists?

---

[4] *Entsiklopedicheskii slovar'Brokgauza i Efrona*, (St. Petersburg, 1890–1907), Vol. 3, p. 148.

[5] *Entsiklopedicheskii slovar'Brokgauza i Efrona*, (St. Petersburg, 1890–1907), Vol. 4, pp. 782–83.

[6] A.A. Polovtsov, ed., *Russkii biograficheskii slovar'* (St. Petersburg: Izdanie Imperatorskago Russkago istoricheskago obshchestva, 1896–1918), Vol. 3, pp. 371–83.

[7] *Entsiklopedicheskii slovar'Brokgauza i Efrona*, (St. Petersburg, 1890–1907), Vol. 20A, p. 935.

The *Salvator Mundi* Motif

The concept of depicting a powerful ruler as one who holds a sphere in one hand (and often a sceptre in the other) dates to Greco-Roman times. A large sculpture of Zeus/Jupiter holding a sceptre and orb can be found in the Hermitage. In Hellenistic mythology, the Muse of Astronomy, Urania, was a daughter of Zeus.[8] In sculpture, she is shown holding a small sphere in her left hand and a ruler's wand in the right, as seen in Roman copies of the muses for Hadrian's Villa in Tivoli, Italy (now at the Prado in Madrid). With the spread of Christianity in the fourth and fifth centuries CE, a cross was added to the globe to portray divine approval and Christ's guidance of emperors. The motif was so popular by the time of the Middle Ages that the globe-with-cross-atop became known as a Globus Cruciger, and when the ruler was replaced by Jesus himself the name became *Salvator Mundi*. By the time of the Renaissance, it was a standard image for Church leaders and Catholic rulers alike – celebrating Christian domination of the world via heaven's anointed popes, kings and queens. Artists were delighted for such commissions since they were inevitably to be very large paintings placed in prominent locations.

In some artistic depictions (FIGURE 17.3), *Salvator Mundi* was clearly holding a terrestrial globe – as in the version by Domenico Fetti (1591–1623) now in the Metropolitan Museum of Art in New York. In other cases, Christ holds a clear glass globe, as Andrea Previtali (ca. 1480–1528) chose for his painting, now in London's National Gallery. The Previtali sphere has two bright spots – one in each hemisphere – showing reflections of an unseen light, but they certainly are not stars. In northern Europe, the *Salvator Mundi* theme was also popular. Albrecht Dürer (1471–1528) started such a painting in the early years of the 1500s, but never completed it. The partial painting (now at the Metropolitan Museum of Art in New York) is still remarkable for the portions he finished. Most appealing are the texture and shadows of the folded red and blue robes worn by Jesus that can be seen through a transparent glass globe. In Antwerp, a version by Joos van Cleve (1485–1540), now in the Louvre, depicted a glass sphere containing a nautical seascape – an appropriate image painted in a city linked to the North Sea by a river.

What was the inspiration for the version of *Salvator Mundi* in Saint Isaac's Cathedral? While the paintings in FIGURE 17.3, or similar ones throughout western Europe, were likely known to the academicians in St. Petersburg, they clearly had one easily accessible source: Titian's masterpiece on that theme – a painting newly

---

[8] Hesiod, *Hesiod's Theogony*, ed. Richard S. Caldwell (Newburyport, MA: Focus Information Group), p. 32, lines 75–79.

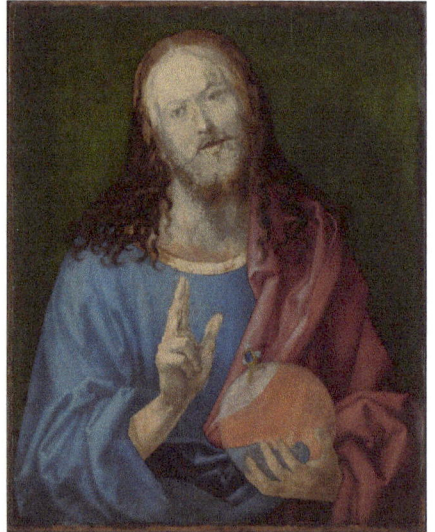

FIGURE 17.3. Examples of paintings of *Salvator Mundi*: (top left) Domenico Fetti (1591–1623) Metropolitan Museum of Art in New York, (top right) Andrea Previtali (ca. 1480–1528) National Gallery in London; (lower left) Joos van Cleve (1485–1540) the Louvre, (lower right) Albrecht Dürer (1471–1528) Metropolitan Museum of Art in New York.

acquired by the Tsar's art collectors and housed in the Hermitage.[9] The Hermitage was in its own formation phase at the time of construction and decoration of Saint Isaac's. Many professors of the Imperial Academy of Art involved with the cathedral also served as expert consultants on the specific works of art to include in the Hermitage. Neff, Bruni and others surely knew of this famous painting by one of the giants of the Italian Renaissance.

Titian (ca. 1488–1578) was eighty-two when he portrayed his Saviour of the World, and his approach was considered innovative. Christ's left hand can be seen through the crystal sphere he holds, a tour de force for depicting details influenced by the optical properties of glass. There are also two bright spots of light reflected from an unseen source on the left of the painting, and these dominate the side of the glass sphere facing the viewer. With the appearance of Titian's *Salvator Mundi*, the classic motif completed its evolution from the solid opaque spheres of medieval times to actual terrestrial globes, and then to solid glass globes, with some of them being translucent glass spheres.

The source of inspiration for Titian's painting in the Hermitage (and thus the ultimate source of the translucent glass sphere held by Jesus in Saint Isaac's) may well have been Leonardo da Vinci. Experts believed that Leonardo had painted a *Salvator Mundi* sometime between 1506 and 1513 – there was just confusion about where it was, or if it actually survived to modern times. Given Leonardo's reputation of conceptual creations far exceeding his actual productivity, the preparatory sketches now in the Royal Collection at Windsor do not guarantee that the actual painting existed. There were many copies of the *Salvator Mundi* identified as Leonardo's throughout Europe, including a detailed etching by Wenceslaus Hollar (1607–77) done in 1650 for Henrietta Maria, wife of Charles I of England. The modern verdict was that the painting's physical existence was assured, but just not its physical location. All of that changed in 2005 when a 'version' in dreadful physical shape was acquired by a group of art dealers in New York and given secretly to restoration experts. In just six years, certifications of its authenticity by scholars became so definitive that the National Gallery in London included it as a bona fide painting in its 2011–2012 blockbuster exhibition 'Leonardo da Vinci: Painter at the Court of Milan'. FIGURE 17.4 presents the da Vinci, Titian and Saint Isaac versions.

Leonardo's achievements, always spectacular, are especially on view in his rendition of the transparent crystal sphere in *Salvator Mundi*. For relevance to the mosaic

---

[9] Irina Artemieva and Giuseppe Pavanello, eds., *Masterpieces from the Hermitage* (Milano: Electa, 1998).

FIGURE 17.4. *Salvator Mundi* paintings by Leonardo da Vinci (left), private collection, and Titian (centre), the Hermitage, in comparison to the mosaic in Saint Isaac's Cathedral (right).

in Saint Isaac's Cathedral, there are several aspects to consider.[10] First is the size of the glass sphere, which Leonardo depicts as about six inches in diameter – a size that very comfortably fits into an open hand.

The globe appears to be solid glass and thus one of considerable mass. A much larger solid glass sphere would be uncomfortable to hold, even in a Saviour's hand. The globe in Saint Isaac's mosaic has a diameter at least twice that in Leonardo's, and thus a factor of eight in volume. Yet in the mosaic it appears easily balanced in Christ's left hand. We can speculate that it is a hollow glass sphere with multiple reflections. Leonardo's globe is clearly not a Globus Cruciger – there is no cross on top of it. Nor are there brass circles for an equator or meridians to portray either geographical or celestial coordinate systems. It could well be seen as a perfect crystalline sphere consistent with the Aristotelian, Church-approved version of celestial bodies, an aspect of astronomy that Leonardo accepted. But Leonardo clearly chose to avoid perfection in preference to reality. The glass sphere strikes the viewer as such a naturally-looking object because, in part, of Leonardo's decision to give it imperfections – many air bubbles appear, particularly so on the right side of the painting.

A second physical property portrayed within Leonardo's *Salvator Mundi* is the optical effects of glass. We see the Saviour's finger tips in front of the sphere and thus they are unaffected visually by it. The palm of Christ's left hand, however, is seen through the globe and it appears unusually wide. This is not a mistake by the artist, and certainly not for a man so expert on the optical properties of lenses. Most intriguing are the three bright spots that form a right triangle on the surface of the globe. The vertical points of light fall almost along the central meridian, while the two horizontal spots are slightly above the equatorial plane. There is no indication

---

[10] Vitushkin, *Glavnyi iconostas Isaakievskogo sobora.*

that they represent stars, or components of a constellation. They could be particularly prominent air bubbles made visible by glints of light coming from the left. Leonardo is clearly putting on prominent display his understanding of how light interacts with glass. For the human anatomy component of the painting, Leonardo offered his unique approach to light and shadow – most famously here with the hand and fingers that Christ raises in blessing. This depiction of the hand was, in fact, one of the pivotal characteristics that experts used to identify *Salvator Mundi* as a true, surviving painting by da Vinci. [11]

Can we see a connection from Leonardo to Titian to St. Isaac's version of *Salvator Mundi*? Certainly in theme, but not in detail. The wall-size mosaic in Saint Isaac's is enormous in comparison to Titian's painting. As in da Vinci's image, and in many others, Christ is portrayed from the waist up; Titian does the same. Yet in the Saint Isaac mosaic the Saviour is seen in full body. The traditional red robe and blue cloak worn by Christ are consistent with the official standards set by the Church. Jesus appears as a robust young man without a beard – an imposing presence upon the iconostasis. The chiaroscuro effect of shading on Jesus's face, and the intensity of his glance towards the viewer, are remarkable achievements for a mosaic.

Surely the artisans who assembled the mosaic did not decide to add bright pieces of glass in an unauthorized attempt to insert star patterns upon the globe. For such a crucially-placed mosaic within the iconostasis – at the very entryway into the sanctuary – such a detail would not have been missed by the review-and-approval bureaucracies of the Academy and the Tsar. Someone with the official clout of a Timothy A. Neff, Professor at the Academy, was needed for such an obvious departure from the standard theme of a *Salvator Mundi*. Might Professor Neff have owned a celestial globe, or had seen one depicted in a celestial atlas? The appearance of a celestial globe upon a desk, if viewed from the viewpoint of one standing in front of it, can be made to coincide (more or less) with the set of constellations portrayed. With Ursa Major (big dipper) in upper right, and Orion in lower left, it is fairly easy to see how Professor Neff could have handed the mosaicist a sketch to solidify upon the iconostasis in Saint Isaac's Cathedral.

Other renditions of *Salvator Mundi* with celestial content

In addition to the readily available and popular *Salvator Mundi* by Titian in St. Petersburg, there are some less well known versions that Professor Neff and colleagues might have seen. To counter the iconoclast phase of the Protestant Revolution in

---

[11] Luke Syson, ed., *Leonardo da Vinci: Painter at the Court of Milan* (London: National Gallery of Art, 2011). Exhibition catalogue.

sixteenth-century northern Europe, the Jesuits embarked upon a series of sponsor-ships of new religious images to re-introduce Catholic doctrine in pictorial modes. In Antwerp and Brussels, the Wierix family comprised a dynasty of printmakers and engravers spanning the sixteenth and seventeenth centuries. They produced few original works, but rather specialized in making engravings of existing works for sale as single prints or for illustrations in books. Hieronymus Wierix (1553–1619) produced over 600 prints, and he was particularly fond of making copies of Albrecht Dürer's works. For his *Salvator Mundi*, however, he (or his customer) chose to copy a version by Maarten de Vos (1532–1603) shown in FIGURE 17.5 (left). This variation is unusual for several reasons – it has Christ looking to the left, his head surrounded by a cloud containing angels, and with the Globus Cruciger component so large it looks like a weathervane. The most dramatic departure is the unambiguous replace-ment of a terrestrial globe with one containing celestial objects. The Sun is given prominence, perhaps in reference to the Son of God holding the sphere. The Earth and Moon could be two of the three secondary circular disks shown, leaving the third one ambiguous. The mosaic in Saint Isaac's Cathedral is clearly not the first ever to have astronomical content in a *Salvator Mundi*.

A second example comes from the German painter and print-maker Ludolph Büsinck (ca. 1600–69), shown to the right in FIGURE 17.5. Büsinck introduced the chiaroscuro method of wood cut prints to France, and one of his most impressive accomplishments was his series *Christ and the Twelve Apostles* now at the Rijksmu-seum in Amsterdam. His depiction of Jesus in the *Salvator Mundi* motif uses the classic above-waist style, but with the right hand fingers used in blessing far closer than usual to the sphere held in the left hand. The globe appears so heavy that Christ rests it against his chest – providing a shadow to exhibit the engraver's chiaroscuro talents. The Sun is prominent atop the globe, taking the place of a Globus Cruciger ornament. The Moon and nearly twenty stars provide evidence for this being a ce-lestial sphere. Yet, the stars are just ornamental with not so much as a hint that they might form a constellation.

Returning to Saint Isaac's Cathedral and its grand mosaic of *Salvator Mundi*, can we find any evidence that the Wierix or Büsinck engravings played any role in the final design adopted? Certainly Professor Neff did not hand the mosaicists these prints to copy. The styles are so different from the final mosaic, and the astronomical content is unrelated in detail. The style is best matched by Titian's painting in the nearby Hermitage. Perhaps Neff or his associates simply suggested making the glass sphere more scientifically acceptable, whether or not they were aware of the prints by Wierix or Büsinck. The instructions could have been nothing more than 'Make it a

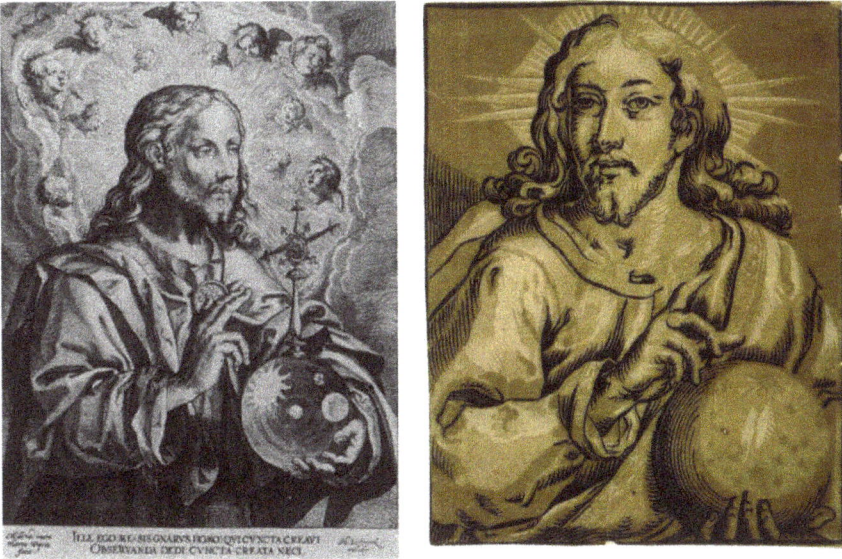

FIGURE 17.5. Prints of the Salvator Mundi motif with astronomical content; (left) version by Hieronymus Wierix, (right) version by Ludolph Büsinck. See text.

full-size Christ, and get the astronomy more representational of reality'. Finally, there is the possibility that the mosaic we see today within Saint Isaac's was not actually completed during the initial decorative phase. In Theophile Gautier's *A Winter in Russia* (New York, 1881) the author describes Saint Isaac's from the point of view of a mid–nineteenth century traveller.[12] Gautier describes the 'globe' as being made of the deep blue semi-precious stone known as lapis lazuli. He suggests that lapis was used for 'pictures upon gold backgrounds imitating mosaics, and to be replaced by veritable mosaics as soon as the latter are completed'. A lapis-lazuli celestial sphere may have been a decorative 'place holder' that ultimately became the celestial globe we find in the mosaic today.

ACKNOWLEDGEMENTS: We are grateful for discussions with Dr Yakov Dimant at Boston University. We very much appreciate the assistance of Eugenia M. Dimant, Library Assistant at the Widener Memorial Library at Harvard University, for providing a copy of the publication Glavnyi iconostas Isaakievskogo sobor and for digital copies of the images used in FIGURE 17.1. Discussions with George Beke Latura pointed us towards the *Salvator Mundi* prints by Wierix and Büsinck shown in FIGURE 17.5. We acknowledge the assistance of Ms Joei Wroten at Boston University with manuscript preparation and for her photography of the celestial globe in FIGURE 17.2. This work was supported, in part, by research funds made available through the Center for Space Physics at Boston University.

---

[12] Theophile Gautier, *A Winter in Russia*, 2nd rev. ed., trans. M. M. Ripley (New York: Holt, 1881).

A SELF-PORTRAIT BY GALILEO?

Paolo Molaro

ABSTRACT: An intriguing reference to the existence of a self-portrait by Galileo Galilei is contained in the biography of the scientist by Thomas Salusbury dated ca. 1665, of which only one incomplete and inaccessible copy exists. Galileo grew up in a Renaissance atmosphere, acquiring an artistic touch. He was a musician, a writer and also a painter, as reported by Viviani and documented by his watercolours of the Moon and drawings of solar spots. Recently a new portrait with a remarkable similarity to the portraits of Galileo Galilei by Santi di Tito (1601), Domenico Tintoretto (ca. 1604), and Furini (ca. 1612) has been found and examined using sophisticated face recognition techniques.[1] If the identity could be confirmed, other elements, such as the young age of Galileo or the seam in the canvas revealed by infrared and X-ray analysis, may suggest a possible link with the self-portrait mentioned by Salusbury.

Galileus Galileus, his life by Thomas Salusbury

On 6 June 1660, a few months before the Royal Society was founded at Gresham College, Thomas Salusbury gave the first manuscript of his life's work to the editor.[2] The whole project was divided into two Tomes, each consisting of two parts and with each part containing several translations and scientific books. The first part of the First Tome contains an English translation of Galileo's *Dialogues* (which at the time was still banned in Italy), Galileo's letter to the Grand Duchess Christina Lorena, and a text by Kepler and one by Foscardini. The second part consists mainly of a collection of Benedetto Castelli's works on the dynamics of fluids.

The first part of the Second Tome contains Galileo's *Discourses on Two New Sciences*, which turned out to be quite important for Newton's work – the first *Discourse* on the things that move in and upon the water, and other works by Cartesius, Archimedes and Tartalia. The second part contains a work by Torricelli, a book by Salusbury himself and his biography of Galileo, 'Galileus Galileus, his life, in five books'. This structure suggests that the book was possibly meant to be published separately. The first part of the second Tome, printed in 1665, burnt almost completely during the Great Fire of London of 1666 and only seven copies are presently known, five of which are in the libraries of the British Museum, Cambridge University, Oxford University and University College of London, with one in a private collection in

---

[1] R. Srinivasan, C. Rudolph, and A. K. Roy-Chowdhury, 'Computerized Face Recognition in Renaissance Portrait Art: A quantitative measure for identifying uncertain subjects in ancient portraits', *IEEE Signal Processing Magazine* 32, no. 4 (2015): pp. 85–94, here p. 85.

[2] Nick Wilding, 'The Return of Thomas Salusbury's "Life of Galileo" (1664)', *The British Journal for the History of Science* 41, no. 2 (2008): pp. 241–65, here p. 241.

London.[3] Only one copy of the second part of the second Tome survived the Great Fire, in the private library of the Earl of Macclesfield. We know of its existence since it was quoted by several authors in the early nineteenth century; the final reference to it is found in Drinkwater's *Life of Galileo Galilei*, published in 1829. Afterwards the book disappeared from Macclesfield's library and only re-appeared two hundred years later in 2006 when it was found by Nick Wilding at Sotheby's in London, when the library of the Earl of Macclesfield was being auctioned. The book was sold to an anonymous buyer for £153,600 and, regrettably, is not available. Nick Wilding, who is one of the few who saw the book, made a detailed report in his paper 'The Return of Salusbury'.[4] The copy examined by Wilding shows several annotations and probably is a proof copy of the book. This is the first published biography of Galileo since the one by Viviani in 1654 and was printed in 1717. The index of the book – anticipated in the previous volumes – is provided here in the appendix to show the complexity and richness of information of Salusbury's work, which is 180 pages long. Very little is known about Thomas Salusbury and the way he collected his sources but we know he travelled through Italy and could speak Italian. We know very little about his life, not even the date of his death, which most probably occurred in early 1666 just before the Great Fire.

Although the book is missing, we know an entire passage from a quote made by Drinkwater in his own biography, published in 1830. On reading this passage in Salusbury's book, Drinkwater was so impressed that he transcribed the text literally to avoid any misunderstanding. Salusbury seems in the following passage to describe a self-portrait of Galileo:

> He did not contemn the other inferior arts, for he had a good hand in sculpture and carving, but His Particular care was to paint well. By the pencil he described what His telescope discovered; in one he exceeded art, in the other, still. Osorius, the eloquent bishop of Sylya, esteems one piece of Mendoza the wise Spanish minister's felicity, to this, That he was contemporary to Titian, and That by His hand he was drawn in a fair tablet. And Galilaeus, lest he Should want the same good fortune, made so great a progress in this curious art, That he Became His Own Baonarota; and Because there was no other copy worthy of His pencil, drew himself.[5]

Drinkwater added a comment of disbelief at the end: 'No other author makes the slightest allusion to such a painting; and it Appears Likely That Salusbury Should be

[3] S. Drake, 'Galileo Gleanings II a kind word for Salusbury', *Isis* 49, no. 1 (1958): pp. 26–33.
[4] Wilding, 'The Return of Thomas Salusbury's "Life of Galileo" (1664)'.
[5] E.J. Drinkwater, *Life of Galileo Galilei with Illustrations of the Advancements of Experimental Philosophy* (London: printed by William Clowes, 1829), p. 103.

mistaken Than That so interesting to portrait Should Entirely lost sight of'.[6]

Drinkwater could be right but the perspective of his epoch is quite different from that of the seventeenth century. In those days it was quite common not to sign or mention paintings and we should not be much surprised if this painting was never mentioned in the short biography of Galileo by Vincenzo Viviani. De' Nelli could have missed it in absence of clear documentation, as was the case for many things in the first half of Galileo's life. For instance, we know very little of the other Galileo portraits painted when he was not yet famous, nor of the portrait which was the source of the engravings of Galileo's image in several of his books.

The Tuscan Artist

Galileo, before becoming the first modern scientist, was a son of the Italian Renaissance. All his biographers agree that he experienced all the arts, from music to literature and painting.[7] According to Viviani, Galileo was a talented musician who could compete in playing the lute with his father Vincenzo, one of the most important musicians of the time, and was also an excellent writer, as recognized by Lepardi and Calvino who considered him 'the greatest Italian writer of all time'. John Milton, who visited Galileo when he was imprisoned in his house in Arcetri, calls him the 'Tuscan Artist' in his *Paradise Lost*:

> […] the Moon, whose Orb
> Through Optic Glass the TUSCAN Artist views
> At Ev'ning from the top of FESOLE,
> Or in VALDARNO, to descry new lands,
> Rivers or Mountains in her spotty Globe.[8]

Galileo's interest and talent for painting is carefully documented by Viviani, who wrote:

> He greatly enjoyed drawing which he pursued with admirable profit, demonstrating such genius and talent in it that he himself would often say to his friends, that if at the time it had been in his power to decide his own profession, he would certainly have chosen painting. And in truth he had such a natural inclination towards drawing, and over time acquired such an exquisite taste, that his opinions on paintings and drawings were more highly appreciated than those of the lead-

---

[6] Drinkwater, *Life of Galileo Galilei*, p. 103.

[7] E. Panofsky, 'Galileo as a Critic of the Arts. Aesthetic Attitude and Scientific Thought', *Isis* 47 (1956): pp. 3–15.

[8] J. Milton, *Paradise Lost* (London: Samuel Simmons, 1667), 1.297–91.

ing masters by the masters themselves, such as Cigoli, Bronzino, Passignano and Empoli, and other famous painters of his time, all of whom were great friends of his. Indeed they frequently turned to him for advice about how to organise the narrative, how to arrange the figures, about perspective and colours and all other aspects that contributed to the perfection of painting. They did so because they realised that in this noble art Galileo had the most perfect taste and supernatural grace, such as it was impossible to find in the work of any others, even the masters. So much so, that the illustrious Cigoli, whom Galileo reputed to be the foremost painter of his time, attributed most of the best work that he did to the excellent teachings of Galileo himself, and in particular he felt himself honoured to say that in the matter of perspective, he alone had been his master.[9]

FIGURE 18.1. Noble gentleman. Is he Galileo?

[9] V. Viviani, 'Racconto Istorico della Vita di Galileo, 29 aprile 1654' (Firenze: Fasti consolari dell'Accademia fiorentina, 1717), p.2.

In 1613 Galileo became a member of Vasari's Florentine Academy of Design where he distinguished himself for his taste in supporting and protecting, *inter alia*, the young Artemisa Gentileschi. Disappointingly, as all biographic sources agree, few records survive of Galileo's artistic ability. Among these are the watercolours he made during his first telescopic observations of the Moon in November 1609. The line of the terminator and the white spots in the dark area need to be drawn very precisely since from their distance Galileo computed the height of the mountains on the Moon. Similarly, the drawing of the solar spots published in the *Istoria e Dimostrazioni sulle Macchie Solari* (1613) are extremely accurate in shape and position and allow us to compute the solar activity in that period, the first one ever recorded by man.[10] The shape of the solar spots near the solar limb was essential in the Galileo's understanding that they were on the solar surface and were not intervening spherical bodies, as proposed by Scheiner. It is worth noting that one drawing is dated 26 June 1612, which is the date of Galileo's famous letter to Cigoli on the 'Comparison of Arts' where, requested by his friend, Galileo compares the relative merits of painting and sculpture and argues that painting is superior because it is more distant to reality.[11] A quite modern view indeed. A few other interesting sketches were found by Horst Bredekamp in Galileo's notes on waste paper, dating to around 1584, certainly something that Galileo would have never shown to anyone.[12]

A new Galileo portrait?

A few years ago an antique painting showing a nobleman has been connected to Galileo.[13] The identification of the gentleman as Galileo is principally based on the resemblance to other portraits of Galileo. The painting shown in Fig. 18.1 is a half-length portrait of a man dressed in black, turned slightly to the left, but looking directly at the observer. The man has short hair and an incipient red beard. The painting – oil on canvas 87 x 69cm – was classified by the experts as a portrait of the Italian school of the seventeenth century. The painting has been analysed with UV-Wood, X-ray and Infrared frequencies. The latter was performed at the National Optical Florence INO-CNR by Raffaella Fontana and collaborators using the tech-

---

[10] Galileo Galilei, *Istoria e Dimostrazioni intorno alle Macchie Solari e loro accidenti* (Roma: Appresso Giacomo Mascardi, 1613).

[11] Galileo Galilei, 'Letter to Cigoli, 26 June 1612', OG, XI, p. 340, n. 713.

[12] Horst Bredekamp, 'Gazing hands and blind spots: Galileo as draftsman', *Science in Context* 14, no. S1 (2001): pp. 153–92, here p. 153; and Horst Bredekamp, *Galilei der Künstler: Der Mond, die Sonne, die Hand* (Berlin: Akademie Verlag, 2007).

[13] Paolo Molaro, 'Possible Portrait of Galileo Galilei as a young scientist', *Astronomische Nachrichten* 333 (2012): pp. 186–193.

nique of infrared reflectography. Twelve images were taken between 0.952 and 2.265 microns of the optical band. A representative image is shown in FIG. 18.2. The pigments are partially transparent to infrared radiation and allow one to see that the preliminary drawing of the face was done with no second thoughts in the execution, while probably the body and hands were sketched up, probably directly with a brush. Interestingly the X-ray image reveals a different and simpler collar.

A horizontal seam at about 24cm from the bottom edge of the picture combines together two pieces of canvas. At the bottom we can see a re-painting, probably made at a later time and by a different and less skilful hand. X and IR images reveal a different original drawing of the right arm, which holds some objects. These seem to include two rolls of paper tied with two bows and other unidentified tools. This could have been done deliberately to prevent the identification of the sitter and a careful restoration could be very useful in this respect.

The first known portrait of Galileo is that of 1601 by the painter Santi di Tito, as reported by De' Nelli, who writes:

> The most illustrious Italian artists wished to have the honour of painting Galileo's portrait. Santi di Tito portrayed him in a small painting in 1601 at the age of thirty-eight, not long before he himself passed away. [...] This portrait is the one conserved in my private Library, and I have placed the derived engraving executed by Mr. Giuseppe Calendi at the opening of this History.[14]

The painting by Santi di Tito was then lost, although it may perhaps have been recently rediscovered, as reported by Tognoni in 2013; also see my paper of 2016 on the subject.[15]

A second portrait is attributed to Domenico Tintoretto, son of the famous Jacopo Robusti, and executed around 1605–1606. The attribution to Tintoretto derives from the similarity to an engraving executed by Natale Schiavoni and Giuseppe Bossi, which bears the inscription 'from a lost portrait by Tintoretto'.[16] More recently in

---

[14] G. B. De' Nelli, *Vita e commercio letterario di Galileo Galilei* (Losanna: 1793), pp. 872, 873.

[15] F. Tognoni, ed., *Le opere di Galileo Galilei: Iconografia galileiana, Appendice*, Vol. 1 (Florence: Giunti Editore, 2013); and F. Tognoni, 'L'immagine di Galileo:tra iconoteche e biografie illustrate', in *Atti e Memorie dell'Accademia Galileiana di Scienze, Lettere ed Arti già dei Ricovrati e Patavina. Parte III: Memorie della Classe di Scienze Morali, Lettere ed Arti* CXXVII (2014–2015). Also see Paolo Molaro, 'Su ritratto perduto di Galileo ad opera del pittore toscano Santi di Tito', *Giornale di Astronomia* 42, no. 1 (2016): pp. 10, 14.

[16] A. Favaro, 'Studi e Ricerche per una Iconografia Galileiana', in *Atti del Reale Istituto Veneto di Scienze, Lettere ed Arti, Anno accademico 1912–1913, Tomo LXXII Parte seconda* (Venice: C Ferrari, 1913), pp. 995, 1051; and A. Favaro, 'Nuove contribuzioni ad una Iconografia Galileiano', in *Atti del*

2008 Mancini proposed, on stylistic grounds, a different attribution of the portrait to the painter Apollodoro di Porcia (ca. 1531–1612), who was a portraitist greatly in vogue in Venice in the early seventeenth century, particularly among the circle of intellectuals that surrounded Pinelli.[17] There is no certain documentation of the painting, and it is not mentioned in the historic biographies of Galileo or in his private correspondence. It should be noted that the first account of the iconography of Galileo, contained in the biography by G. B. Clemente De' Nelli (1793), makes no mention of it.[18]

The very dating of the painting is fairly approximate, having been proposed by Fahie on the basis of the double letters 'LL' in the legend on the painting, which appears to be an element detected only in documents of that part of the Venetian period.[19]

Another portrait is the one possibly painted by Filippo Furini, known as *lo Sciamerone* in 1612 and used for the engraving of the frontispiece of the *Istoria e dimostrazioni intorno alle macchie solari* of 1613 and in *Il Saggiatore* of 1623.[20] The painting is at the Vienna Kunsthistorisches (INV GG 7976) where, however, it bears a different attribution to Tiberio Titi, son of Santi di Tito.

Then there are three identical pencil portraits, dated and signed, which were executed by the papal portraitist Ottavio Leoni in 1624 during Galileo's visit in Rome to the new Pope, Urban VIII. Of the same year is the oil painting on a plate of silver attributed to the painter Domenico Passignani, now part of a private collection in Helsinki. The two most famous portraits of Galileo which are always mentioned in the various biographies are those by the Flemish painter Justus Susterman, court painter of the Medici, dated 1636 and 1640 and now conserved in Florence in the Palatine Gallery and in the Uffizi. These show Galileo late in life, and represent the paradigm of the Galilean icon. Copies of them are in the Bodleian libraries, sent by Viviani in 1661.

Reale Istituto Veneto di Scienze, Lettere ed Arti, Anno accademico 1914–1915, Tomo LXXIV (Venice: C Ferrari, 1915), p. 310.

[17] V. Mancini, 'Contribution to the iconography of Galileo Galilei', *Padova e il suo territorio* 131 (2008): pp. 24–26.

[18] De' Nelli, *Vita e commercio letterario di Galileo Galilei.*

[19] J. J. Fahie, *Memorials of Galileo Galilei (1564-1642), Portraits and Paintings, Medals and Medallions, Busts and Statues, Monuments and Mural Inscriptions* (Leamington and London: Courier Press, 1929), pp. 10–12.

[20] Galilei, *Istoria e dimostrazioni intorno alle macchie solari*, and in Galileo Galilei, Il *Saggiatore nel quale con bilancia esquisita e giusta si ponderano le cose contenute nella libra astronomica e filosofica di Lotario Sarsi Sigensano scritto* (Roma: Appresso Gioacomo Mascardi, 1623).

FIGURE 18.2. Left: X-ray. Right: IR 2265nm in false colours performed at the INO-CNR in Florence by Raffaella Fontana.

The identification of the gentleman shown in FIG. 18.1 as Galileo is based on the resemblance to the portrait by Tintoretto and Santi di Tito, obviously taking into consideration the different age. It is true that the characteristic mole above the left cheekbone is missing, but this might only have appeared later. Detail of the face is shown in FIG. 18.3, where the particular of the face is set in direct comparison with Tintoretto's portrait. In the Tintoretto the face appears slightly thinner, but the essential proportions are the same. The most significant difference concerns the nose, which is slightly shorter in Tintoretto's portrait, while the shape and the blue colour of the eyes are extremely similar. The painting is also very similar to the engraving by Giuseppe Calendi. An overlay shows that the shape of the head coincides perfectly, while there is a slight difference in the shape of the tip of the nose and the lobe of the left ear, which is missing in the painting where it is perhaps covered by the beard.

The painting was recently studied within the framework of the project FACES (Faces, Art, and Computerized Evaluation Systems) coordinated by Conrad Rudolph of the Department of the History of Art, University of California, Riverside, in which the most sophisticated face recognition technology is applied to works of art. Their conclusions regarding the portraits of Anne Boleyn received extensive coverage in the media. Indeed, they rejected the only portrait believed to be of Anne Boleyn, but found a new identification of Anne in a portrait previously believed

Figure 18.3. On the left Galileo Galilei by Domenico Tintoretto, and on the right is the anonymous.

to be of Jane Seymour. When tested against eight other known portraits of Galileo Galilei, they found consistently decreasing match scores for the three paintings chronologically closest in age (1601–ca. 1612), 'no decisions' for the two dated 1624, and non-matches for the final three done when Galileo was old.[21] More specifically, the matches with the portraits by Santi di Tito (via Calendi), Tintoretto and the Furini have maximum probability (probability 0.724, 0.718, 0.708 respectively, with the peak of probability of the procedure at 0.725), whereas they are ambiguous with the Leoni and the Passignano (0.672, 0.650) and are not consistent with those of Susterman (0.639 and 0.590). The latter has to be attributed to the process of ageing that altered the physiognomy of the scientist's lineaments.

It should be noted that this analysis does not consider colour. Indeed, elements such as the colour of the eyes and hair and of the flesh tones which play a significant role in the processes of human recognition, are not taken into consideration in the FACES procedure. This confirms that the anonymous figure portrayed displays a remarkable similarity to Galileo, even though – in the absence of certain documentation – it is not possible to distinguish between Galileo and a perfectly identical double. If he is Galileo, he must have been aged between 20 and 25, which corresponds to the period during which he became a professor at Pisa or in the early years of his teaching in Padua. The portrait offers us a unique image of the young Galileo, with his intense and captivating gaze, fully aware of his own capacities and almost as

---

[21] C. Rudolph et al., 2016 in press; Srinivasan et al., 'Computerized Face Recognition in Renaissance Portrait Art', p. 85.

if he had a premonition of the important role he was to play in the history of science.

The author of the painting could have been one of the many painters belonging to the circle of his artistic acquaintances, and may have been executed as an exercise. There is also documentation of other portraits executed by the painters of the time which are now lost. For example the one by Ludovico Ciardi – known as Cigoli – who had been Galileo's friend since the time they both took lessons of perspective from the mathematician Ostilio Ricci and who portrayed the physical Moon of the *Sidereus* in the fresco in Santa Maria Maggiore in Rome. The reference to the portrait by Cigoli is contained in a letter from Luca Valerio to Galileo dated 4 April 1609 and is probably too late to be the one we are considering here.[22]

However, since at that time Galileo was anything but famous, and that portraits were painted only for figures of a certain status, the hypothesis of the self-portrait executed by Galileo in his youth, mentioned by Thomas Salusbury, might not be entirely implausible. Certainly it is possible that Salusbury is wrong, as Drinkwater concludes, but it should be not too surprising that no mention to this self-portrait is present in the earliest biographies of Galileo. Viviani's is very concise and De' Nelli's appears very late, when many sources were compromised. These biographies or later ones, including that of Drinkwater, lack references to other portraits of Galileo. Moreover, many famous portraits of Galileo are not signed and can hardly be dated, following the customs of the time. It should not be a surprise that a work probably made in his youth was overlooked. We recall that the whole of Galileo's life when he was in Padova is very little documented.

In the sixteenth-century portraits were commissioned only by important people, but this contrasts with this painting, made in a 'cheap' way, which shows a degree of economic difficulty of the author or of whoever commissioned the work. Thus, the possibility that Galileo made a self-portrait is an intriguing one which is certainly unlikely but that cannot be excluded *a priori*.

ACKNOWLEDGEMENTS: It is a pleasure to thank Federico Tognoni, Pier Andrea Dandò, Franco Zotto, Raffaella Fontana, Marco Barucci, Ferruccio Petrucci, Conrad Rudolph, Paolo di Marcantonio and Francesco Palla.

Appendix
Index of the Thomas Salusbury's *Galileus Galileus, his life, in five books*. The book belongs to an anonymous private collector. We strongly pray its owner in honour of the science of making it accessible.

---

[22] Luca Valerio, 'Letter to Galileo of 4 April 1609', the National Library in Florence, Mss. Gal., P. VI, T. VII car. 93 (autograph), 1609.

III. GALILEUS GALILEUS, his LIFE: in Five BOOKS

BOOK I. Containing Five Chapters.
Chap. 1. His Country.
    2. His Parents and Extraction.
    3. His time of Birth.
    4. His first Education.
    5. His Masters.
BOOK II. Containing Three Chapters.
Chap. 1. His judgment in several Learnings.
    2. His Opinions and Doctrine.
    3. His Auditors and Scholars.
BOOK III. Containing Four Chapters.
Chap. 1. His behaviour in Civil Affairs.
    2. His manner of Living.
    3. His morall Virtues.
    4. His misfortunes and troubles.
BOOK IV. Containing Four Chapters.
Chap. 1. His person described.
    2. His Will and Death.
    3. His Inventions.
    4. His Writings.
    5. His Dialogues of the Systeme in particular, containing *Nine Sections.*
    Section   1. Of Astronomy in General; its Definition, Praise, Original.
        2. Of Astronomers: a Chronological Catalogue of the most amous of them.
        3. Of the Doctrine of the Earths Mobility, *&c.* its Antiquity, and Progresse from *Pythagoras* to the time of *Copernicus.*
        4. Of the Followers of *Copernicus,* unto the time of *Galileus.*
        5. Of the severall Systemes amongst Astronomers.
        6. Of the Allegations against the *Copern.* Systeme, in 77 Arguments taken out of *Ricciolo,* with Answers to them.
        7. Of the Allegations for the *Copern.* Systeme in so Arguments.
        8. Of the Scriptures Authorities produced against and for the Earths mobility.
        9. The Conclusion of the whole Chapter.
BOOK V. Containing Four Chapters.
Chap. 1. His Patrons, Friends, and Emulators.
    2. Authors judgments of him.
    3. Authors that have writ for, or against him.
    4. A Conclusion in certain Reflections upon his whole Life.

EINSTEIN, GALILEO, AND KEPLER:
THE SCIENTIST PORTRAIT OPERAS OF PHILIP GLASS

David L. Morgan

ABSTRACT: Over the course of his career, contemporary composer Philip Glass has written three operas based on the characters and biographies of physicists and astronomers: *Einstein on the Beach* (1976), *Galileo Galilei* (2002) and *Kepler* (2009). While these three works are stylistically diverse, and range from almost entirely abstract to straightforwardly narrative, they share common themes and stylistic techniques, and may be viewed within the context of Glass's career-spanning exploration of law-like mathematical algorithms as a generative tool for musical expression.

With the possible exception of a handful of film composers, Philip Glass is arguably the most well-known and popular living composer of orchestral music.[1] He has been incredibly prolific well into his later years and continues to perform his compositions in both solo piano concerts and with his Philip Glass Ensemble. He has composed more than two dozen operas and musical theatre works, ten symphonies, seven string quartets, and countless chamber works and compositions for solo piano and other instruments. Along with contemporaries such as Steve Reich and Terry Riley, he is considered one of the founders of the stylistic movement often referred to as 'minimalism', although Glass himself has never been fond of the term. Whatever label we attach to Glass's aesthetic, it is impossible to deny that his compositional style remains incredibly influential, and imitations of his distinctive rhythmic and harmonic structures can be heard in everything from TV commercials to film scores.

Philip Glass's 1976 opera *Einstein on the Beach* remains one of the pivotal works of the so-called 'minimalist' genre. In this work, which can last up to five hours with no intermission, Glass and his collaborator Robert Wilson eschew a narrative structure and instead present a theatrical meditation in which Einstein is not so much a character as he is a visual metaphor. The creators of the opera take advantage of the fact that Albert Einstein's name and image is synonymous with 'genius' in a way that needs no explanation, and there are few individuals who could fulfil the role of both thematic focus and free-associative nucleus that Albert Einstein plays in Glass and Wilson's work. It would not be quite accurate the say that *Einstein on the Beach* is 'about' Albert Einstein in any meaningful way, and yet the presence of Einstein does

---

[1] Philip Glass Biography, http://philipglass.com/biography/; Philip Glass, *Words Without Music: A Memoire* (New York: W.W. Norton, 2015).

provide content and meaning to the opera that it would otherwise lack, and the piece would be reduced to a work of pure abstraction in his absence. The creators made no excuses for the free-associative and sometimes plodding nature of the opera. Choreographer and collaborator Robert Wilson said of the work, 'It's something you can freely associate with... It's ok to get lost'[2] In fact, viewers of the original productions were even encouraged to get up and move about during the performances. And yet, the recurring visual image and oblique textual references to the figure and ideas of Einstein encourage the viewer to find meaning in the churning musical structures, frantic dances, and slowly unfolding set pieces.

It is, of course, possible to find explicit references to the physics of Einstein in *Einstein on the Beach* if one looks hard enough. The sharp vertical beams of light and slowly advancing train seen during the *Train* movement call to mind Einstein's *Gedankenexperiment* about the relativity of simultaneity that were based on the premise of observers on a moving train witnessing a strike of lightning. In some movements, the leaping bodies of the dancers, their bodies at a striking angle to vertical, suggest that the choreographer Lucinda Childs might have been riffing on Einstein's idea of the Equivalence Principle – that the effects of gravity and of an accelerating frame of reference are indistinguishable from one another. And the climactic *Spaceship* movement invites us to recall the relativity of time aboard a spaceship traveling close to the speed of light – both in terms of its set design and in the increasingly propulsive pace of Glass's music as it approaches its destination. This metaphorical sense of momentum and energy was, to Glass, an important part of the structure of the work. When discussing the absence of a slow movement in the work, Glass said, 'In classical music there are allegros and prestos in all kinds of pieces... But with *Einstein*, the idea of unstoppable energy was all there was. There was no need for a slow movement.'[3]

But it is the concept of the relativity of time that finds its clearest expression within the musical structures of *Einstein on the Beach*. From the opening notes of the work, the listener is presented with a musical motif that consists of a descending 3-note pattern of four beats, then six beats, then eight beats. This is a very simple example of a compositional technique that Glass employed often in his early works – an 'additive' structure in which a musical motif is established via repetition and then lengthened by adding notes or repeating phrases. (More will be said about these

[2] Robert Wilson, in M. Tarnopolsky, 'Robert Wilson, Philip Glass and Lucinda Childs discuss Einstein on the Beach', (28 October, 2012), https://www.youtube.com/watch?v=k8iLOGPm7AY [Accessed 2 January 2016].
[3] Glass, *Words Without Music*, p. 259

mathematical structure later in the chapter.) As the viewer of *Einstein on the Beach* is watching images that are explicitly referencing Albert Einstein, he or she may be reminded of Einstein's notions of time as expressed in his theories of relativity. Time is not, as was thought before the twentieth century, a universal constant that proceeds apace in the Cosmos with no relation to anything external. Rather, the measurement of time is malleable and flexible, and can depend upon the relative motions of observers making the measurements. Relative to a 'stationary' observer, time slows down for an individual moving at speeds approaching the speed of light – a phenomenon known as time dilation. Likewise, for the observer of *Einstein on the Beach*, Glass presents a sort of musical time dilation, in which the distance between musical events – bar lines, chord changes, phrase repetitions – gradually expands the more we listen.

Glass is certainly conscious of this notion of the role of the observer in the performance, and has spoken about the role of the listener in terms that bring to mind Einstein's notions of relativity and motion in a very explicit, albeit metaphorical way. Speaking of the role of the observer in Einstein, Glass said,

> Let's say that you're sitting there and the piece is up here. You have to travel a certain distance... in your effort to apprehend it, to understand it, to see what it is. It's the traveling of that distance that personalizes the work. [...] If you don't make that journey, it doesn't happen.[4]

In contrast to Glass's somewhat more abstract character operas, of which *Einstein on the Beach* is perhaps the most abstract, *Galileo Galilei* is a relatively concrete biographical work exploring the actual life events of Galileo. While the narrative is presented in reverse-chronological order, the work touches on recognizable events from the life of Galileo, including his trial and recantation at the hands of the Inquisition, his scientific work with the inclined plane, his *Dialogue Concerning the Two Chief World Systems*, and his presentation of the telescope to his patrons. The opera also examines the relationship between Galileo and his daughter, Maria Celeste – a topic explored by many authors in recent years, most notably by Dava Sobel in her book *Galileo's Daughter*.

The music of *Galileo Galilei* is representative of Glass's more mature style. Compared to *Einstein on the Beach*, the scope of the opera is much more modest. The work has only a single act, and a production of *Galileo* only runs 90 minutes or so in length. While the cyclical harmonic structures are immediately recognizable as

---

[4] Philip Glass, in Tarnopolsky, 'Robert Wilson, Philip Glass and Lucinda Childs discuss Einstein on the Beach'.

Glass, gone are the extensive experimentations in minimalist aesthetic and repetition for its own sake. And yet metaphorical connections can still easily be drawn between Glass's stylistic hallmarks and the subject matter of Galileo's science – the sense of forward motion (the inclined plane), repetition (the swinging pendulum), and circularity (the orbits of the planets) being the most obvious connections.

Galileo Galilei also includes a nod to the origins of the form of opera itself, via a reference to Galileo's father, Vincenzo Galilei, at the end of the story. The elder Galilei, as a member of a group of Italian intellectuals known as the 'Florentine Camerata' is sometimes credited with the introduction of *recitative*, a stylistic form somewhere between talking and singing, which was a stylistic element crucial to the early evolution of opera.[5]

The 2009 opera *Kepler*, commissioned by the Linz Landestheater in Austria, is a somewhat biographical work about Johannes Kepler that shares more with *Galileo Galilei* than with *Einstein on the Beach*, although the work is not strictly narrative in structure. The libretto includes not only the words of Johannes Kepler himself, but texts from the Old Testament, and poems by German lyric poet and dramatist Andreas Gryphius (1616 – 1664)[6]. Both Gryphius and Kepler lived through the Thirty Years' War, and with the inclusion of these poetic works, Glass and librettist Martina Winkel are attempting to say something about the society and the time that Kepler found himself living in…

> Kepler's times – the sixteenth century – this remote period has a lot in common with today's. Back then, there was the Counter-Reformation, the Thirty Years' War, the Catholics against the Protestants . . . Kepler said the that blood was flowing down the streets. He was watching the world around him. It was horrible! He said 'The only thing that we can do is to pursue our studies and melt into eternity.' [7]

As was the case in *Einstein on the Beach*, there are points during *Kepler* where one can find references to Johannes Kepler's scientific ideas in the staging of the opera and chorography of the performers. The connection between the subject matter and the musical devices is again easy to make. There is a scene in *Kepler* during which the chorus repeats the phrase 'Number… quantity… circular motion…' over and over. They could just as easily be referring to Glass's own musical structures as to the planetary orbits of Johannes Kepler.

---

[5] Piero Weiss and Richard Taruskin, *Music in the Western World* (Cengage Learning, 2007), p. 139.
[6] Kepler, http://www.philipglass.com/music/recordings/kepler.php
[7] Jean-Luc Clairet, 'Interview With Philip Glass', (11 March, 2014), http://www.resmusica.com/2014/03/11/interview-with-philip-glass/ [Accessed 2 January 2016].

Speaking as a scientist, it is possible to criticize 'Kepler' on the basis that it doesn't do nearly enough to directly explore the scientific ideas of Kepler in interesting ways. In their review of the opera in *Nature*, INSAP IX attendees Jay M. Pasachoff and Naomi Pasachoff wrote, 'We wished that astronomers or historians of astronomy had been consulted at an early stage of writing and been given an opportunity to make suggestions.'[8] If one went into 'Kepler' expecting to learn a great deal about the details or the impacts of Johannes Kepler's work, one would likely come away disappointed.

This raises the question which motivated my INSAP IX presentation in the first place. Looking back at these three very different operas, spanning more than three decades of Glass's career, the most obvious question to ask is – 'Why?' Why scientists, and why these astronomers and physicists in particular? So little of the science itself enters Glass's works in a narrative role, and even the metaphorical value of the scientific concepts sometimes feels only cursory. So why Einstein, Galileo, and Kepler?

One facile answer, and one that is perhaps not at all fair to Glass himself, is that all of these scientists were, in their own times, 'revolutionaries' of a sort, and that Glass, being one of the pioneers of the twentieth-century movement that came to be known as minimalism, counts himself among the pantheon of revolutionaries and visionaries that includes Einstein, Galileo, and Kepler. (One might also include Gandhi, Akhenaten, and Walt Disney – the subjects of Glass's other operatic character studies – in this list.) At the very least, one might conclude that Glass's apparent preoccupation with revolutionary genius derives in part from a desire to understand and situate the role of the radical, rebel, and revolutionary within the culture at large. But since I can find nothing in Glass's own writings about his operas, or any of his works, to suggest any knowing self-identification between himself and his subject matter, I will set aside this psychoanalytical approach in favour of one that examines the structure of the music itself and the guiding philosophy behind some of the compositional techniques that define Glass's characteristic style.

Philip Glass's music has always had not just a mathematical aspect to it, but an algorithmic one. What I mean by this, is that many of Glass's works are based on taking a simple melodic or harmonic structure and either expanding upon it or permutating it according to a simple rule. In many of his early works, this rule was simply additive. A four-note phrase would be repeated a certain number of times, then expanded into a five-note, then a six-note, then a longer and longer musical

---

[8] Jay M. Pasachoff and Naomi Pasachoff, 'Third Physics Opera for Philip Glass', *Nature* 462 (2009): p. 724.

idea. This technique can be seen very clearly in early Glass works such as *Two Pages* (1969) or *Music with Changing Parts* (1970), but it can also be heard in *Einstein on the Beach*, for example in the aria of the *Bed* scene in Act IV, where a simple four-chord harmonic structure grows from a starting length of 22 beats, and as it repeats itself over the course of the next eight minutes, slowly expands to a length of 84 beats.[9] While the technique is simple, the effect on the listener can be profound – demanding that they become aware of musical events and changes that occur on timescales of several minutes.

Another algorithmic technique used by Glass is the repetition of a single harmonic phrase within different rhythmic structures. For example, in opening of the fourth 'Knee Play' in *Einstein on the Beach*, a four chord pattern is arpeggiated in groups of nine (6+3) eighth notes, then eight (4+4) sixteenth notes, then nine triplets (3+3+3), and then six (2+2+2) eighth notes. Each pattern repeats itself a number of times, then the whole cycle reverses itself. The musical impression on the listener after this algorithm reaches its conclusion is very much that of having traversed a circle, and the feeling of resolution is satisfying, albeit a very different sense of musical resolution than the harmonic resolutions or melodic destinations we might be accustomed to experiencing when listening to more traditional classical music.

Finally, a technique employed in many of Philip Glass's compositions is the rhythmic device of polyrhythm – patterns with different rhythmic content played at the same time. The simplest polyrhythm is a 'two against three' pattern, where an eighth-note pattern of two alternating notes is played against a triplet pattern of two or three notes. Several of Glass's works for solo piano, such as *Openings* (1982) and *Mad Rush* (1989) are built around such a structure. In both of these piano works, the left hand plays eighth-notes and the right-hand triplets. (A feat that presents a challenge in hand-independence for the novice pianist, and which took the author some time to master and automatize.) In some of Glass's works, this simple rhythmic structure underlies the entire piece. In *Openings*, for example, the two hands play a four-chord pattern that repeats itself four times. There are three such four-chord patterns, and the sequence of three four-chord patterns itself repeats three times, and then simply ends. There is no melodic or harmonic unfolding toward some destination or resolution; there are only cycles of twos, threes, and fours. Out of this raw material of cycles and repetition, Glass constructs works in which the listener is challenged to perceive musical events that are not moments of melodic or harmonic resolution, but rath-

[9] Kyle Gann, 'Intuition and Algorithm in Einstein on the Beach', (6 March, 2013), http://www.newmusicbox.org/articles/intuition-and-algorithm-in-einstein-on-the-beach/ [Accessed 5 January 2016].

er the unfolding and layering of cycles on much larger timescales. It is this sort of compositional technique that earns Glass's music the label of 'boring' or 'repetitive' by his detractors. But the larger musical question that Glass (and many of his fellow so-called 'minimalist' contemporaries) seem to be addressing in such pieces is–how rich a musical experience can one create by combining the simplest rhythmic and harmonic units according to law-like mathematical rules?

Philip Glass's music can often be seen as an exploration of form for its own sake. Speaking in a 2009 interview, Glass himself said that 'With the [Philip Glass] Ensemble music... the issue was always about form and content. The thesis of the music was that the structure and the material were identical, in the same way that a Jasper Johns painting of a flag is identical with the flag.'[10] Glass has also not been silent on the topic of the aesthetic beauty of mathematics and the elegance of mathematical forms. In the same interview, he adds,

> Mathematicians are subject to the same kinds of enthusiasms as everybody else. The beauty of mathematics is something that mathematicians talk about all the time, and the elegance of a mathematical theorem is almost as good as its proof. Not only is it true, but it's elegant. So you get into almost aesthetic questions.[11]

It is in this formal exploration of layered hierarchies of structure, and the aesthetic value of simplicity and unity in which I find the clearest connection between the music of Philip Glass and the history of physics and astronomy. The larger goal of science is to unify and simplify – to explain the phenomena of nature using the fewest number of elements and the smallest number of rules. Galileo unified the heavens and the Earth when his telescope revealed the Moon to be nothing more than a rock covered in mountains and valleys. Kepler unified the motions of the planets when he discovered a triplet of mathematical rules that they all obeyed. And Einstein unified our notions of energy and mass, motion and light, and space and time with his theory of relativity.

Glass has even borrowed the language of physicists to describe some of the musical ideas that he has employed in these works. For example, speaking of the climax of *Einstein on the Beach*, Glass suggests that 'The spaceship at the end represents the culmination of this 'unified field' of harmony, melody, and rhythm.'[12] Glass's appropriation of this terminology is clearly no accident, and reveals at least a passing

---

[10] Philip Glass, in Corinna Da Fonseca-Wollheim, 'Where Music Meets Science'. Updated (24 November, 2009), http://www.wsj.com/ [Accessed 8 January 2016].

[11] Philip Glass, in Da Fonseca-Wollheim, 'Where Music Meets Science'.

[12] Glass, *Words Without Music: A Memoir*, p. 258.

knowledge of the goals of contemporary physicists and a recognition of the parallels between their goals and his own. In the twos-and-threes polyrhythms, one can even see something of one the ultimate goals of the particle physicists: To describe the entire structure of matter as nothing more than simple combinations of quarks, bound to one another in triplets (protons and neutrons) and doublets (mesons).

Indeed, one can view the entire history of science as a series of unifications, and one can trace these unifications through the works of pivotal scientists who were able to recognize patterns in complexity. Like these revolutionary physicists and astronomers, the compositional practices of Philip Glass challenge us to imagine how vast a universe it is possible to create using the smallest handful of elements obeying the simplest of mathematical rules.

John Bevis's Eighteenth-Century Uranographia Britannica and the Atlas Celeste: Oft-Overlooked Treasures

Jay M. Pasachoff and Kevin J. Kilburn

ABSTRACT: One of the most beautiful star atlases of all time was prepared by John Bevis, who had earlier discovered the Crab Nebula, but was never formally published because of the bankruptcy of his printer. Several of the printed pages were bound in 1750 while further copies were released in 1786. We discuss these unpublished atlases, especially the copy in the library of the Manchester Astronomical Society and a newly discovered atlas in the library of the Duke of Devonshire. We maintain a list of the approximately 30 copies of the atlases, both the 1750 and 1786 editions, which are now known.

In November 1997, Mike Oates and Tony Cross and one of the present authors, Kevin J. Kilburn, all members of the Manchester Astronomical Society (c/o the God-lee observatory, University of Manchester, England), discovered, with the help of Peter Hingley, librarian at the Royal Astronomical Society, that an old star atlas that had been in their library for over fifty years was an original, bound set of star charts from Dr John Bevis's unpublished atlas, *Uranographia*, dating to ca. 1750.[1] William B. Ashworth Jr, then Assistant Professor of History at the University of Missouri, Kansas City, first described the strange history of this atlas in 1981.[2] By the time of Ashworth's article only twelve copies were known; three others have since become known and the Manchester atlas is the sixteenth to be identified.

This discovery started a long quest to learn more about this rare atlas and to compare and rank known copies with the one at Manchester.[3] In 2003, by which time 23 atlases had been identified, and in collaboration with Professor Jay M. Pasachoff, who had bought a Bevis atlas the previous year, and Professor Owen Gingerich, who had also carried out research about its history, a paper was published in the *Journal for the History of Astronomy*.[4] Our aim was to bring the story of the atlas up to date, to bring it to the attention of astronomy historians and, we hoped, to locate other copies.

---

[1] This is the technically correct description.

[2] William B. Ashworth, 'John Bevis and his "Uranographia" (ca. 1750)', *Proceedings of the American Philosophical Society* 125, no. 1 (1981): pp. 52–73..

[3] The definitive list of all known copies of Bevis's atlas can be found on the Manchester Astronomical Society website, available at http://www.manastro.org/bevis.html.

[4] Kevin J. Kilburn, Jay M. Pasachoff, and Owen J. Gingerich, 'The Forgotten Star Atlas: John Bevis's Uranographia Britannica', *Journal for the History of Astronomy* 34 (2003): pp. 125–44.

To put the atlas into context, we need to understand what Bevis's intentions were and what *Uranographia* and *Atlas Celeste* actually are:

The first mention of *Uranographia Britannica* is in a newspaper advertisement placed by Thomas Yeoman in the *Northampton Mercury* of 11 April 1748.[5]

> URANOGRAPHIA BRITANNICA BEING an exact Survey of the Heavens, on fifty large Copper-Plates; wherein are represented, in their Places to the present Time, all the fix'd Stars, which have hitherto been observed in any Part of the World, with their proper Asterisms or Images, each accompanied by an explanatory Index, containing both the ancient and modern Catalogue, and curious Remarks pertinent thereto. At the End will be added two Hemispheres, with Ptolemy's Stars; and to the Whole will be prefix'd an Introduction, containing an Historical account of the Asterisms, and the whole Astronomy of the Fix'd Stars from the earliest Antiquity to the present Time. This will be followed by a general Alphabetical Index of all the Stars on the whole Uranographia.
>
> Proposals, and the Plan at large, may be had gratis at Mr. Professor Bliss's in Oxford, the Rev. Dr. Hoopers's at Trinity College in Cambridge, Mr. Thomas Yeoman's in Northampton, and at the Undertaker's own House in Leadenhall street, London; at all which Places most of the Copper-Plate Prints may be inspected.
> N.B. The newest and most curious Experiments in ELECTRICITY will be exhibited, during the present Week, at Mr. Yeoman's Experiment Room in Gold Street, Northampton.

In about 1745, Dr John Bevis F.R.S., a London doctor and amateur astronomer, had proposed to compile a modern British star atlas, *Uranographia Britannica*.[6] He enlisted the help of John Neale, variously described as an instrument maker or toy maker of Leadenhall Street, as the undertaker of the project, to finance the work with advanced subscriptions from those wealthy enough to afford a new, expensive, star atlas.[7] Other collaborators, mainly London scientific instrument makers, helped publicise it.

Uranographia was based on Flamsteed's star positions. His star catalogue, *Historia Coelestis Britannica*, was published posthumously in 1725, edited by his widow, Margaret. This contained a catalogue of 2,935 stars to much greater accuracy than any other previous work. Halley's contemporary observations of the southern hemisphere, together with additional stars from Bevis's own transit observations made

[5] *The Scots Magazine* noted the intended publication of *Uranographia Britannica* by Mr Neale on Friday 1 April 1748. This is of course following the Old Style Julian Calendar.
[6] Kevin J. Kilburn, 'Bevis [Bevans] John', *Biographical Encyclopedia of Astronomers* 1 (2007): pp. 118–19.
[7] The 1746 list of subscribers is held at Glasgow University Library. Sp Coll. f465.

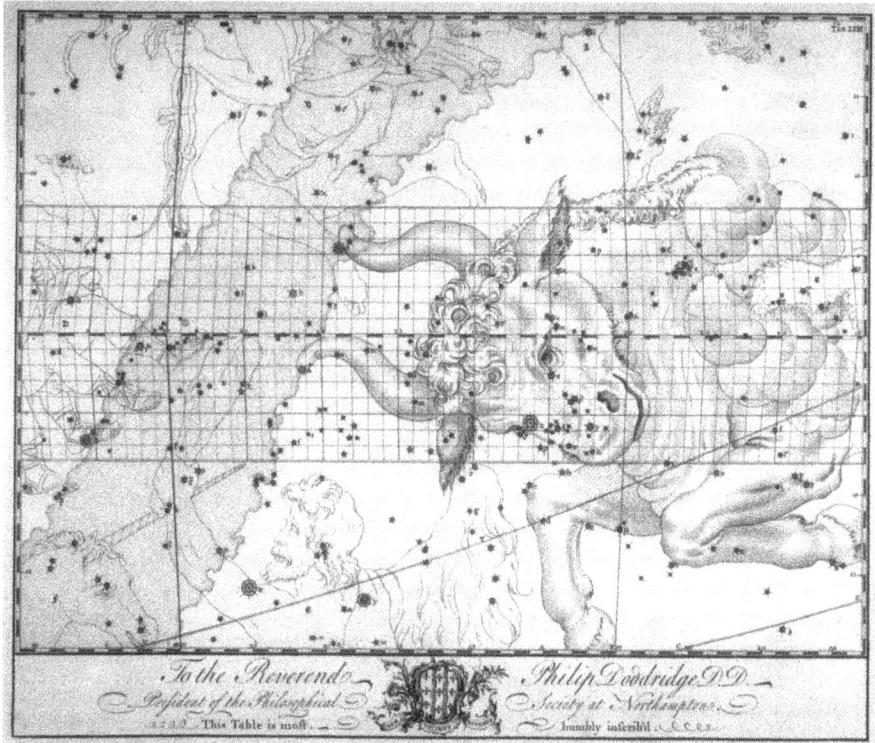

Figure 20.1. The Taurus plate XXIII showing the ecliptic and dedication to the Reverend Philip Doodridge.

between 1738 and 1739 from Stoke Newington, helped to expand the catalogue for Bevis's new atlas.

In 1731, Bevis had been the first to notice what we now call the Crab Nebula, something acknowledged by Messier in a second version (1780) of his still-widely-used list of non-stellar objects, which had been first issued in 1771 without mention of Bevis. The Crab Nebula was the first object on Messier's list; it is still known as Messier's M1.[8] Bevis's atlas was to be the first of the 'classical' star atlases based on Bayer's *Uranometria*, Augsburg 1603, and Flamsteed's *Atlas Coelestis*, 1729, to include non-stellar objects.[9]

In 1750, as the atlas was still in the process of being printed and compiled, disaster struck. Neale was declared bankrupt, the copper plates were sequestered by the Lon-

---

[8] Jay M. Pasachoff, 'Messier, Copernicus, Flamsteed: The SAF Rare-Book Collection in Paris', in *American Astronomical Society Meeting Abstracts#* 223 (Vol. 223), id.107.07 (2014).

[9] About 125 examples of Bayer's atlas are still extant. Flamsteed's atlas is not so rare.

don Courts of Chancery and the project was abruptly terminated. It was not until 1785, long after the death of Neale and of Bevis, that Bevis's library was auctioned by the widow of his executor, James Horsfall F.R.S. According to the auction catalogue, which still survives in the Whipple Museum, Cambridge, three near-complete atlases were sold together with an unknown number of pre-printed star charts. Ashworth lists the relevant part of the catalogue, lots 765–770 as followed:

765   Dr. John Bevis's *Uranographia Britannica, being an Exact survey of the Heavens, on Fifty-One large Copper-Plates with a double Nomenclature and tables of all the fixed Stars: A work never published. Some Sheets of the Nomenclature are wanting.*
766   *Another copy.*
767   *Another copy, interleaved.*
768   *Several duplicate impressions, and Sheets of the Nomenclature, and complete Catalogues of the fixed Stars.*
769   *Ditto*
770   *Drawings and Proof Sheets of several of the Plate.*

The surplus, finished star charts were later compiled into an unknown number of atlases and offered for sale cheaply in 1786 as *Atlas Celeste* by an anonymous seller. It is this, *Atlas Celeste,* which forms the bulk of the currently identified Bevis atlases. A typical *Atlas Celeste* comprises an elaborate frontispiece showing the Muse, Urania, offering a star atlas to a seated man (Frederick, Prince of Wales), with Greenwich Observatory in the background and 51 star charts, Tab I to LI, each carrying a dedication to institutions or individuals who subscribed to the project. A few atlases have a simple index thought (by Kilburn) to have been printed in 1786 together with an exceedingly rare title sheet advertising it as *Atlas Celeste.* Manchester's *Atlas Celeste* is still, as of 2016, considered to be the most complete and original in its format.[10]

Of the three, nearly-finished *Uranographia,* one of them was bought at the Sotheby & Wilkinson sale in London on 21 Jan 1856 by the American Philosophical Society (APS) of Philadelphia. It was formerly owned by Sir George Shuckburgh-Evelyn (1751–1804). This is the atlas on which Ashworth based his 1981 seminal description. Another, also identified by Ashworth, is at St John's Library, Cambridge. In poor condition and dirty, it is not quite as complete as the APS atlas. The third atlas went unidentified until November 2011.

As described in the 1785 auction catalogue, what sets these three *Uranographia* atlases apart from *Atlas Celeste* is that Bevis intended to include descriptive tables

---

[10] The *Atlas Celeste* in the British Library is folded. Their reference is C.21.c.5.

ATLAS CELESTE;

OR, THE

CELESTIAL ATLAS.

Being the moſt correct, copious, and ſuperb Work of the Kind,
that has ever been offered to the Public.

The Expence of the Engravings was *immenſe*, as the moſt Capital Artiſts in *Europe* were
employed in executing them, and the learned and ingenious Delineators and Directors of the
Work had determined to ſell it by way of Subſcription at Five Guineas the Set. The heavy
Charge attending it, rendered ſome of them Inſolvent, others were removed by Death, which
with divers adverſe Occurrences were the Means of retarding the Publication until the preſent
Period 1786. Many of the Copies have been deſtroyed by Fire and Removals; *the few that
remain are now offered at One Guinea and a Half each Set.*

This elegant and uſeful Work is not, nor ever has been in the Hands of any Bookſeller. The Copies ſaved
are all of the firſt Impreſſion, and will be an Ornament to any Library, and highly worthy the Notice and
Patronage of every Lover of the Sciences.

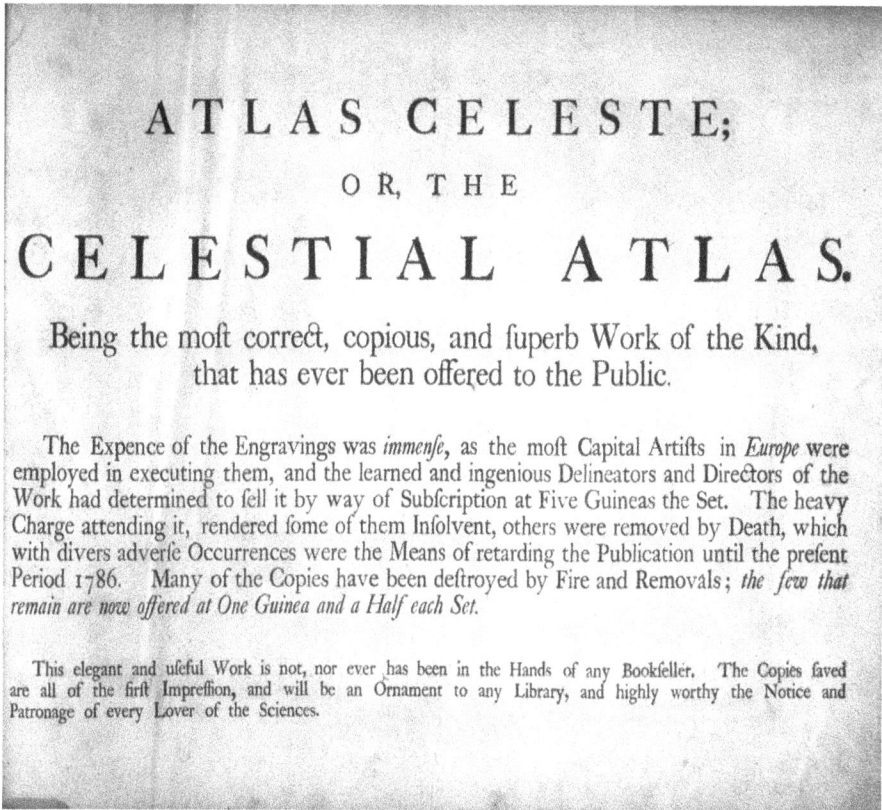

FIGURE 20.2. The title sheet in the atlas owned by Manchester Astronomical Society advertising *Atlas Celeste* in 1786.

opposite the star charts. The last two charts, unlike the majority that depicted specific constellations, show wide-field southern hemisphere regions or planispheres.
The tables list Ptolemaic and Bayer's stars whose magnitudes and zodiacal positions,
corrected to the epoch 1746, are shown together with additional explanatory notes.
Bevis included a separate fourteen page catalogue of the main stars in each constellation and also intended a comprehensive index and additional notes. The American
Philosophical Society *Uranographia* has fifty-one star charts, thirty-two tables (the
nomenclature) and fourteen pages of star catalogue. The Cambridge *Uranographia*
has forty-nine star charts, twenty-nine tables and fourteen pages of star catalogue.
None of Bevis's surviving *Uranographia* atlases have the intended index or additional notes, which were probably never printed.

In September 2011, a visit by Kilburn to an autumn country fair at Chatsworth,
Derbyshire, brought to mind that the eighteenth-century scientist, Henry Caven-

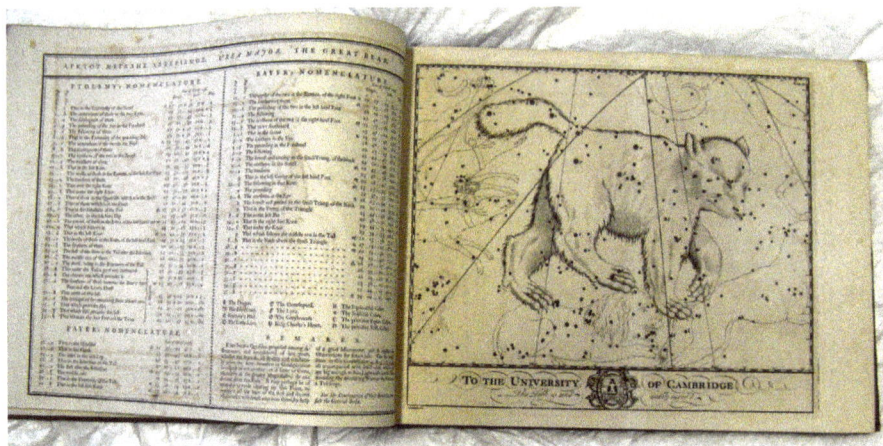

FIGURE 20.3. The Chatsworth *Uranographia Britannica*, one of only three copies offered for sale by auction in 1785. Photograph by permission of Chatsworth Settlement Trustees.

dish F.R.S., had connections with the Chatsworth estate.[11] He was the son of Lord Charles Cavendish, the youngest son of William Cavendish, 2nd Duke of Devonshire, whose family seat was at Chatsworth. Henry Cavendish was an avid book collector and was contemporary with the auction of Bevis's library in 1785. His father, Lord Charles Cavendish, was a subscriber to the intended Bevis atlas but had died in 1783. Chatsworth soon confirmed that they indeed had a Bevis atlas in the Devonshire Collection.[12] It was subsequently identified as being the missing *Uranographia*, having the frontispiece, fifty-one star charts, twenty-nine pages of tables (the APS atlas has thirty-two pages of tables) and fourteen pages of star catalogue. An inked book stamp unambiguously identifies it as having belonged to 'H. Cavendish'.[13]

All three copies of Bevis's intended *Uranographia* have now been identified. We also know the whereabouts, as of 2016, of twenty-seven of the twenty-nine described copies of *Atlas Celeste*: most are in the UK or US while two are missing, presumed to be in private collections. Other than a copy at the Lund University library in Sweden, none has yet been identified elsewhere in continental Europe. There is one copy in Australia.[14] There may be others in libraries, university collections or in private own-

---

[11] Henry Cavendish is noted for the discovery of hydrogen or what he called 'inflammable air'. His experiment to measure the density of the Earth has come to be known as the Cavendish experiment.

[12] The Devonshire Collection may be accessed by prior arrangement with the Archivist Librarian. Details can be found on the Chatsworth website http://www.chatsworth.org/.

[13] Kevin J. Kilburn, 'The Chatsworth Uranographia Britannica', *Astronomy & Geophysics* 53, no. 1 (2012): pp. 22–23.

[14] The Astronomical Society of Victoria owns an *Atlas Celeste* lacking the frontispiece. Old Mel-

ership elsewhere that we would like to know about.

In July 2013, the present authors were invited to examine a bound set of star charts owned by the Science Museum, London, and thought to be from Bevis's atlas. This document was subsequently examined by Kevin J. Kilburn at their Wroughton collection near Swindon and found to be an incomplete set of proof prints from Bevis's *Uranographia* taken before the copper plates were finished and probably dating to 1747–48. This set comprises 36 proof sheets of constellation figures, many of them lacking the dedication to individual subscribers and in some cases missing quite large areas of stars and the ecliptic grid lines that are present in the finished versions. In addition is one extremely rare example of a single sheet of the nomenclature, *THE LITTLE BEAR*, which would have faced the appropriate star chart.

This collection may be Lot 770, or a part thereof, from the 1785 auction that was salvaged and bound post-1786 and later mis-identified in the Science Museum catalogue and on the binding as being 'Flamstead's$_{(sic)}$ Celestial Atlas'.[15] It was also incorrectly described in their catalogue as being John Flamsteed's *Atlas Coelestis,* 1729.

ACKNOWLEDGEMENTS: Our sincere thanks go to James Towe, Archivist Librarian, the Devonshire Collection, for permission to research their Bevis atlas and to Diane Naylor, Photo Librarian, for permission to publish the picture of their Bevis atlas, supplied © Devonshire Collection, Chatsworth, and reproduced by permission of Chatsworth Settlement Trustees. Also thanks to Mr John Underwood, Site Librarian, Wroughton, with prior permission from Mr Nick Wyatt, Acting Head of Library, facilitated via Mr Kevin Johnson, Science Museum, London.

---

bourne Observatory, Melbourne, Victoria, Australia.

[15] Item reference (Wroughton) OB FLA FF Flamstead$_{(sic)}$. The Science Museum's proof set of Bevis's plates was bought ca. 1925. Its provenance is as yet unknown.

SIR CHRISTOPHER WREN: ARCHITECT-ASTRONOMER

Valerie Shrimplin

ABSTRACT: Sir Christopher Wren is better known in many circles as an architect rather than an astronomer, having designed St Paul's Cathedral and another fifty-one churches in London following the Great Fire of London in 1666. However, from the time of his appointment as Gresham Professor of Astronomy in 1657 at the age of 25, his proficiency in science and astronomy was evident. This paper will aim to bridge the gap between these two major aspects of Wren's career by focussing on the astronomical elements that he transferred over to his later architectural career, looking in particular at the astronomical symbolism and significance of St Paul's Cathedral and other selected Wren buildings in London.

A contemporary, Robert Hooke, wrote of Sir Christopher Wren that '*Since the time of Archimedes there scarce ever met in one man in so great perfection such a mechanical hand and so philosophical mind*.'[1] As a visionary – yet with a practical bent for physics and engineering – Wren is best known perhaps as the architect of St Paul's Cathedral in the City of London (FIG. 21.1). But during his long life (1632–1723) he was also responsible for a very large number of other architectural designs, as well as for having a notable career as Professor of Astronomy, firstly at Gresham College London (1657–60), and subsequently as Savilian Chair of Astronomy at Oxford. As well as being such a renowned architect, Wren should be remembered as a leading scientist of his age, and instrumental in the founding of the Royal Society and Royal Observatory.

In fact, had Wren died young, as it were, instead of reaching the age of ninety, he would doubtless have been remembered as a great astronomer, geometer and mathematician, regardless of his architectural work. A great deal has been written about Wren as an architect but clearly he would need to have known a great deal about mathematics to enable his great edifices, such as St Paul's, to stand. This paper aims to bring these two aspects together and, in the light of the theme of the INSAP conferences, to consider how his earlier scientific career informed his later astronomical work and whether 'astronomical phenomena' inspired some of his greatest architectural works, such as St Paul's Cathedral, The Monument, the Royal Observatory and other of his City churches.

Wren's prowess as astronomer was first recognised in his appointment as Gresham Professor of Astronomy (the oldest Chair of astronomy in Europe), so the

---

[1] Robert Hooke, *Micrographia* (London, 1665).

FIGURE 21.1. Johan Clostermans, *Sir Christopher Wren*, ca. 1690. Photo: Wikimedia Commons.

context of Gresham College is highly significant. Space does not allow extensive discussion of the significance and importance of Gresham College in London since the sixteenth century, but Wren's involvement with the College as the hub of intellectual development in London at the time clearly had a profound effect in the early stages of his career.[2]

Wren's background and early life are also significant for an understanding of these dual aspects of his career. His father was Rector at East Knoyle in Wiltshire until he became Dean at Windsor, where the young Christopher would have been immersed in contemporary theology and the workings of the liturgy, as well as befriending members of the Royal family, namely King Charles I and his young sons (subsequently to become Charles II and James II). Whilst still at school, the young Wren

[2] The first Professor of Astronomy in England was appointed at Gresham College in 1596, predating the Chairs in astronomy at Oxford and Cambridge by more than twenty years. An unbroken line of Gresham Professors of Astronomy has continued to the present day and the collective contribution of Gresham Professors to the study of astronomy is immense. Briefly, Sir Thomas Gresham (1519–79) was a financier, merchant and philanthropist who served as Royal Agent for Henry VIII, Edward VI, Mary and Elizabeth I in Antwerp, Flanders, and Spain. Trading and negotiating loans, he saved the monarchy from bankruptcy, formulated 'Gresham's Law', acquired vast estates in London and Norfolk and built (and owned) the 'Royal' Exchange, forerunner of the modern Stock Exchange. Sir Thomas's son died in 1564, so he left his vast fortune in Trust to found a College with Professors in Divinity, Astronomy, Musick, Geometry, Law, Physick and Rhetorick. The College has continued to provide free public lectures to the public for over four hundred years, presenting and recording lectures in a wide range of subjects, and making them available worldwide via the internet (www.gresham.ac.uk).

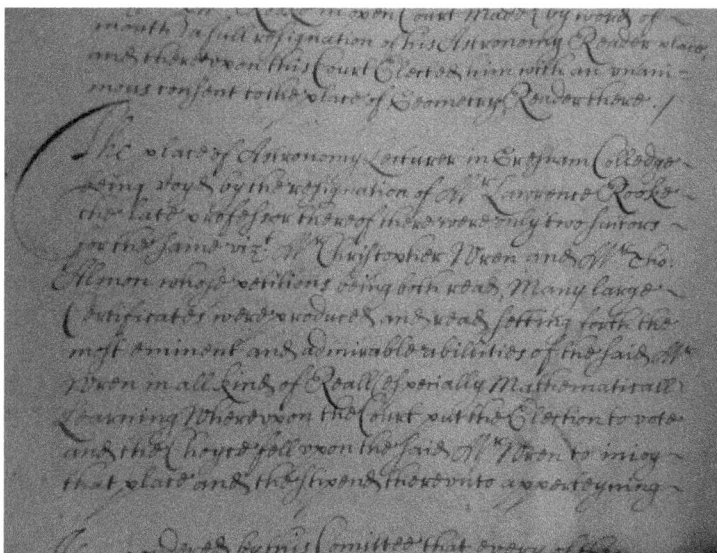

FIGURE 21.2. The Joint Grand Gresham Committee, appointment of Mr Christopher Wren. Photo: Valerie Shrimplin.

made models of the moon and solar system. He then studied at Oxford, becoming a Fellow of All Souls and, at this time, pursuing some early observational work and theories of terrestrial magnetism, as well as searching for a method to calculate longitude. He also carried out medical experiments on blood circulation and transfusion. He was familiar with the classics, especially Vitruvius, who significantly specified in his *Ten Books on Architecture* that it is essential for all architects to 'be acquainted with astronomy and the theory of the heavens'.[3]

Wren was appointed Gresham Professor of Astronomy in 1657, yet Sir Thomas Gresham had endowed the Professorship of Astronomy much earlier, in 1575, at a time when the subject was not taught in any school or university in England, when its rudiments were almost unknown and when many of the leading figures of the age believed in astrology. As well as his personal interest in astronomy (reflected in his coats of arms), Gresham's idea seems to have been based on his aim to ensure a practical approach to academic study, emphasising the functional rather than theoretical use of subjects like astronomy, used so much in navigation at the time. The minute books of the Joint Grand Gresham Committee have survived since the sixteenth century and record the appointment of 'Mr' Christopher Wren (FIG. 21.2).

---

[3] Vitruvius, *The Ten Books on Architecture*, trans. M. H. Morgan (New York: Dover, 1960), p. 6. Knowledge of astronomy is particularly significant architecturally in the history of the construction, development and symbolism of the dome.

Duties were specified: '… to read the principles of the sphere and the theories of the planets… to explain the use of common instruments for the capacity of mariners… to use by reading the area of navigation'.

Considerable detail is known about Wren's time at Gresham College. From 1657–1660 he lectured (on Wednesdays) on telescopes, the Moon, Saturn and the Satellites of Jupiter. He was interested in the prediction of solar eclipses and also worked on astronomy as an aid to navigation, including the search for a means of calculating longitude. Optics, meteorology and mechanics were of interest at this time as he worked with other notables at the College including Robert Boyle and Robert Hooke. Wren was said to have influenced Newton's *Mathematica Principia* by disputing Hooke's explanation of motion round the sun; the transcript of Wren's inaugural Gresham lecture, 1657 ('*Oratio inauguralis, habita Londini in collegio Greshamensi per Christophorum Wren, astronomiae professorem electum, ann. 1657, aetatis suae 25*') survives. The lecture contains extensive classical references and an emphasis is laid (as Gresham would have wished) on astronomy for practical purposes such as navigation ('*Astronomia navigantium tutelam puma suscepit*'), with references to Christopher Columbus and circumnavigation, as well as Copernicus, Galileo and Kepler's elliptical orbits.

Perhaps even more important than his work as Professor of Astronomy was Wren's role in the foundation of a 'College for the Promoting of Physico-Mathematical Experimental Learning', which took place following one of Wren's own lectures at Gresham College on a damp November evening in 1660 when Wren, Hooke, Boyle and others set up a society, later named as the Royal Society by decree of Charles II in 1662. Wren undoubtedly played a major role in the early life of the Royal Society since his wide expertise in so many different subject areas helped the exchange of ideas between colleagues. The Royal Society is the oldest scientific academy in continuous existence, and has led to some of the most fundamental and significant discoveries in the history of science. As a fellowship of the world's most eminent scientists in all areas of science, engineering and medicine, it continues 'to recognise, promote, and support excellence in science and encourage the development and use of science for the benefit of humanity' and has led to whole new branches of science, with profound theories on atoms, evolution, gravity and motion. The importance of its foundation at Gresham College, 'the focus of scientific life in the capital', should not be underestimated and until very late in his life Wren acknowledged and supported the College.[4]

---

[4] Michael Hoskin, *Cambridge Concise History of Astronomy* (Cambridge: University of Cambridge Press, 1999), p. 132.

In 1661 Wren was appointed Savilian Professor and moved to Oxford, where he was largely based until 1668, although he did make frequent trips to London to attend the Royal Society (then lodged in a corner of Gresham College). Wren's scientific achievements are recorded in some detail at the Royal Society, his work ranging from astronomy, optics, the problem of finding longitude at sea, cosmology, mechanics, microscopy, surveying, medicine and meteorology.[5] Wren continued his work with the Royal Society during this period, but it was around this time that he began to turn his attention to architecture. His scientific interests seem to have gradually waned as his architectural and official duties absorbed more time. A turning point seems to have come in 1665 when Wren's interest in architecture increased dramatically as a result of a visit to Paris where he met Bernini (architect and sculptor famous for his contribution to St Peter's in Rome) and saw his designs, as well as the architectural projects and rebuilding promoted by the 'Sun King', Louis XIV. Strongly influenced by the French and Italian baroque styles, Pevsner often quipped that this resulted in the beginning of an English 'Wrenaissance'.

Wren's first major architectural design was a chapel at Pembroke College, Cambridge, commissioned in 1665 by his uncle, the Bishop of Ely. Around the same time he worked on a design for the Sheldonian Theatre, Oxford, completed in 1668 and demonstrating real mathematical research. This was Wren's first attempt to design a dome, for which he had studied drawings of Michelangelo's great dome at St Peter's in Rome and also developed his study of engineering. At a time when architecture was considered to be a part-time interest for wealthy and educated gentlemen, Wren was one of the few architects to have a sound knowledge of the structure of buildings. Architecture, with its technical challenges and delivery of both functional and imposing spaces, enabled Wren to show off his architectural talent, and the Restoration of the monarchy in 1660 brought increased opportunities to those who, as mentioned above, had previously known and been loyal to Charles I and his family. The family's Royalist sympathies had caused difficulties under the primacy of Oliver Cromwell but caused Wren to experience special favour when the Civil War and Commonwealth came to an end with the Restoration of Charles II in 1660.

Apart from St Paul's Cathedral, Sir Christopher Wren was responsible for a wide range of architectural projects (many by Royal patronage), including fifty-one other 'Wren' churches in London replacing those destroyed by the Great Fire of London 1666: The Sheldonian, 1668; The Monument, 1671–76; the Royal Observatory

---

[5] Wren presented a coded manuscript outlining possible longitude methods to the Royal Society in November 1714 – the year the first Longitude Act offered a £20,000 reward for a solution. See Lisa Jardine, *On a Grander Scale* (London: HarperCollins, 2002), p. 460.

Greenwich, 1675–76; the Library at Trinity College Cambridge, 1676–84; Chelsea Hospital, 1682–92; Kensington Palace, 1689–96; South front of Hampton Court Palace, 1689–1700; and the Winchester Palace designs, modelled on Versailles. However, it was really St Paul's (built 1675–1711) that dominated his architectural work.

At the same time as Bernini and Borromini were changing the face of Rome, the rebuilding of London was also taking place, since the redesign of St Paul's Cathedral had already been underway before the Great Fire of London in 1666 made total rebuilding necessary. Sir Christopher Wren had already been consulted about St Paul's, which was in danger of collapsing, and proposed replacing the tower by a dome according to his designs of 27 August 1666. By chance, the Great Fire of London was to start just a few days later on 2 September 1666, resulting in the vast majority of the buildings of the City of London being burned to the ground during the four days that the fire raged until 5 September (Wren was in Oxford at the time...). While it had first been decided to improve the medieval cathedral with a tower and spire, Wren now opted for a domed structure. This may well have been influenced by Wren's knowledge of the late sixteenth-century architecture of Palladio (such as the Villa Rotunda) or by his contact with Bernini in Paris in 1665–66 (suggesting an emulation of St Peter's). But it could also be the result of his astronomical knowledge relating to the celestial and cosmic implications of large, especially domed, structures. Although the concept of the flat earth covered by the dome of heaven had been superseded and no educated person (let alone a professor of astronomy) seriously believed the earth to be flat, the idea of the earth covered by the dome of heaven still persisted in theological thinking and was reflected in architecture. The celestial implications appear to have acted as an inspiration for the final design of the dome of St Paul's (1675), which can hardly be doubted by anyone able to visit the Cathedral and experience its 'cosmic' proportions. The later nineteenth-century mosaic decoration of the creation cycle demonstrates how securely this cosmic aspect was felt by succeeding generations. Bearing in mind that Wren was first a mathematician and astronomer and afterwards an architect, and that he was also the architect for the Greenwich Observatory, lying on the prime meridian, the astronomical implications seem clear. Domed architecture continued to be reserved for schemes of special significance.

The first plans, 27 August 1666 (pre-dating the Great Fire) made provision for the correction of the dangerous bulging of the walls and the addition of a dome. But after the Fire (2–5 September 1666) Wren was given a rare chance to redesign a major Church and the capital city that surrounded it. Invited then to dismantle and rebuild, rather than to restore the great Church under the 1667 Rebuilding Act, Wren's initial design, called the 'First Model' (winter 1669–70), was not approved by

FIGURE 21.3. Christopher Wren, *The Great Model*. Photo: Wikimedia Commons.

the City Council. They thought it was insufficiently grand, so his Greek Cross design then followed (1671–72), eventually known as the 'Great Model' (1674) since a large model of his design was made in wood, six metres long (FIG. 21.3). This was also rejected, this time by the clergy who did not like its Greek-influenced plan. Wren was at first upset and then angry. His third design, called the 'Warrant Design', was for a Latin Cross plan with a relatively small dome surmounted by a spire, and this design was approved by Royal Warrant in 1675 (FIG. 21.4). However, Wren was determined to arrange things as he wished as building progressed, and it was significantly written in to the agreement that Wren was allowed 'to make variations… as from Time to Time he should see proper.' The finished work (FIGS. 21.6 and 21.7) was a far cry from the approved 'Warrant' design as a result. The Foundation stone was laid on 21 June 1675, the cathedral opened and the first service held in 1697, with the dome completed in 1711 thirty-five years after the laying of the Foundation stone.

Domed architecture is traditionally linked with astronomy and cosmology, being imitative of natural eye observation of the earth covered by the dome of heaven, according to scriptural references. So there are various indications that, in addition to this basic concept, Sir Christopher Wren used his astronomical training and career to lay a basis for, and influence, his later architectural career, and his building of St Paul's exemplified this in several ways. Domed architecture was unknown in

FIGURE 21.4. Christopher Wren, *The Warrant Design*. Photo: Wikimedia Commons.

England until this time and the continental influences on Wren can be traced, apart from his visit to Paris, to prints and drawings of other examples from Rome to Constantinople. St Paul's significantly measures 365 feet high to the top of the cross, and the South West Tower was known to operate as a scientific instrument for telescopic observations and for pendulum experiments. More significant, perhaps, is the question of the orientation of the new Cathedral, which was changed from the medieval building. Orientated to the East, as is normal in Christian places of worship, the medieval Cathedral lay on a true east-west axis. Wren's Cathedral was however rotated to lie circa 6 degrees north of due east (FIG. 21.5).

Various reasons have been put forward to explain why Wren should have made this change. All Wren's expertise as a mathematician had to be put to use as he had to devise immensely strong foundations on London's soft clay soil, based on the use of complex catenary arches, but this does not explain the change in orientation. It could, it has been argued, have simply been to fit the awkward shape of the site which was unsuitable for Wren's original grandiose scheme. Or perhaps Wren still had hopes of implementing his design for an entire new City for which a change in orientation of the Cathedral might be needed? Although the Warrant design was only finally approved on 19 May 1675 and the Foundation Stone laid on 21 June 1675, there is some evidence from the building accounts (April–September 1673) that scaffolding for a domed design was already being set up in 1673, and also that in May-September 1674 there was some evidence of staking out of the building. The Greek Cross design, known as the Great Model design, would have been the chosen option at that time, although later abandoned for the Warrant design. So the argument that

FIGURE 21.5. Ordnance Survey Map of London, 1890, overdrawn to show orientation of St Paul's Cathedral. Photo: Valerie Shrimplin.

FIGURE 21.6 & FIGURE 21.7. Christopher Wren, St Paul's Cathedral: West façade (left) and Dome (right). Photos: Valerie Shrimplin.

the change in orientation was simply to fit the site, is based on the idea, as Schofield proposes (following Crayford), that it was the large symmetrical design that needed to be accommodated. However, given the authority accorded to Wren as early as 1667 by the Re-building Act, and the fact of the destruction of surrounding areas, it is here argued that this simple explanation seems unlikely. More importantly, with a tradition of orienting Christian Church buildings to the east, and with Wren's well known prowess as an astronomer, it is argued that the reason for the change had an astronomical basis. Whilst recent work has shown that the notion of the orientation of Christian buildings to the direction of sunrise on the Saint's Day of the dedication of the church is less likely than previously thought, the idea was supported in the late seventeenth century. Hence, the orientation of a major building such as St Paul's by the Astronomer-Architect Wren, must surely have had significance.[6]

The answer seems to lie in looking at when sunrise was 6 degrees north of due east in the years 1673–75, assuming an unobscured horizon – which would be reasonable given that most of the City had burned down. It clearly was not the precise orientation on the Saint's Day of the church, since the saint's day for St Paul falls on 25 January when sunrise would be south of east for the latitude of London. Nor would it likely have been related to the day chosen for the laying of the Foundation Day (21 June 1675) because at the time of the summer solstice sunrise would, at the latitude of London, be around 50 degrees (and as mentioned, some laying out had already started in 1673).

The current Table of the Declination of the Sun Mean Value for the Four Years of a Leap-Year Cycle, giving a positive sign (+) where the sun lies north of the Celestial Equator, and a negative sign (-) where the sun lies south of Celestial Equator should be considered at this point.[7] An orientation of circa 6 degrees north of east yields a declination of ca. 3.5, that is, for 31 March. This would have been eleven days earlier in 1673 (since the Julian, not the Gregorian, Calendar was used in England until

---

[6] For new work on the orientation of Christian churches see Peter G. Hoare, 'Orientation of English Medieval Parish Churches', in *The Handbook of Archaeoastronomy and Ethnoastronomy*, ed. Clive L. N. Ruggles (New York: Springer, 2015) , pp. 1711–18; and Stephen C. McCluskey, 'Orientation of Christian Churches', in Ruggles, ed., *The Handbook of Archaeoastronomy and Ethnoastronomy*, pp. 1703–10. However, McCluskey does point out (p. 1708) that the notion was popular and widely believed in the late seventeenth century, as cited by Wren's contemporaries and fellow members of the Royal Society, John Aubrey ('Customs and Manners of the English', 1678), Silas Taylor and Robert Plot. I am grateful to the anonymous referee for drawing the volume to my attention.

[7]   https://www.starpath.com/blog_files/Table%20of%20the%20Declination%20of%20the%20 Sun.pdf [accessed 26 September 2016].

Figure 21.8. Christopher Wren, St Paul's, West Tower. Photo: Valerie Shrimplin.

1752).[8] This indicates a likely actual date of 21 March, which indicates the Vernal Equinox, also a likely symbol of rebirth appropriate to the building (by a trained astronomer) of the most important ecclesiastical edifice in the country – which furthermore was quite literally rising from the ashes. The idea of the orientation being changed to fit with the vernal (spring) equinox also seems even more appropriate when it is considered that Easter sunrise occurred on 2 April in 1673. It also fits in with the idea of choosing the summer solstice for the laying of the Foundation Stone. These seem to be far more plausible reasons for the change in orientation than simply a way of fitting a building onto an awkward site. Wren appears to have chosen the calendrical dates over the empirical observable dates for the vernal equinox and summer solstice. Because of the shift in the Julian calendar by 1673 (as mentioned), the calendar was off by eleven days which means that the calendar marked 21 March (the Spring equinox) eleven days after the empirical observable equinox. The orientation to the calendrical date instead of the empirical one was certainly a careful and

[8] Assistance from Dr Fabio Silva, UCL Archaeology and UWTSD, for advice on, access to and use of declination tables is gratefully acknowledged (personal communications October 2014 and June 2015).

considerate choice by Wren, implying perhaps an opposition to the new calendar.

As mentioned, the South West Tower of St Paul's was also used for scientific and astronomical work (FIG. 21.8), demonstrating again how Wren's astronomical career influenced and went over into his later architectural one. The South West Tower was very different, standing out as having a suite of separate rooms, together with an aperture, probably used for observational experiments, since it was known that the building would take many years and problems like stellar parallax could be investigated with such a great height involved. Pendulum experiments were also said to have been carried out and Ward (in his *Lives of the Gresham Professors*, written soon after in 1740) also speaks of a planned telescope on the south side between the nave and transept. The minutes of the Royal Society, February 1704, record that 'Sir Christopher Wren proposed that the Telescope given by Mr Huygens to the Society should be set up in [St] Paul's and astronomical observations made' – providing substantial evidence of the interaction between Wren's astronomical and architectural careers.

The Great Fire devastated the City of London with the loss of eighty-six parish churches, as well as St Paul's. The replacement churches by Sir Christopher Wren similarly, but perhaps to a lesser extent than St Paul's, also reflect his earlier interests in astronomy in their overall designs, for which Wren was responsible, rather than in close supervision and detail – Wren was clearly above such niceties as details of stars on finials and so on. Twenty-three of the original fifty-one other Wren Churches in the city of London survive and are in use, the other 28 having been destroyed (many of them during the bombings of the Second World War) or having been very much rebuilt. In comparison with St Paul's, the other fifty-one parish churches for which Wren was appointed to oversee the construction have been somewhat overlooked. This resulted in a unique set of contemporary churches, even though some have not survived. Working with Robert Hooke (also a Gresham Professor) and Nicholas Hawksmoor, Wren also made use of the famous wood carver, Grinling Gibbons. Not all are masterpieces, like St Stephen Walbrook or St Brides, but they all have merit and several show astronomical features. For example, St Michael Cornhill has a vaulted ceiling with star features in the glass roundels in the south aisle. The ceiling is now painted blue which no doubt reflects the contemporary idea, alluded to above, of the vault as a symbol or reflection of the heavens. St Mary at Hill features a dome with an aperture to let in the sunlight, whilst a star is featured in the roundel situated above the altar. St Mary Abchurch unusually consists of a 40-foot dome supported only by brick walls with the heavens (painted by Thornhill in 1708) significantly depicted on the dome. Wren had more control and input into St Stephen Walbrook (FIG. 21.9), since it was his own parish Church, nearby where he lived

Figure 21.9. Christopher Wren, St Stephen Walbrook. Photo: Valerie Shrimplin.

at 15 Walbrook. The dome again reflects the heavens in a demonstration of light, weightlessness and illumination – achieved by the creation of the dome by the use of lath and plaster over a wooden frame, rather than stone.

Amongst the other surviving Wren churches, further examples, such as St Edmund King and Martyr, St Margaret Pattens, St Michael Paternoster, St Magnus the Martyr, St Mary le Bow and St Lawrence Jewry are built in the same tradition of airy, light weightlessness, with many having domes or vaulted ceilings. Even the flat ceiling of St Michael Paternoster continues to be painted blue.[9] Although rarely containing direct astronomical allusions, there is an overall feeling that Wren's scientific background was indeed carried over to his architectural work, other than in the purely practical aspects of the mathematics needed for the completion of such enterprises. On entering St Paul's Cathedral itself, it is clear that it is nothing if not cosmic – in scale, proportion and feeling.

---

[9] The other surviving churches are St Andrew by the Wardrobe, St Andrew's Holborn, St Anne and St Agnes, St Benet's Paul's Wharf, St Brides Fleet St, St Clement Eastcheap, St James Garlickhythe, St Martin within Ludgate, St Margaret Lothbury, St Mary Aldermary, St Nicholas Cole Abbey, St Peter upon Cornhill and St Vedast Foster Lane.

Turning to some of Wren's other works, The Monument, 1673–79 (FIG. 21.10) shows clear astronomical features. Built as a memorial to the Great Fire with the assistance of Robert Hooke, it was actually conceived as a telescope, with the idea of replacing the symbolic urn of ashes at the top with a lens made by Constantine Huygens. Over 200 feet (62 m) high, it was known to have been used for astronomical experiments, such as the attempt at detection of Stellar Parallax (for proof of the earth's orbit) and also gravity and pendulum experiments, with an underground laboratory beneath. Like St Paul's, it was both a national Monument and an over-sized scientific instrument.[10]

Other significantly astronomical works by Wren included the Royal Observatory (1675–76) where John Flamsteed, the first Astronomer Royal, lived until his death in 1719. Wren was amongst those who advised Charles II that an observatory was needed in order to undertake observational work and mapping of the sky. It seems to have been his suggestion to construct the observatory at Greenwich where there had been an important Royal palace and with good river transport and communication. Much may have been left to his Royal Society Colleague Robert Hooke and the builders themselves, but Wren was clearly a great influence and driving force. The project was supported by Charles II, with the idea that a detailed study of the moon and stars would assist with navigation at sea – a concept still very much in line with Sir Thomas Gresham's ideas about using astronomy for practical purposes. The use of telescopes and other scientific instruments would enable the recording of the moon's position relative to specific stars which would, in turn, enable navigators to set their positions and routes more accurately so as to avoid loss of life and trade. Wren was behind and supportive of this approach, demonstrating that he never forgot about astronomy in his later career, even when he later became immersed in other major Royal projects such as the Royal Hospital for soldiers at Chelsea, Christ Church Oxford, Kensington Palace, Hampton Court, and as Surveyor General for Repairs to Westminster Abbey.

In his later career, Wren was elected as Member of Parliament for Old Windsor, near where he had lived and grown up, in 1680, 1689 and 1690, but he did not take up his seat. He was President of the Royal Society (1681–83) and continued always to be active in scientific circles, serving on a range of committees and commissions relating to scientific, astronomical and even medical interests. He referred to astronomy as 'The trade I was once well acquainted with', acknowledging perhaps that his main interests had moved on, although he never dismissed it completely. He also never forgot Gresham College, as reflected in his continued links with Robert Hooke

---

[10] Jardine, *On a Grander Scale*, p. 318.

Figure 21.10. Sir Christopher Wren and Robert Hooke, The Monument, 1673-79. Photo: Valerie Shrimplin.

and his involvement in 1702 in plans to expand the Royal Society at Gresham College (still based, until 1768, in Gresham's mansion in Bishopsgate). A crater on Mercury, 'Wren', was named after him in honour of his astronomical work, but finally his greatest achievement can perhaps be considered as St Paul's Cathedral – of truly cosmic proportions. His memorial (and tomb) beneath the dome, record '*Lector, si monumentum requires, circumspice*' (Reader, if you seek his monument, look around you). St Paul's Cathedral has become a lasting national icon, its importance summed up by Sir Winston Churchill when it became threatened during that second period of great fire in London, the Blitz of 1940. Rather than seeing a further opportunity to rebuild, Churchill simply insisted, 'St Paul's must be saved at all costs'.

Junking Astronomy Jargon

Roberto Trotta

ABSTRACT: My book, *The Edge of the Sky* tells the story of the All-There-Is (the Universe) as seen through the eyes of a Student-Woman (a female scientist) spending a night of solitude with Big-Seer (a giant telescope), trying to unlock the mystery of dark matter. All of this using only the most common 1,000 words in English. In this paper I reflect back on why and how I chose this format and I present the lessons I learnt in trying to explain the entire All-There-Is this way.

A Momentous Discovery... or is it?

In 1965 Arno Penzias and Robert Wilson published a short article in the *Astrophysical Journal*.[1] Barely over a page long, it was entitled: 'A measurement of excess antenna temperature at 4080 Mc/s'. Behind this cryptic title lurked one of the greatest discoveries of all times: they had found the cold radiation left over from the Big Bang, evidence that the Universe had a beginning in time. Not something you would have guessed from the title.

Fast-forward to a more recent occasion, when on July 4th, 2012, Joe Incandela, the spokesperson for the CMS experiment at CERN, announced to a packed auditorium: 'If we combine the ZZ and gamma-gamma, in the region of 125 GeV they give a combined significance of 5 standard deviations!'[2] As everybody cheered (and Peter Higgs shed a few tears), it was not immediately obvious to anybody but the particle physicists in the room what the significance of this was. What Incandela was saying was that they had discovered the Higgs boson, the 'God particle' that gives mass to all other particles.

For the public at large to partake in these momentous discoveries, nothing short of a translation would do. Fundamental science often deals with very esoteric concept, far removed from our everyday experience. The technical language (often imbued with mathematics) that scientists use to describe their object of study is unhelpful in connecting with our understanding of the world. It would be better if precision was somewhat eschewed in favour of better communication: a translation

---

[1] A.A. Penzias and R.W. Wilson, 'A Measurement of Excess Antenna Temperature at 4080 Mc/s', *Astrophysical Journal* 142 (1965), pp. 419–21.

[2] CERN Press Release, 'CERN experiments observe particle consistent with long-sought Higgs boson', (4 July 2012), online: http://press.cern/press-releases/2012/07/cern-experiments-observe-particle-consistent-long-sought-higgs-boson [accessed 4 Jan. 2017]; Joe Incandela, 'Status of the CMS SM Higgs Search', Talk given at CERN on 4 July 2012: https://cms-docdb.cern.ch/cgi-bin/PublicDocDB/RetrieveFile?docid=6125&filename=CMS_4July2012_Incandela.pdf [accessed 4 Jan. 2017]

of difficult ideas into their basic elements that can better put them in contact with our experiences at the human scale.

Public Communication Enemy No 1

The obvious enemy to a clear communication with the public is jargon.[3] As scientists, we are guilty of slipping back into it all too often, sometimes involuntarily. The above two examples were cases of scientists writing for or talking to their colleagues, and so in fairness they might not necessarily be expected to use language that a non-specialist would understand. But fundamental science being funded with taxpayers' money, it is a duty for the professional scientist to engage the public in a two-way discussion about their work, its objectives and the very reason of its being. The first obstacle to this aim is jargon.

As an astrophysicist with a passion for communicating with the public, I have been looking for novel ways of engaging new audiences with my science. For over a decade, I gave public lectures to a wide variety of audiences; I worked with filmmakers, artists, designers and architects to create videos, artwork and installations inspired by cosmological ideas; most recently, as part of a Science and Technology Facilities Council public engagement fellowship I was awarded, I have been using cookery and food to approach astrophysics and cosmology with young audiences in a hands-on way. My partners at Kitchen Theory and myself are now working on taking this notion one step further and are developing an idea for an astronomy lecture for visually impaired people entirely based on food and sensorial experiences other than sight.

All this time – I now realise – what I was searching for was a language to translate in a more pictorial, immediate way the often complex and abstruse cosmological concepts my research is about: dark matter, dark energy, the Big Bang and the fundamental nature of the universe. A language that would speak not only to people's minds, but most importantly to their hearts.

Inspirational examples of this can often be found in astronomy, perhaps because its breath-taking beauty lends itself well to engaging our emotions as well as our brains. Andrew Fraknoi gives a surprising survey of how much astronomy has seeped

[3] Aviv J. Sharon, A. Baram-Tsbari, 'Measuring Mumbo Jumbo: A Preliminary Quantification of the Use of Jargon in Science Communication', *Public Understanding of Science* 23, no. 5 (2014): pp. 528–46; K. Hinko, J. Seneca & N. Finkelstein, 'Use of scientific Language by University Physics Students Communicating to the Public'. Paper presented at the Physics Education Research Conference in Minneapolis, MN, 30–31 July 2014, *PER Conference Series*, P.V. Engelhardt, A. Churukian and D.L. Jones, eds., pp. 115–18.

into fiction, drama and poetry.[4] Some of my personal favourites are the poetry of Simon Barraclough (Poet in Residence at the Mullard Space Science Laboratory), Nick Payne's play *Constellations* (aptly mixing the multiverse, love and the meaning of life) and the cosmologically inspired sculptures of Josiah McElheny (produced in collaboration with astrophysicist David Weinberg).[5] In trying to devise a similarly touching way of talking about my subject, my hope was to tear down the barrier to a genuine two-way dialogue that is the technical knowledge gap between the science professionals and the public.

Less is more

In the back of my mind I had the apocryphal story I once heard about a bet made by Ernest Hemingway. It is recounted that one night, around a dinner table, his friends challenged him to write a novel with only six words. After a moment of reflection, the great novelist grabbed a napkin and on it he wrote: 'For sale: baby shoes, never worn'.[6] His friends readily conceded the bet.

I loved the immediacy of what would later be called 'flash-fiction': its economy of words left space for the readers' imagination to fill in the gaps – indeed it demanded it. Somehow, this seemed to promote a stronger, more active engagement from the part of the reader. Today, the genre has evolved in a number of different ways, including the Six-Word Story (in homage to Hemingway's six-worder) the drabble (word count limited to one hundred), and of course, Twitter-based storytelling (David Mitchell's 'The Right Sort' short story was narrated over 280 tweets sent in the course of a week[7]).

'Was it possible to achieve something similar with science?', I asked myself. It was clear to me that the flash-fiction format would require to pare down ideas to their absolute essence, thereby hopefully helping their quintessential meaning to shine through more brightly.

Then one day in January 2013 I stumbled on the web on the Ten-Hundred Words of Science challenge – a website collecting people's descriptions of their job written

---

[4] Andrew Fraknoi, 'Interdisciplinary Approaches to Astronomy: Cosmic Fiction, Drama and Poetry', *Communicating Astronomy to the Public Journal*, 18, Sept 2015): pp. 7–11.

[5] Simon Barraclough, *Sunspots* (London: Penned in the Margins, 2015); David Weinberg, 'The Glass Universe: Where Astronomy Meets Art', *New Scientist* (Dec 6th 2008): p. 2685.

[6] Josh Jones, 'The (Urban) Legend of Ernest Hemingway's Six-Word Story: "For sale, Baby shoes, Never worn."', *Open Culture*, 24 March 2015, http://www.openculture.com/2015/03/the-urban-legend-of-ernest-hemingways-six-word-story.html [accessed 4 Jan. 2017].

[7] David Mitchell, 'The Right Sort': https://www.theguardian.com/books/2014/jul/14/the-right-sort-david-mitchells-twitter-short-story [accessed 4 Jan. 2017].

using only the most-used 1,000 words in English.[8] The format had come from a cartoon by Randall Munroe, the creator of the XKCD website. This is a humorous site with original, geeky stick-like cartoons, often revolving around physics, maths, computer science and other technical subjects. Randall had drawn a picture of the Saturn V moon rocket (or 'Up-Goer Five') , and labelled its parts using only the 1,000 words list.[9] For example, in this spartan vocabulary the escape pod had become 'Thing to help people escape really fast if there is a problem and everything is on fire so they decide not to go to space'. I could see that this could be fun.

I spent a frustrating hour writing up my job with the 1,000 words lexicon, and I found it harder than I had imagined. I posted a copy on my website, then forgot about it. The next month I gave a public talk at the White Building, an art venue in East London. The person who introduced me mentioned that he had found this unusual description on my website, and a member of the audience brought this up at the end – what was this business with the 1,000 words about, exactly?

I read out the couple of paragraphs I had written:

I study tiny bits of matter that are all around us but that we can not see, which we call dark matter. We know dark matter is out there because it changes the way other big far-away things move, such as stars, and Star Crowds. We want to understand what dark matter is made of because it could tell us about where everything around us came from and what will happen next.

To study dark matter, people like me use big things that have taken lots of money, thought and people to build. Some of those things fly way above us. Some are deep inside the ground. Some are large rings that make tiny pieces of normal matter kiss each other as they fly around very, very fast – almost as fast as light. We hope that we can hear the whisper of dark matter if we listen very carefully. We take all the whispers from all the listening things and we put them together in our computers. We use big computers to do this, as there are lots and lots of tiny whispers we need to look at.

I go to places all over the world to talk to other people like me, as together we can think better and work faster. Together, perhaps we can even find new, better ways to listen to dark matter. Most of them are good people, and after we talked we go out and have a drink and talk some more.

I was surprised by the unexpectedly strong, positive reaction of the audience. That got me thinking: perhaps this was the new language that I had been looking for! And

---

[8] Ten Hundred Words of Science: http://tenhundredwordsofscience.tumblr.com [accessed 4 Jan. 2017]

[9] http://splasho.com/upgoer5/phpspellcheck/dictionaries/1000.dicin [accessed 4 Jan. 2017]; R. Munroe, 'Up Goer Five', 12 Nov., 2012: https://xkcd.com/1133/ [accessed 5 April 2016].

perhaps it could even be used to talk about everything in the Universe, not just my job. My book, *The Edge of the Sky* is the result of that small Eureka moment.[10]

A New Language

Over the next three months, I dutifully sat down at my desk at the University of California Santa Barbara (where I was free of teaching duties during a research stay) and spent some time every day wrangling with the difficulty of talking about the Universe using only the most common 1,000 words.

The first hurdle was to find a new word for 'Universe', which was not in the list. So it became the 'All-There-Is'. A 'planet' became a 'Crazy Star'; a telescope, a 'Big-Seer'; scientists were 'Student-People'; our galaxy 'the White Road', the 'Big Bang' the 'Big Flash' (after my editor vetoed my earlier choice, the 'Hot Flash'!) and other galaxies became 'Star Crowds'.

As this new language started to emerge, little by little a new voice took over. A voice that I had not anticipated, and that was created by the poetic straitjacket imposed on me by my chosen format.

Not only did I find that limiting my lexicon to the most-used 1,000 words swept the table clean of jargon (as I was sure it would): it also forced me to think afresh about seemingly familiar concepts, and how to describe them in a more pictorial, metaphorical way. It gave me a fresh, childlike perspective on the Universe. I realised that this is particularly important when talking about concepts that might be familiar to us, the professional practitioners of our discipline, but that are very far-removed from the everyday experience of the general public. We tend to get lulled into a false sense of comfort, by using terms that we mistakenly believe non-scientists understand the same way we do, like 'galaxy', or 'electron' or 'black hole'. So why not get rid of all those words and use instead simple language that everybody can understand?

The All-There-Is in 707 different words

In the *The Edge of the Sky*, I've tried to follow the apocryphal advice attributed to Einstein, who, who reportedly once said: 'You do not really understand something unless you can explain it to your grandmother' – and a mere 707 words from the 1,000 words list is all that I ended up using to do that. The book tells the story of a Student-Woman who spends a night observing far-away Star Crowds with the help of Big-Seer, looking for dark matter:

---

[10] Roberto Trotta, *The Edge of the Sky: All you Need to Know about the All-There-Is* (New York: Basic Books. 2014).

She steps outside in the cold night, holding her cup of hot coffee with both hands. The White Road is beautiful in the dark, clear sky, and, once again, she can not help but be amazed by it all.

It does not matter how many times she has seen this before, or how much she knows about what is out there. The sight of the stars is enough to make her gasp.

'It all seems so still and yet it's changing all the time,' she whispers to no one.

It is hard to believe that everything out there past the White Road and its stars is running away from us.

Yet, like Mr Hubble found long ago, the Star-Crowds are running away from each other, as the space between them gets bigger and bigger. The All-There-Is is growing with time.

From sunset to sunrise, we follow her as she reflects about our Home-World and the other Crazy Stars around the Sun, and the many more that go around far-away stars; the way the All-There-Is grows, and how it began in a Big Flash; and all the questions we still have on it, like dark matter, the Dark Push and the existence of other kinds of All-There-Is.

Whether or not *The Edge of the Sky* succeeded in its goal is a question that only my readers can answer. From my perspective, writing the book has been a fascinating challenge that brought up questions I had never considered before. That meant re-thinking my understanding of concepts I thought I grasped – I had to see them with new eyes, and I hope this perspective shines through in the book. Perhaps the hardest part was how to talk about the early Universe, when none of the words I would normally have used (such as 'particle', 'energy', 'speed' or even 'soup'!) were available. Or indeed at times I felt that abiding by my own rules was generating verbosity rather than the concision I was after, for example in the following passage, where there is a table 'with a large number of small gray round pieces on it – of the type that you can use to buy a coffee, or a paper, or to pay for parking. The ones with one head on one side and some other picture on the flip side'. In short, coins. With some distance now, I have come to see the language of the book as a sort of post-apocalyptic tongue. One could imagine a bunch of survivors recounting stories around a fire, at a time when all the details have been forgotten, and only the core images remain.

If the book will help some of my readers connect with some of the complex ideas of modern cosmology and generate curiosity and enthusiasm for fundamental science, my aim will be achieved.

SOLARGRAPHY:
MAKING THE INVISIBLE VISIBLE

Tarja Trygg

ABSTRACT: Astronomical phenomena have provided inspiration for many artists, stimulating our curiosity and desire to see more than our eyes are capable of seeing. It is human nature to try to understand our position in the universe. In my research on solargraphy, the most fascinating and meaningful photographic power is in its ability to make the invisible motions of sun trails visible in diverse ways. In this presentation, I will start to speak first about the curiosity and desire to see more than our eyes are capable of. Second, I highlight some examples of how photography can show us something we have never seen before. Sometimes we cannot believe what we see. It is quite common for us to try to find some explanation. What motivates people to get intrigued by some unknown novelty? Some photographs awake our suspicion: is it true or false, real or unreal, a joke or a fake? The last – but not the least part – of this presentation is to demonstrate some artistic ways to better understand the rotation of the Earth and how to make the invisible visible in works of art.

Astronomical phenomena have provided inspiration for many artists, stimulating our curiosity and desire to see more than our eyes are capable of seeing. In my research, the most fascinating and meaningful photographic power is its ability to make the invisible visible in diverse ways. Since 2002 I have been fascinated with solargraphy. My research on solargraphy focuses on catching the movements of the Sun at various latitudes from a human being's viewpoint. This kind of astronomical phenomena can be made visible using pinhole photography, which creates solargraphs – photographs that people have not seen before the year 2000.

In my licentiate thesis work, I interviewed over ten photographers whose artworks make the invisible visible.[1] When I became interested in solargraphy in early 2002, I was suspicious of the statement that sun trails go horizontally in the northern hemisphere and vertically on the equator. Is it the whole truth? I don't believe it unless I see it with my own eyes. I wanted to verify it in an experimental way. So I started to do solargraphs and found I needed these kinds of photographs from all over the world. This is why the website of the Global Photographic Project was set up in 2005 – it was the best way to ask volunteers to participate in this project.[2] The

---

[1] Tarja Trygg, 'The Fascination of A Photograph. In the pragmatical frame of reference – Art As EXPERIENCE' (Licentiate thesis, University of Art and Design Helsinki, 1999), http://www.solargraphy.com/index.php?option=com_content&task=view&id=2&Itemid=3 [accessed 6 September 2016].

[2] Targa Trygg, The invitation to participate to the Global Photographic Project of solargraphy, available at http://www2.uiah.fi/~ttrygg/project.html [accessed 27 September 2016].

FIGURE 23.1. Summer in Helsinki. 60°N, 24°E. Solargraphy by Tarja Trygg 2014.

goal of this project was to collect solargraphs from all over the world and to build a world map of solargraphs for this research. In 2006 I started to publish solargraphs on a new website.[3]

Solarigrafía is a quite recently invented technique for recording long-term weather variations by using long exposure pinhole camera photography. Photographers Diego López Calvín, Sławomir Decyk and Paweł Kula invented it. Their first project of *Solaris* started in 2000 with strong relations with photography, astronomy, climate and agriculture. My first touch to this technique started 2002 in the Poland International Photographic Workshop PROFILE ´02 organised by the Academy of Fine Arts in Poznan (nowadays it is the University of Fine Arts). The workshop was held on the theme of solarigrafía at Skoki, and Paweł Kula was a leader there. The resulting image registers the trail of the sun during its apparent movement on the ecliptic, the sun's path across the sky. Using black & white photographic paper in a pinhole can camera, with long exposure times, where no developing chemicals are allowed, creates an image with many colours. How is it possible? When we cannot explain something, it is easy to say it's 'magic'. But 'it is not magic, it is technology', said one young astronomer to me at INSAP VII, the seventh international conference on *The Inspiration of Astronomical Phenomena*, held in Bath, England, in 2010. Later I started to use the name *solargraphy* in writing English and in my presentations after I had tested the term in international conferences as to how American professors understand these variations of *solarigrafía, solarography, solarigraphy, and solargraphy*. All these variations mean the same thing. ((Solar = Sun) + Photography = Solargraphy).

[3] Tarja Trygg, Solargraphy Gallery of The Global Project of Solargraphy, available at http://www.solargraphy.com [accessed 6 September 2016].

FIGURE 23.2. The Flammarion engraving (1888) depicts a traveller who arrives at the edge of a flat Earth and sticks his head through the firmament. Flammarion's book reads: 'A missionary of the Middle Ages tells that he had found the point where the sky and the Earth touch...'.[4]

1. Curiosity is a human trait

The invisible universe offers new findings continually. It is human nature to try to understand our position in the whole universe. How can we understand the huge space surrounding our planet? I have chosen the Flammarion engraving for the first picture in this chapter for being characteristic of human curiosity.

2. Photography has the power to convey something we have never seen before; it makes the invisible, visible.

Before the birth of photography, scientists drew pictures to convey their observations. Since its invention, photography has been very useful for catching sharper and sharper images in different scales of the macro and micro cosmos, for example in finding the hidden universe or in medical examinations. The only part of the electromagnetic spectrum that human eyes are sensitive to is called the visible region. But by using photography, it is possible to make something visible from the non-visible region as well. For example, in infrared photography, the film or image sensor used is sensitive to infrared light.

FIGURE 23.3. Wavelengths of light (non-visible and visible). The light spectrum, showing the

[4] Nicolas Camille Flammarion, *L'atmosphère: météorologie populaire* (Paris: Hachette, 1888), p. 163, available at http://gallica.bnf.fr/ark:/12148/bpt6k408619m [accessed 7 September 2016].

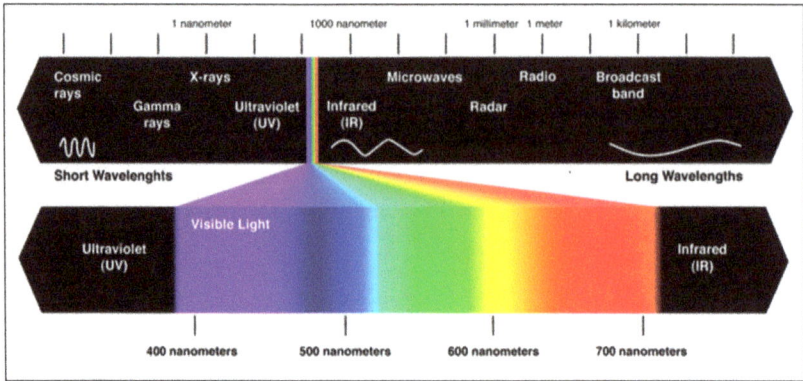

small segment we call visible light.[5]

Infrared, or "IR" photography, offers photographers of all abilities and budgets the opportunity to explore a new world – the world of the unseen. Why "unseen"? Because our eyes literally cannot see IR light, as it lies just beyond what is classified as the "visible" spectrum – that which human eyesight can detect. When we take photographs using infrared-equipped film or cameras, we are exposed to the world that can often look very different from that we are accustomed to seeing…[6]

Visible and Invisible Light

The electromagnetic spectrum consists of all the different wavelengths of light (also known as electromagnetic radiation).
The only region in the entire electromagnetic spectrum that our eyes are sensitive to is the visible region.[7]

In this paper, I argue that photography is meaningful in its ability to make the invisible visible in different ways. Using photography, astronomers want to see beyond the visible region and record discoveries from the cosmos. Nowadays it is possible to observe the universe with technologically advanced telescopes and attain increasingly sharp pictures from the hidden universe. But it is also feasible to photograph the movements of the sun over landscapes or city views using the most low tech and simplest 'camera obscura' method using, for example, a small black film canister as a

[5] LucidScience, Tutorial: Invisible Light Basics, available at http://www.lucidscience.com/tut-invisible%20light%20basics-1.aspx [accessed 20 September 2016].
[6] Bob Vishneski, 'Introduction to Infrared Photography', (12 March 2012), available at https://photographylife.com/introduction-to-infrared-photography [accessed 22 September 2016].
[7] Hubblesite, 'What is the electromagnetic spectrum', available at http://hubblesite.org/reference_desk/faq/answer.php.id=70&cat=light [accessed 22 September 2016].

Figure 23.4. The simplest pinhole can camera. Photo Targa Trygg.

pinhole camera, as you can see on my website www.solargraphy.com; more information about the project of Astronomy & Camera Obscura can be read in the Sunrise project by the EAAE in 2011.[8]

3. How to feel the rotation of the Planet Earth — Rotation / Motion

Intervention artist Juliane Stiegele started her workshop with students concentrating on one of her silence exercises. Its duration was five minutes. The statement of Globe was as follows:

Find a comfortable position in space you are.
You can sit, lay, lean or stand.
Don't move at all for 5 min, besides breathing.
Afterwards: while you have not been moving at all, you were moving in a gigantic circle line around the globe with speed. It depends on the position of respective location – around 1000 km/h.
Within the duration of the exercise, you might have covered a distance of around 80 km.[9]

---

[8] 'Astronomy & Camera Obscura, EAAE, Catch a Star, Winners 2013, available at http://www.innovativeteachersbg.org/EAAE_CAS_2013/3_project_astronomy_and_camera_obscura2.pdf [accessed 22 September 2016].
[9] Juliane Stiegele, Munich, 11 February 2014, personal communication.

Globe
Performance Exercise
Juliane Stiegele

FIGURE 23.5. Juliane Stiegele: Globe.

In this case how far did they move when sitting still for five minutes? We cannot feel the rotation of the Earth, but we can count the DURATION of the journey.[10]

> The movement of the Earth rotation in five minutes in the latitude 60°N is 830 km/h x (5/60) h = c. 70 km. It can be calculated or configured in the formula cos (latitude) x pi x the diameter of the Earth) / 24 x 60 min) x 5 min.
> Another case from the latitude 55°N the count is like that: cos 55 x 3.14 x 12760 km / (24 x 60 min) x 5 min = n. 80 km.[11]

A very well known experience happens, for example, when you are in sitting in a train in a railway station. When the next train starts moving but it feels like the static train in the station is moving.

How Earth Moves is the title of the science and visual presentation where it is possible to get facts about astronomical knowledge and see movements and Earth rotation around a moving Sun. It is a YouTube video created by Michael Steven in 2010. [12]

---

[10] 'Earth Rotation & Revolution around a moving Sun', YouTube video, 2 December 2008, available at http://www.youtube.com/watch?v=lkWyM-M8o0c [accessed 23 September 2016].

[11] Personal communication with Sakari Ekko, who carried out this math problem for me, Sakari Ekko, 'Sunrise project', available at http://eaae-astronomy.org/sunrise/images/documents/sunriseproject.pdf [accessed 29 September 2016].

[12] Michael Stevens, *How Earth Moves*, video created in 2010 and published on 13 June 2016, available at https://www.youtube.com/watch?v=IJhgZBn-LHg [accessed 23 September 2016].

4. How to make the invisible visible by artistic methods

A Cyclic movement during one year is visual artist Eeva Karhu's recent artwork. She has three times walked along the different routes of the old pilgrimage route, Camino de Santiago, ca. 800km long, in Spain. After the last route she wanted to do a series from those walks. She walked eight km route, which formed an imaginative circle, starting from her home and ending back there. Her statement of the series is from the Gallery Taik Persons/Helsinki School page on the Internet. Here it is as follows:

> A spiral lies on its side. It grows by forming counterclockwise moving circles on top of each other, linearly, from left to right. Years are following one another; winter solstice comes every year at the same time at the highest point of the circle. Years consist of smaller rings, days, spinning inside this line. Curves of this fractal structure have already been specified and the present is a twirling hint of light in this perimeter. The camera is my eye. It captures moments between what I'm viewing and me. I record the time that I will soon pass through while I experience the timelessness of its passing. I walk a circle that path has neither beginning nor end. I photograph this path where each beginning is the horizon of the last one. By layering all these photographs together, they form one image that documents my journey. In a sense, I record time and in so doing, I continue its movement forever. I study this cyclic movement. I take part in it and imitate it, by walking the same circle route during one year. While walking, my feet find the rhythm of the way. A monotonic beat is unleashing my thoughts. Knowledge of the past and the future is dropping onto the path; bit by bit the present reveals its timeless essence.[13]

I asked Eeva Karhu to choose one image of her *Path* series for my presentation, and she chose the photograph *Path 15*. Here it is her experience with her words:

> This is a picture of my favourite artwork. It reminds me of Werner Holmberg's painting Old Country Road. The atmosphere in both of them is of the heat of a summer day, with a strong smell of pine needles in the air and the feeling of an adventure ahead. (There are no pines along my path but the colours in the painting bring them to mind.)

And for technical details see below:

> I have walked the old pilgrim voyage, Camino de Santiago in Spain three times along different routes. After the latest pilgrimage, I wanted to make a series of pho-

---

[13] Eeva Karhu, Statement, Gallery Taik Persons, available at http://gallerytaikpersons.com/artists/eeva-karhu/statement/24 [accessed 7 September 2016]; Andrea Holzherr, and Timothy Persons, *The Helsinki School. Vol. 4, A Female View* (Ostfildern: Hatje Cantz, 2011).

Figure 23.6. Eeva Karhu, *Path 15*, 2010–2011.

tographs based on those walks. So I chose a path of eight kilometres that makes an imaginary circle, starting from my doorstep and coming back home. I focus on eye level at the horizon and then walk there, generally to a turning of the road. I repeat this until the last photo which concentrates on the place where I took the first photo. I am walking a circle at the same time as the earth makes a circle around the sun. The structure in the background of the pictures is of circles, or rather spirals because they never close. I study the changes of the light in the same place during a year.

The picture consists of about 90 photos on top of each other.[14]

Karhu's photographs look somewhat like impressionist paintings and are also like the traces of solargraphs without the movements of the Sun paths. There are some commonalities, for example, long exposures.

In solargraphy, the light-sensitive material as black and white photographic paper is exposed continuously – for example for half a year. Eeva Karhu's photograph of *Path 15* consists of ca. 90 photographs that she has taken in walking this path (see Fig. 23.6).

Figure 23.7, *Rowing*, 2006, is an example of the photographer Pekka Luukkola's way of showing traces of movements in his book, *Sense of Time*.[15] The photograph is

---

[14] Eeva Karhu, Helsinki, 15 February 2014, personal communication.

[15] Pekka Luukkola, *Maiseman Aika (Sense of Time)*, trans. Malcolm Hicks (Helsinki: Aalto Arts Books, 2012), p. 24–27,

FIGURE 23.7. Pekka Luukkola, *Rowing*, 2006, Kolovesi, Finland.

from Kolovesi, Finland. The exposure was five minutes. He wrote his description of the event of this photo in his book as follows:

> We often sense that someone has been there in the landscape. But we cannot see that person at all. In this photograph flares attached to the oars of a boat have been made to remain visible in the twilight landscape. A long exposure can allow the whole history of a brief event to be recorded in one picture.[16]

FIGURE 23.8, *Moving Earth*, 2008, is Pekka Luukkola's photograph; I asked his permission to share these two pictures in my presentation. He photographed this in Kilpisjärvi, the fell Pikku-Malla. Luukkola's photograph shows traces of stars moving over a period of 20 minutes. He wrote about his feelings and thoughts like this:

> We are hurtling around in a vast cosmic system. Day gives way to night as the Earth spins on around its axis. My camera is firmly on the ground and is recording traces of the stars, representing motion and the passage of time in a visible form. By

---

[16] Luukkola, *Maiseman Aika (Sense of Time)*, p. 26.

FIGURE 23.8. Pekka Luukkola, *Moving Earth*, 2008.

FIGURE 23.9. Miska Knapek: Sunshine light during a voyage from Helsinki to Stockholm (June 2007).

monitoring the stars with the camera, I was able to create a picture of the Earth´s rotation.[17]

The next two pictures are samples of time-lapse photography by the visualization designer Miska Knapek.

FIGURE 23.9 is one of his travel images. These two artworks deal with sunlight. The first one showed sunshine light during a voyage from Helsinki to Stockholm in 2002. He was curious to see what long distances look like, as a whole. Miska Knapek told me that the travel images are his reflections on memory journeys and information representation. Often we remember fragments of passages. This project tries to investigate what a trip would like if one could see it all in one go. Each journey was photographically 'scanned', allowing the viewer to see its visual entirety all at once. He uses the word spatiotemporal when he describes his travel images, scanning views over long distances: catching and showing journeys.[18]

His travel images are images of long journeys. How do you do this kind of travel images?

> Typically I photograph out the side of the vehicle I'm traveling in, taking 3–5 images per minute, for the whole route. Then, to make it possible to get a visual sense of the landscape that's passed by, I've written some software that takes the hundreds or thousands of resulting images, and puts them side by side. Following, the software compresses the images horizontally together, so all the long line of images can be seen at in a once glance.
>
> It's a bit like using a scanner to catch the whole route, instead of a camera. Quite fun, in a way, of creating one very wide-angle lens, seeing hundreds | thousands of kilometres at a glance.[19]

Miska Knapek calls his work *Sun/Shade, Time-light calendar, Physicalization of a Helsinki year's sunlight (measurements)*. His idea is to make interfaces that help us understand, navigate, and learn from the world in both traditional and experimental ways.[20]

Let's see solargraphs from different latitudes.

---

[17] Luukkola, *Maiseman Aika (Sense of Time)*, p. 52.

[18] Miska Knapek, Travel Images, available at http://knapek.org/travel_images_overview01.html [accessed 7 September 2016).

[19] Miska Knapek, Travelimages, available at https://www.flickr.com/photos/miska_too/collections/72157603780184048/ [accessed 10 February 2016].

[20] Miska Knapek, About me, available at http://knapek.org/aboutMe_page02.html [accessed 7 September 2016].

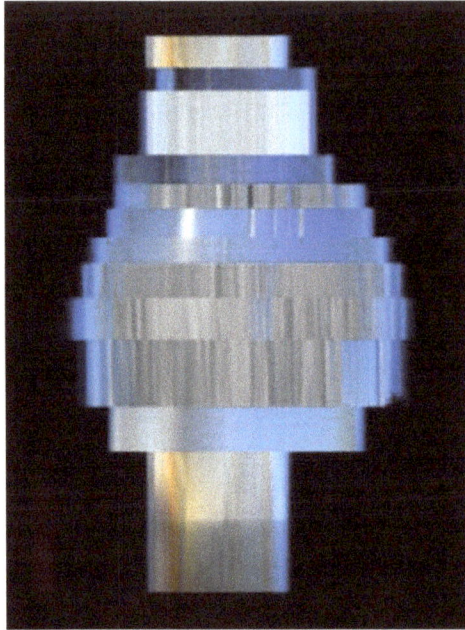

FIGURE 23.10. Miska Knapek, Sun/Shade, Time-light calendar, Physicalization of a Helsinki year's sunlight (measurements).

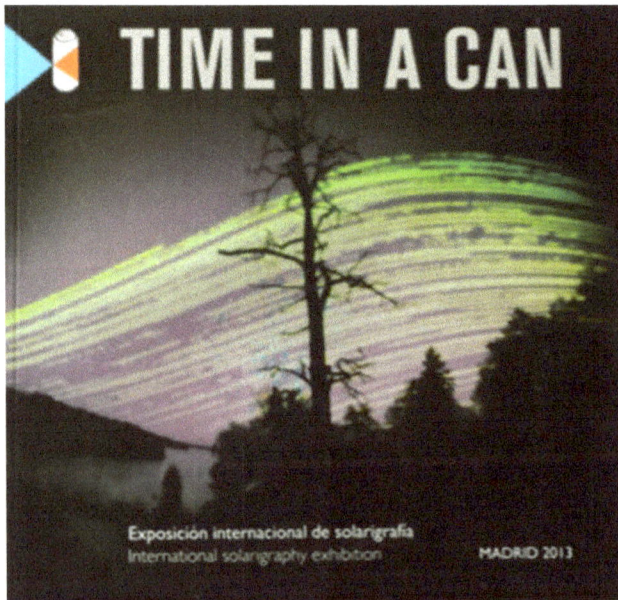

FIGURE 23.11. The catalogue of the project: Time in a can.

FIGURE 23.12. Tarja Trygg's can exposure shows the view for half a year in Suomenniemi, Finland, for the project *Time in a can*.

FIGURE 23.11 is the solargraph by T. Trygg of Helsinki, Finland 60°N, 24°E, that was chosen for the cover of the catalogue of the *Time in a Can* Project organised by the Spanish team at Estudio Redondo in Madrid.[21] Diego López Calvín tells about the project as follows:

> Some years ago I started as a partner in Estudio Redondo (a creative platform in Madrid), and we decided to relaunch the original idea of Solaris, so we looked for pinhole artists around the world who wanted to work with us. We designed and built the best solargraphy camera ever thanks to the support of our sponsor Asociación de Latas de Bebidas (Association of Beverage Cans) and we sent five cameras to 40 different pinhole photographers spread around the globe, from the Equator to both Hemispheres. We were working together during six months, the period between summer and winter solstice 2011. All the participants were using the same cameras and keeping the same parameters during all the procedures to achieve and compare the solar phenomena in the same period.[22]

Solargraphy records the weather of several months in one image. During half a year, the Sun seems to rise more from the east; in June during the summer solstice the Sun is at that time at its highest point in the sky. The Sun seems to move towards the south in December at the winter solstice. In fact, the Sun does not rise at all over

---

[21] Estudio Redondo, *The Project Time in a Can*, available at http://www.estudioredondo.com/project/proyecto-time-in-a-can/ [accessed 10 October 2016].

[22] International Analog Advocates, 'Interview with Diego López Calvín about the project *Time in a Can*', available at http://www.analogadvocates.com/page/time-in-a-can/Analog+review [accessed 7 September 2016].

FIGURE 23.13. The solargraph from the can exposed at Suomenniemi for half a year. Tarja Trygg.

FIGURE 23.14. *Solarigrafía Temporal.* Solargraph in Almeria, Spain, by Diego López Calvín: the weather of half of the year 2008. Available at http://solarigrafia.com/solarigrafia_solargraphy/Temporal.html [accessed 7 September 2016].

FIGURE 23.15. Almería, Spain, Location: 36°N, 2°W (Google Earth).

the horizon between December and the 15th of January in Lapland, over the Polar Circle. But in the summer time, the Sun does not at all set on the horizon. The Midnight Sun or the Nightless Night is a local speciality in the north, too much sun and light. For recording that phenomenon, it is better to point a pinhole can towards the north. According to the Earth's rotation, we are moving around the Sun, yet we see the tracks of the movements of the Sun paths in solargraphs.

Conclusions

1. Photography has an ability to make the invisible visible. It is possible to achieve sharper photographs and get information from the universe, discover new things from outside and inside for different purposes (including space research and medical research).

2. Photography can make the invisible visible for a better understanding of the astronomical phenomena from the human perspective. Photography helps humankind to see into deep space and even go deep inside our minds and bodies, through observing the sky with the simplest of low-tech tools such as a small pinhole camera.

3. According to the Earth's rotation, we are moving around the Sun. Although we do not feel it, we see the tracks of the movements of the Sun paths in solargraphs instead of the rotation of the Earth.

The purpose of this chapter is to emphasise human curiosity as a significant catalyst or power to show experience, the movement of time and all that is not visible. Our curiosity and desire is to see further than the visible light area can offer. The purpose is to find out explanations for astronomical phenomena. Unknown and non-visible events inspire artists to create new alternative ways to study them in their art and to try to make non-visible phenomena visible. Art and science together add to our knowledge for better understanding the hidden universe. Movements and the rotation of the Earth have their influence to help interpret sharper the images we not see before. Interdisciplinary resources create new approaches to discover the world and make non-visible astronomical phenomena visible and understandable.

CITIZEN SCIENCE ON THE ISS:
STE[+A]M IT UP!
PRELIMINARY RESULTS OF A
STORYTELLING EXPERIMENT USING BIOSENSORS

Elizabeth Forbes Wallace

ABSTRACT: 'What's it like'? we ask astronauts about their view of Earth from space. We know how we feel when looking down from a mountaintop or the top floor of a skyscraper. But how it feels to gaze at Earth from orbit, more than 200 miles away, traveling 17,500 mph, billions of us are dying to know. We cannot, however, all board the International Space Station (ISS) nor buy a $200K ticket for a commercial spaceflight 100km above the Earth. Sometimes we wait years for an astronaut's memoire to reveal a special moment that gives us goosebumps while reading in a gravity-bound armchair. However, an almost instantaneous way to uniquely share the experience is now possible through technological advances in neuroscience and communications; citizen science, i.e., scientific research conducted, in whole or in part, by amateur or nonprofessional scientists; and just plain, old-fashioned storytelling.

Imagine: An astronaut floats into the ISS cupola. Seven windows arranged in a hemisphere reveal a 180 degree field-of-view of Earth. For the next 90 minutes as she orbits Earth, biosensors track her reactions, not just her facial expressions but also the changes in her heart rate, her brain waves, and even the secretion of sweat on her palms. Software instantly captures, syncs, analyzes and live-streams the results in real-time directly to you. A 3-D video displays not only the grand scope of Earth but also exactly where she's looking. You, too, are wearing biosensors, along with dozens of other citizen scientists back on Earth. Simultaneously you compare, contrast, discover and report "what's it like" for her, for you, for each other. But the journey started, you realize, long before you ever entered the room. Now it's time to craft your own personal epic.

Our first three goals for STE[+a]M It Up!, a unique Education and Public Outreach (EPO) experiment to invigorate public interest in the ISS mission and Earth stewardship, were to test the iMotions platform and equipment; discover if there is a statistically significant difference in emotions and engagement of subjects while they look at still photos of Earth from space compared to images accompanied by astronaut narrative or story; and find at least one astronaut who might volunteer for the job. We succeeded in all three and then set more goals.

The Overview

In 1961, more than half a century ago, cosmonaut Yuri Gagarin made the first orbit of Earth by man and exclaimed 'I see Earth. It's so beautiful'.[1] Nine days later *Life Magazine* reported on his journey.

The mysteries of outer space for the first time seemed to lie within the measure

---

[1] Anon., 'International Space Hall of Fame at the New Mexico Museum of Space History: Yuri Gagarin, USSR, Inducted 1976' at http://www.nmspacemuseum.org/halloffame/detail.php?id=8 [accessed 10 October 2016].

of human instruments…but (Gagarin's) words, describing what no human being had ever seen before, outsoared the shadow of the cold war and touched the hope and imagination of all men. 'I saw for the first time the Earth's shape. I could easily see the shores of continents, islands, great rivers, folds of the terrain, large bodies of water. The horizon is dark blue, smoothly turning to black. The feelings which filled me I can express with one word – joy'.[2]

In 2015 within seconds of astronaut Samantha Cristoforetti's arrival on the ISS, she located the closest porthole, grabbed two handles to anchor herself and smiled wide, her face bright with sunlight reflected off the Earth's surface as well as with utter joy.[3] The view is always engaging, according to astronaut Joseph P. Allen:

It's endlessly fulfilling. You never see quite the same thing as you are orbiting. There is a different ground track every time. The time of day is different, the clouds are different. The cloud patterns show different colors. The oceans are different. The dust over the deserts is different. It doesn't get repetitive.[4]

Cosmonaut Boris Volynov reported a change in his relationship to other human beings:

During a spaceflight, the psyche of each astronaut is reshaped. Having seen the sun, the stars, and our planet, you become more full of life, softer. You begin to look at all living things with greater trepidation and you begin to be more kind and patient with the people around you. At any rate, that is what happened to me.[5]

The first Arab and Muslim astronaut Sultan bin Salman bin Abdulaziz Al Saud began to change his perception of the concept of home. 'The first day or so we all pointed to our country. The third or fourth day we were pointing to our continent. By the fifth day, we were aware of only one Earth'.[6] Astronaut Michael J. Massimino said it's like looking at Paradise.[7] Astronaut Edgar Mitchell admitted having a spiritual experience: 'Instead of an intellectual search, there was suddenly a very deep

[2] Hugh Sidey, 'Soviet Traveler Returns from Out of this World: the Story of Yuri Gagarin and his RemarkableAchievement. How the News Hit Washington', *Life Magazine* 50, no. 16 (1961): p. 21.

[3] Toni Myers, *A Beautiful Planet* (2016).

[4] Frank White, *The Overview Effect: Space Exploration and Human Evolution* (Boston: Houghton Mifflin, 1987), p. 234.

[5] Kevin W. Kelley, *The Home Planet* (Reading: Addison-Wesley, 1988), p. 88.

[6] Kelley, *The Home Planet*, p. 81.

[7] Mike Massimino, 'A View of the Earth From the Hubble Space Telescope. Which I nearly broke.' (2013) at http://www.esquire.com/news-politics/news/a24509/a-view-of-earth/ [accessed 10 October 2016]

gut feeling that something was different… I suddenly experienced the universe as intelligent, loving, harmonious.[8]

Joy and awe. Fleeting and unpredictable on Earth, yet, perhaps, predictable nearly every time an astronaut looks out an ISS porthole toward our fragile blue planet. The experience changes some astronauts. They later describe having had what Frank White, author of several books on space exploration and the future, calls 'the over-view effect', that is, a paradigm shift so that they return home with a 'different philosophical view as a result of having a different physical perspective'.[9]

But most of us would like to know what it is like without having to go on the dangerous journey of escaping Earth's gravity. To date, experiments with humans in LEO on the ISS have remained solely in the purview of national space agencies. Most have been focused on making sure astronauts survive the journey and live healthfully in space. How best can we include citizen scientists of all ages in the exploration of what Dr Shaun Gallagher calls 'the inner space of experience, while traveling in outer space'?[10]

We Can STE[+a]M IT UP!
By Putting the Art of Story into the STEM of Neuroscience and Space

STE[+a]M refers, in this case, to engaging the Art of storytelling in the STEM of space exploration and Earth observation. Our experiment, STE[+a]M IT UP!, proposed at the 2015 ISS R&D Conference, is a chance to facilitate a shared reality between an astronaut in Low Earth Orbit (LEO) and citizen scientists on Earth; that is, a chance to facilitate a process whereby citizen scientists on Earth might adopt an astronaut's perspective regarding the effect of seeing Earth from space.[11]

While amazing virtual reality (VR) opportunities for the public are on the late 2017 horizon with companies like SpaceVR, citizen science projects are opportunities for non-scientists to meaningfully contribute to scientific research. Like the ISS *Ants in Space* program which 'allows K-12 students to participate in near real-time space research, learn about collecting and analyzing research from NASA scientists and gather their own parallel, ground-based data to analyze', STE[+a]M

---

[8] Kelley, *The Home Planet*, p. 138.

[9] White, *The Overview Effect*, 4.

[10] Shaun Gallagher, Lauren Reinerman-Jones, Bruce Janz, Patricia Bockelman and Jorg Trempler, *A Neurophenomenology of Awe and Wonder: Towards a Non-Reductionist Cognitive Science* (Hound-smills-Basingstoke: Palgrave Macmillan, 2015), 3.

[11] Gerald Echterhoff, E. Tory Higgins and John M. Levine, 'Shared Reality:Experiencing Commonality with Others' Inner States About the World', (2009), http://pps.sagepub.com/content/4/5/496. abstract, [accessed 16 August 2016].

IT UP! would as well.[12]

First, at least two ground-based venues, named *Your Brain on Space Camp,* would train citizen scientists of many ages and backgrounds in basic neuroscience, life on the ISS, Earth observation and the art of storytelling.

Second, on entering the cupola for a full ninety minute orbit, an astronaut would be equipped with biosensors to monitor "what it's like" using neuroscience and bio-metrics. She would be equipped with:

- An electroencephalogram (EEG) to measure the electrical activity of her brain.
- Galvanic skin response (GSR) sensors to measure sweat secretion, a sensitive marker for emotional arousal.
- An electrocardiogram (ECG) to measure the electrical activity of her heart.
- Eye tracking sensors to ascertain where she focuses her gaze synced with a video camera pointed at Earth.
- Micro facial expression sensors to detect subtle subconscious emotions.

iMotions Global[13] software would sync and analyze the metrics in real time to identify the focus of the astronaut's attention as well as her engagement, relaxation and excitement. During the astronaut's transmission, the reactions of citizen scientists would also be monitored by biosensors as they view Earth through the astronaut's cameras and follow her biometrics. Results would eventually be shared in scientific reports, but more importantly, and almost immediately, the participants would craft and share their neuroscience (STEM) findings via their own personal narrative using the dramatic art of storytelling.

In order to be able to replicate the experiment without incurring prohibitive costs in space, we would follow the lead of the *Ants in Space* program. The *Your Brain on Space Camp* could use footage from the first astronaut's experience and replicate the experiment in multiple classrooms and venues over time with innumerable students and lifelong learners.

On the Shoulders of Giants

The first documented neuroscience experiments performed in space took place during the flight of the third Russian Vostok spacecraft in 1962.[14] These experiments,

---

[12] Charles Moore, 'BCM Center for Educational outreach Ants in Space Project Lifts Off For ISS', (2014) at https://bionews-tx.com/news/2014/01/10/bcm-center-for-educational-outreach-ants-in-space-science-project-lifts-off-for-iss/, [accessed 16 August 2016].
[13] Anon., 'iMotions Biometric Research Platform: One Stop Shop for Human Behavior Research', (2016) at https://imotions.com/products/, [accessed 16 August 2016].
[14] Gilles Clement and Millard F. Reschke, *Neuroscience in Space* (New York: Springer Science and

the first of many by RosCosmos, the Russian Federal Space Agency, and those conducted by NASA starting with Gemini 5 in 1965, were initially focused on gathering information to address the issues of nausea and spatial disorientation in weightlessness. After research and conversations with several NASA representatives, We have not uncovered evidence of any experiment performed by NASA using an EEG and other biosensors to track positive emotions while looking at Earth from LEO. Cameras and other tools are used in other parts of the ISS to monitor changes in astronauts' moods and productivity.

The first scientific study of the feelings of awe and wonder experienced by astronauts during spaceflight was actually ground-based and led by Dr. Shaun Gallagher. Volunteers looked at pictures of Earth from space in a simulated ISS environment. His study employed methods from neuroscience, psychology, hermeneutics and phenomenology. Their working definitions of awe and wonder follow.

> **Awe**: a direct and initial experience or feeling when faced with something amazing, incomprehensible, or sublime.
> **Wonder**: a reflective experience motivated when one is unable to put things into a familiar conceptual framework – leading to open questions rather than conclusions.[15]

If feelings of awe on the ISS are directly connected to the experience of nature on a grand scale, it is important to consider the results of ground-based experiments which explore the effect of nature on the human psyche. Some non-space related experiments show that 'experiences of awe bring people into the present moment, and being in the present moment underlies awe's capacity to adjust time perception, influence decisions, and make life feel more satisfying than it would otherwise'.[16] People become more willing to help others and learn to value experience over possessions.

The Overview: Beyond Borders

Perhaps because of my own experience as a visual artist and photographer, when I look at photos of Earth taken from LEO, I see patterns, textures, colors, and contrast, that is, I see them as abstract paintings. I don't see them as a botanist, geologist

---

Business Media, 2008), pp. 43–69.
[15] Gallagher, *Neurophenomenology* , 6.
[16] Melanie Rudd, Kathleen D. Vohs, and Jennifer Aaker, *Awe Expands People's Perception of Time, Alters Decision Making, and Enhances Well-Being* (Washington, DC: Psychological Science, 2012), p. 1.

or climate change expert might. The context of the experience is determined by the observer, not the object itself. Because the visuals alone were not enough to inspire awe in me, I produced a six minute film, *The Overview, Beyond Borders* to see if the words of astronauts could.[17] For presentation at INSAP VIII, in a matter of a few days, I worked with several ten-year-old children who chose and narrated quotes by astro- and cosmonauts which reveal the emotional side of their journeys from Earth, to space and return. Photos taken by NASA, National Geographic and individuals worldwide accompany the oral narrative.

I often felt that I was going into a very relaxed mode while watching it and noticed that others did too. Many reported good feelings, goosebumps, and even tears. I wanted to learn what was happening inside viewers' brains and find out which emotions they experience (neurophenomenology).[18]

Goal 1: Equipment Test and Experiment Design

The film became my low-budget ground-based simulation. According to Gallagher's findings from the astronauts' journals, it was the view outside their portholes that contributed to their emotions, not the physical experience of weightlessness[19] so, fortunately, a zero-G simulation was not necessary. Dr. Roxanna Salim, Product Specialist at iMotions Global, arranged a preliminary experiment with eight adults from Copenhagen, Denmark and Boston, US including two females and six males. The age range was thirty to forty-nine years.

A **pre-survey** was completed by participants to gauge their level of interest in space exploration, the ISS and views of Earth. They were then shown several still shots of Earth from space as well as of astronauts on the ISS including one of astronaut Anousha Ansari snapping photographs. This was followed by a viewing of the video, *The Overview: Beyond Borders*. A post survey was given to ascertain if the level of interest in space and/or the ISS had increased.

An **electroencephalogram** (EEG) specifically a mobile Emotiv headset, measured the brain activity by calibrating electrical activity from the scalp. An EEG provides the fastest response of all biometric sensors and it can track several emotional and motivational states including engagement, meditation, and excitement.

---

[17] Elizabeth F. Wallace, *The Overview: Beyond Borders* (2013) at http://bit.ly/1SAH9d3 [accessed 10 October 2016].

[18] Gallagher, *Neurophenomenology*, 11.

[19] Gallagher, *Neurophenomenology*, 14.

The **eye tracker** monitored 600 gaze points per second. A cluster of gaze points indicates a fixation. A *heat map* tracks the cluster of visual attention with red indicating the areas most frequently viewed.[20] A *bee swarm* is an aggregate of all volunteers.[21]

The **galvanic skin response** (GSR) measures arousal or excitement. The hands and feet secrete the highest amounts of sweat. Since it is not under conscious control, GSR is a source of insight to validate verbal reports or interviews or to question them.

Micro facial expressions were tracked to determine subconscious reactions which occur in 100–200 ms. On the other hand, feelings are conscious reactions. Facial expressions from feelings are more visible and occur 100–200 ms after an emotion. The equipment can measure the presence or valence of an emotion or feeling but not its power or arousal.

The iMotions platform synchronized all sensor, stimuli and API (application programming interface) data streams in real time. No manual post synchronization of data sets was required. Other scientists like Dr. Erin MacDonald, Assistant Professor of Mechanical Engineering, Stanford University, testify to is accuracy and ease of use:

> If I compare (iMotions) to other software that we have been using before, then we always designed a study that would fit to the software. Now we just design our study and we know that iMotions can handle it. I can do things so much faster with iMotions and they even understand the budget of an assistant professor.[22]

The iMotions dashboard clearly and simultaneously displays a video of the volunteer, the film and the levels of engagement (blue), meditation or relaxation (green) and excitement (yellow).[23]

Goal 2: Statistically Significant Results

While we were fortunate to be able to use the eye tracking software, the results are of minimal importance for the purpose of this test. The GSR was hard to collect during the stills but detected a marked increase during the video. The facial expres-

---

[20] The iMotions Global Specialist Team, 'Eyetracking Photo of Anousha Ansari', (2015) at http://bit. ly/2dvS86F [accessed 10 October 2016].

[21] iMotions, 'Bee Swarm', (2015) at http://bit.ly/2d3ONfO [accessed 10 October 2016

[22] Anon., 'iMotions: Increasingly Sophisticated Research More Efficiently', (2016) at https://imotions.com/customer-success-story/increasingly-sophisticated-research-efficiently/[accessed   10 October 2016].

[23] iMotions, 'TOBB Gaze Gens', (2015) at http://bit.ly/22TxX2O [accessed 10 October 2016].

sion results were not extremely significant but there was more joy detected during viewing of the film compared to the stills.

What we found most interesting was there were higher, statistically significant levels of engagement and meditation during the video than during the still images. For engagement, the Mean was .37 and the Standard Deviation .17 for the video; the still images were Mean .16 and SD .22. For meditation, the Mean was .36 and the SD was .03; the Mean for the still images was .25 and the SD was .02.

The **post survey results** indicated that the subjects showed an increase in overall interest in learning about space, the ISS and 'the overview effect'. All subjects reported a new interest in googling pictures of Earth and/or outer space. Seven out of eight participants wanted to share the video with others.

Most importantly, they reported having experienced a variety of emotions similar to those of astronauts and participants in Gallagher's study: *peaceful, curious, relaxed, awe, exciting, adventuresome and wonderment*. Respondents to the film at the INSAP IX conference verbally reported having similar feelings as have viewers in informal venues including a film festival. Children who viewed the film and later imagined themselves in space reported 'the whole world is my home', 'I'd be a different person', 'my heart feels like a life of love and hearts', 'amazed', 'dreaming', and 'I appreciate everything the world has given me'.

Goal 3: Astronaut Engagement

Sending an experiment to the ISS is an expensive endeavor. Astronaut time is at a premium. STE[+a]M It UP! may require the volunteer assistance of an astronaut in order to keep the costs down. The equipment is of minimal weight and could be carried in the astronaut's personal gear and the experiment performed in their free time. Time in the cupola is something astronauts look forward to, much more than giving blood or urine samples.

When I asked Astronaut Karen Nyberg her opinion she replied, 'it would have been interesting to have that when I was capturing the ATV because my heart was about to pound out of my chest'.[24] Astronaut Sunita Williams replied,

> I certainly don't think it would ruin the experience of being in the cupola. It might feel a little bit weird to have that on your head but I think you would still get that 'aaaahhhh' feeling when you go into the cupola. So if that's what you're looking for, and see the stress level maybe decrease, and potentially increase when you see

[24] Karen Nyberg, 'ISSRDC 2015 Astronaut Keynote', (presentation, ISS R&D Conference 2105, Boston, MA. July 7-9, 2015). at http://www.iss-casis.org/NewsEvents/OnStation/tabid/113/ArticleID/183/ArtMID/570/2015-ISS-Research-and-Development-Conference-Recap.aspx.

some really amazing great things outside, too. I think you'll definitely get that.[25]

I prefaced my query to Astronaut Rick Mastracchio, after his presentation at the ISS R&D 2016 Conference, with the observation that students are getting more interested in neuroscience and their emotions thanks to animated films like Pixar's 'Inside Out'. When asked if he would be willing to wear and EEG and other biosensors when looking at Earth from the ISS cupola for EPO, he smiled wide and said "Sure!"

Why Teach Citizen Scientists to Tell Stories?

Why not just write a scientific paper and publish that? For the purposes of Education and Public Outreach (EPO), story is the most effective way to communicate new information to an audience and have it stick. TEDTalks, for example, are adult storytelling and STEM education at its finest.

Thanks to neuroscience and master storytellers like Kendall Haven, there are reams of scientific proof that the best way to learn new data is through story. Not only are we descended from a grandmother in the Rift Valley who told stories of the night sky, we're hardwired by evolution with an amazing tool which Haven, author of *Story Smart,* calls the Neural Story Net (NSN).

It's like a scaffolding on which we catch and attach new information, structure it, and connect to it. According to Haven, the NSN facilitates understanding and retention. 'Information is remembered better and longer, and recalled more readily and accurately when it is remembered within the context of a story'.[26] Researcher K. Mallan concluded that emotional engagement of the listener in a story makes it easier to remember the new and key information. 'Stories differ from other narratives (arguments, scientific reports, articles) in that they orient our feelings and attitudes about the story content'.[27]

Therefore, storytelling is the perfect tool for EPO since effective stories, with key ingredients, according to the NSN's rules, can inspire, teach, persuade and more.

Call to Story by a Former NASA Astronaut

From another perspective, astronaut Frank Culbertson asserts it is absolutely necessary that not only students but adults, including NASA employees, to tell their stories in part to help fund NASA missions:

---

[25] Sunita Williams, 'ISSRDC 2015'.

[26] Kendall Haven, *Story Proof: The Science Behind the Startling Power of Story* (Westport, CT: Libraries Unlimited, 2007) 67.

[27] Kerry Mallan, 'Storytelling in the School Curriculum', *Educational Practice &Theory* 19, no. 1 (Hillsdale, NJ: Lawrence Erlbaum, 1997), pp.75-82, as quoted in Haven, *Story Proof,* 63.

It's clear that what (students who sent experiments to the ISS) are doing is spreading into the community what's going on in space. That's really the grass roots... When you leave a conference, or a speech,...you hang out with your neighbors, flipping your burgers... tell them what you do! ... Tell them what's important to you. That way it may become important to them...They may get excited...(and)... talk to their politicians on whom we have to rely for leadership and funding.

But the important thing is that the public and the world, not just the US, knows that we have three to six people in space all the time...doing important work... that will enable the next generation to go beyond. That's the generation that will go to Mars... It's ok to talk about it. Get really excited about it! And shed a few tears because we have to deal with things that are hard, really hard.[28]

Even during the exciting Apollo program, US citizens wanted space exploration but didn't want to pay for it. Currently .5% of the US budget goes to NASA. After we reached the moon and won the space race, interest waned somewhat. But we did and still do reap widespread benefits, called spinoffs, from NASA-funded programs.

NASA wants to inspire all citizens to encourage lawmakers to support its missions., The most effective grassroots skill set the participants of this project could be taught to positively affect NASA outreach is storytelling. If a higher percentage of their mission outreach was in story form, perhaps the budget would increase to much more than 1%.

Putting the Art of Storytelling in STE[+a]M Education

While there is still some discomfort in parts of the scientific and aerospace communities about the evolution from STEM to STE[+a]M, it's the law in the United States.[29] Integration of the arts into STEM (Science, Technology, Engineering, and Math) education is now required in K-12 schools in the United States in order to be eligible for Title 1 funds.[30] Artists have been encouraged and trained for the past several years by esteemed organizations like The Kennedy Center for the Performing Arts in Washington, DC to incorporate the arts throughout curricula.[31]

---

[28] Frank Culbertson, 'ISSRDC 2015 Astronaut Keynote'.

[29] Anon., 'CulturePop: New Education Reform Amendment Adds Arts to K-12 STEM Curriculum', (2010), at http://culturepop.com/art/suzanne-bonamici-education-reform-adds-arts-stem-curriculum/ [accessed 10 October 2016].

[30] Anon., 'US Department of Education: Improving Basic Programs Operated by Local Educational Agencies (Title I, Part A)', (2015), at http://www2.ed.gov/programs/titleiparta/index.html [accessed 10 October 2016].

[31] Anon., 'The Kennedy Center Arts Edge: Growing from STEM to STEAM', (2014) at https://artsedge.kennedy-center.org/educators/how-to/growing-from-stem-to-steam [accessed 10 Oct.2016].

In the K-12 environments, artists and teachers alike are introduced to the Multiple Intelligences (MI) theory of Howard Gardner.[32] The MI theory lists nine types of learning intelligences: verbal-linguistic or *word smart*; bodily kinesthetic or *body smart*; musical or *music smart*; logical-mathematical or *math smart;* naturalistic or *nature smart*; visual-spatial or *picture smart*; interpersonal or *people smart*; and intrapersonal or *self smart.* Use of a few or all of one's intelligences empowers individuals to learn and express themselves through many modalities. Teaching artists are tasked with finding innovative ways to instruct STEM education using the arts of dance, sculpture, painting, drama and more but more importantly to facilitate the students' expression of their knowledge of a STEM subject using an art form rather than, or in addition to, an exam or term paper.

The art of oral narrative uniquely uses all of the intelligences in one art form. Both an astronaut/engineer and a storyteller use logic; one for solving a technical problem, the other a story plot. An astronaut who knows celestial navigation shares nature smarts with a planetarium storyteller who helps listeners locate constellations where ancient tales abide. An astronaut repairing the robotic arm during an EVA (ExtraVehicular Activity) outside the ISS would use body smarts as would a 'teller relating his story using mimicry to imitate wearing cumbersome space gloves to tighten a bolt during an EVA. Most importantly, by using interpersonal and intrapersonal smarts, storytellers introduce emotions associated with ISS experiences from curiosity to frustration, to homesickness and awe.

In the STE[+a]M It Up! project, citizen scientists would be trained to become story- or sciencetellers whose tales could be groomed for K-12 curricula, podcasts, EPO family events and more.

Next Steps

Before going to LEO the STE[+a]M It Up! experiment could be redone with more subjects and a more diverse demographic. Also the film (or a newly edited version) could be shown with or without narrative, music, or video so that the effects of narrative are more precisely measured. And, as Astronaut Nyberg suggested, find out if there is any data on Earth about views from a mountaintop which might correlate.

This begs the question: Can the overview effect be created on Earth using story and visuals alone? The answers will be valuable to those in the process of creating a virtual reality (VR) version of the view from the cupola as well as those selling space tourism flights.

---

[32] Howard Gardner, *Multiple Intelligences: New Horizons in Theory and Practice* (New York: Basic Books, 2006).

Goal 4: Spinoff for Society

Traditionally, NASA spinoffs are technologies originally developed to meet the needs of the space agency's missions 'off the Earth' and subsequently benefit life on Earth in the form of commercial products or services. They include a variety of things from precision GPS to invisible braces to airplane winglets.

The first STE[+a]M It Up! spinoff would not be a commercial product but a sociological process with an environmental and humanitarian effect. Having had a shared reality with an astronaut, the individual perspective of each participant could be shifted toward a global view. That is, many of us could be motivated to work together to protect the Earth, our shared home, and recognize each person as a neighbor. Astronaut Nyberg expressed it this way:

> Of course, everybody's seen pictures from the cupola. Everybody knows that one of our favorite things to do is go to the cupola. Take pictures. Look. It is the most amazing thing to see in the entire world. I wish everybody could go up there for just 90 minutes, everybody in the world, and look out that window. The world would be a different place.[33]

By extending STE[+a]M UP! to a longitudinal study, we could investigate whether or not the perspective gained could be sustained by the ground-based participants as well as by astronauts on return to life on Earth. Astronaut Karen Nyberg loved to go into the cupola during breaks from required workouts:

> I would do a set of squats and then I would float up to the cupola and rest, take a picture or two, come back and do another set. And I did that *every* time. And I remember one time, when it was about 3–4 months into my mission, and I went up and said 'the tip of South America again!' Then I went back to exercising again and then I stopped and I paused and realized that I just took for granted that I am 250 miles above the Earth orbiting! From that point on I made a point to relish it a little more. But it also made me think, how many things that become normal on Earth that are *magnificent* that we just take for granted, that are happening every day...so I tried to take that perspective back with me.[34]

Another spinoff would be 'overview' products, perhaps 3D Earth holograms, to support the mental heath of astronauts suffering from depression in long-duration spaceflight. The interior of the ISS has little in the way of nature except some botanical or biological experiments. According to Bratman the experience of nature reduces rumination and activates the subgenual prefrontal cortex:

[33] Nyberg, 'ISSRDC 2015 Astronaut Keynote', (2015).
[34] Nyberg, 'ISSRDC 2015 Astronaut Keynote', (2015).

More than 50% of people now live in urban areas. By 2050 this proportion will be 70%. Urbanization is associated with increased levels of mental illness, but it's not yet clear why. Through a controlled experiment, we investigated whether nature experience would influence rumination (repetitive thought focused on negative aspects of the self), a known risk factor for mental illness. Participants who went on a 90-min walk through a natural environment reported lower levels of rumination and showed reduced neural activity in an area of the brain linked to risk for mental illness compared with those who walked through an urban environment. These results suggest that accessible natural areas may be vital for mental health in our rapidly urbanizing world.[35]

Participants in one study by Stanford researchers were randomly assigned to take an urban walk or a nature walk, it was found that the experience of nature decreased anxiety and rumination as well as increased cognition for completing a complex working memory span task.[36] While there is no garden path on the ISS, astronauts have reported relief from stressors by catching a quick look outside a window or tending plants.

The same researchers, in another experiment, found that a nature walk can lower the risk of depression. 'Neural activity in the subgenual prefrontal cortex, a brain region active during rumination – repetitive thought focused on negative emotions – decreased among participants who walked in nature versus those who walked in an urban environment.'[37]

Since sleep disorders, cognitive impairment and depression on the ISS are an ongoing cause for concern, perhaps more scheduled downtime at the window would be more effective than pharmaceutical sleeping aids. In another ground-based experiment with persons also living in isolation, as do the ISS crew, Officer Michael Lea at the Snake River Correctional Institution in eastern Oregon reported that calmer behavior is exhibited by prisoners who exercise for forty minutes several days a week in a room where nature videos play compared to those who exercise in a gym with no videos. 'I thought it was crazy at first. There's a lot of yelling really loud. It echoes horribly (in the plain gym). In the (other) room they tend not to yell. They say, 'Hold

---

[35] Gregory N. Bratman, J. Paul Hamilton, Kevin S. Hahn, Gretchen C. Daily, and James J. Gross, Nature Experience Reduces Rumination and Subgenual Prefrontal Cortex Activation', *Proceedings of the National Academy of Sciences of the United States of America* 112, no. 28 (2015): pp 8567–72.
[36] Gregory N. Bratman, Gretchen C. Daily, Benjamin J. Levy and James J. Gross, 'The Benefits of Nature Experience: Improved Affect and Cognition', *Elsevier Science Direct: Landscape and Urban Planning* 138 (2015): pp. 41–50.
[37] Bratman, Daily, Levy and Gross, 'Nature Experience', *Elsevier:Landscape*.

on, I got to watch my video'.[38] The spinoff made for astronauts might also help those suffering from anxiety or depression on Earth and eliminate the need for pharmaceuticals when a guided meditation of Earth from space in a VR headset might be similarly effective.

But does one also need Earth's sounds, breezes and smells? Do you need to be 'in' nature rather than looking at it through a portal or a VR headset? Cosmonaut Georgi Grechko said,

> It's impossible to imagine how much pleasure this green oasis of terrestrial life (a biological experiment on board) gave us. We even said, 'Come on, let's go walk in the grove!' For human eyes accustomed to the greenness of Earth, living plants give tranquility. Even sleeping beside their thin stems was better'.[39]

In Sweden physician Matilda van den Bosch found that after a stressful math task, subjects' heart rate variability—which decreases with stress—returned to normal more quickly when they sat through 15 minutes of nature scenes and birdsong in a 3D virtual reality room than when they sat in a plain room.[40] Perhaps the meditation spinoff for astronauts might need to include nature sounds and smells as well while patients on Earth could wear the hardware outside.

Goal 5: Oxytocin and Bonding with Earth

Another possible experiment tangential to STE[+a]M It Up! would involve measuring oxytocin levels. Astronauts report having a changed feeling about Earth. Perhaps their oxytocin levels increase before and after time spent in the cupola. If so, does an increase help them bond with Earth the way a child does looking at her mother's face? Astronaut Taylor Wang expressed it, inspired by an ancient story: 'A Chinese tale tells of some men sent to harm a young girl who, upon seeing her beauty, become her protectors rather than her violators. That's how I felt seeing the Earth for the first time. I could not help but love and cherish her'.[41]

Conclusion

STE[+a]M It Up! is an important experiment which would collect baseline data from an astronaut looking at Earth from space and citizen scientists on Earth to help create new, shared experiences through storytelling which just might contribute to a

---

[38] Florence Williams, 'This is Your Brain On Nature', *National Geographic 1* (2016) at http://ngm. nationalgeographic.com/2016/01/call-to-wild-text

[39] Kelley, *The Home Planet*, p. 108.

[40] Williams, 'Brain on Nature', *Nat Geo.*

[41] Kelley, *The Home Planet,* p. 60.

worldwide paradigm shift. As Astronaut Scott Kelly said upon return to Earth,

> The more I look at (the Earth)…the more I feel like an environmentalist. (There are) definite areas where the Earth is covered in pollution all the time…and…unexpected storms… This is a human effect… not naturally occurring…We can fix that if we put our minds to it.[42]

---

[42] Anne Ball, 'VOA Science in the News: Astronaut Scott Kelly Talks About His Year In Space', (2016), http://learningenglish.voanews.com/a/scott-kelly-about-a-year-in-space/3208316.html

BALLA'S MERCURY PASSING BEFORE THE SUN AND THE *MODERNIST SUN*

Gary Wells[1]

ABSTRACT: In November 1914, artist and astronomy enthusiast Giacomo Balla observed the transit of Mercury across the face of the sun. He then painted a series of canvases about the event in the abstract and dynamic style of the Italian Futurist movement, of which he was an important member. Balla's interest in this astronomical event was part of a larger Modernist inclination toward the valorisation of new technologies and discoveries, and a celebration of the extended reach of human knowledge. It was also a moment where art began to reimagine and revisit certain astronomical subjects with a new set of expressive and conceptual tools. This paper examines Balla's painting in the context of early twentieth century modern art, particularly in its focus upon the image of the sun. Balla's interest in the developments of modern science was shared by many other artists who explored astronomical themes derived from the discoveries of the day, and were drawn to the long and rich history of the sun as a subject in art.

On the 7th of November 1914, Mercury passed across the visible face of the sun, an event witnessed from a wide swath of the earth's surface. The total transit took 4 hours and 10 minutes from first to last contact. For viewers in Europe, the entire transit was visible during the day with the sun situated favourably high in the mid-day sky.[2] Transits of Mercury are not especially rare, the previous one to 1914 having occurred in 1907 and the following one in 1924. For scientists and astronomers, the event elicited little interest, with brief notices appearing in journals in advance, and scant follow-up afterward.[3] Unlike transits of Venus, which still attracted interest from professional astronomers, those of Mercury were marked by little fanfare.

There was also little in the way of widespread popular interest, although newspapers from New Zealand to Europe made note of the upcoming opportunity for amateurs to observe the event, with proper safeguards. It was also unfortunate that the transit took place a few months into the opening phase of World War I, so that few paid much attention to an astronomical event that was visible only with optical aid.

One person who did witness the 1914 transit was the Italian artist Giacomo Balla (1871–1958), who saw the event from Rome through a smoked glass filter with his small refractor telescope.[4] He used the experience as the source for a series of paint-

---

[1] Ithaca College.
[2] NASA Catalog of Mercury Transits, available at http://eclipse.gsfc.nasa.gov/transit/catalog/MercuryCatalog.html [accessed 16 June 2015].
[3] *International Catalogue of Scientific Literature, (E) Astronomy* (London: Royal Society, 1918), Vol 14, pp. 133–34.
[4] Elica Balla, *Con Balla*, 3 vols (Milan: Multhipla, 1984), Vol 1, p. 346.

FIGURE 25.1. Giacomo Balla, *Mercury Passing Before the Sun (Mercurio transita davanti al sole)*, 1914. Tempera on paper lined with canvas. Gianni Mattioli Collection. Long-term loan to the Peggy Guggenheim Collection, Venice. © Artists Rights Society (ARS).

ings produced almost immediately after the event. Usually titled *Mercury Passing Before the Sun (Mercurio transita davanti al sole)*, these works are rare instances of early modernist art devoted to a specific astronomical event. They are also notable as works that reflect conflicting and divergent ideas about the direction of art in the Modern era, and highlight the role and influence of science upon art in the early twentieth century.

316

FIGURE 25.2. Giacomo Balla, *Mercurio che passa davanti al sole (Mercury Passing before the Sun)*, 1914, opaque watercolor over graphite on textured wove paper adhered to canvas. Philadelphia Museum of Art. Gift of Sylvia and Joseph Slifka, 2004. © Artists Rights Society (ARS).

Balla's *Mercury Passing Before the Sun* exists in several versions across a variety of media, including oil on canvas, tempera on cardboard, pastel, and pencil (FIGS. 25.1 and 25.2). Drawings served as preliminary studies for the paintings. All of the works appear to have been executed within a short span of time shortly after Balla made his observations of the Mercury transit, so they provide an interesting insight into the artist's working process at a moment when the Italian Futurist agenda was interrupted during the opening months of the Great War. What motivated the artist to devote such effort to this relatively minor astronomical event? And what are we to make of the images themselves? How are we to read them, in terms of expressive or symbolic content? Finally, how do these works fit into a picture of modern art's engagement with astronomy and science, and what can they tell us about the inspiration of astronomy at the beginning of the twentieth century?

By the time Balla painted the series *Mercury Passing Before the Sun* in 1914, he had been working as an artist for two decades. The son of an industrial chemist who experimented with nighttime photography, he shifted from the study of music to a job in printing before beginning to learn about art in the later 1880s. His subsequent encounters with Impressionism and Neo-Impressionism in the 1890s strongly influenced his interest in light, colour and atmosphere. While he took up some of the socialist/anarchist agenda of other turn of the century Italian and French Symbolist and Neo-impressionist artists, he was never strongly committed to a political art. Instead, he painted portraits, landscapes, and the occasional urban scene.

Balla's painting style shifted dramatically over these early years of his career, as he absorbed and reinterpreted the many competing modes of modern art that he encountered in Milan, Rome, Düsseldorf, and Paris. Drawn into the circle of Italian artists who sought a new direction for painting in the early years of the twentieth century, he was a co-signer of the 1910 'Manifesto of Futurist Painters' and the 'Technical Manifesto on Futurist Painting', and had become a mentor to some of the younger members of the group. He would embrace the essential conceptual elements of Futurism, with its attraction to Cubist-inspired abstraction, love of modern technology, and celebration of speed and energy. He would develop these fundamental ideas not only in painting, but also, into the later 1920s, through fashion, furniture design, film, theatre and costume, and photography. His 1915 document 'Futurist Reconstruction of the Universe', written with co-author Fortunato Depero, suggested a radical agenda for the role of art in modern society through its application in everyday life, construction and play.[5]

Balla's interest in the transit of Mercury as a subject was based in part on his interest in astronomy and optics, particularly photography. He had done a few previous works about cosmic or astronomical themes, most notably *Orion* of ca. 1910. His friendship with scientists such as Francesco Ghilarducci and photographers like Marius de Paolis (pseudonym for Mario Bellusi) connected Balla with individuals who facilitated his direct contact with science and optics.[6] His personal observation

[5] Umbro Apollonio, ed. and trans., *Futurist Manifestos* (New York: Viking, 1970), p. 197.
[6] Balla, *Con Balla*, Vol. 1, p. 360.

of the 1914 event, however, tells us that his interest in astronomy was significant.

Balla's art had evolved throughout the early 1910s toward a distinctive style of abstract and non-objective forms inspired by technology and industry. His works were not so much depictions of specific objects as the expression of forces. The 'force lines' that characterized his Futurist works, and which were his distinctive contribution to Futurist style, were inspired in part by his study of chronophotography in the movement works of Jules-Etienne Marey and the industrial studies of Frederick Winslow Taylor. Balla was exposed to popular scientific literature in textbooks and journal articles, and he adapted the distinctive abstractions of scientific illustrations to his own purposes as a way of picturing the invisible forces of gravity and energy. Many Futurists had been interested in the writings of Gustave Le Bon, whose book *The Evolution of Matter* went through many editions after 1905 and was translated into several languages. In this work, Le Bon argued that energy was the slow disintegration of matter, that matter slowly transforms into energy over time, and that the material universe is a temporary state in a larger order.[7] This appealed strongly to artists like the Futurists, who had begun to think about light as both energy and substance. One of Balla's most recognizable paintings, *Street Light* (1910–11, dated 1909, Museum of Modern Art, New York), directly visualizes this idea by giving form to the energy of the light rays as tiny parabolic arcs streaming from the bulb. The evolution of these ideas about the visualization of energy, and the symbolic shorthand for representing 'force', would guide Balla's style for much of the rest of his career.

The Mercury transit works are situated near the conclusion of a brief but intense period of development in Balla's career, at the end of his initial Futurist exploration of movement and dynamic forces, but just before he began to shift his attention beyond painting to other areas of visual creativity, including clothing, furniture, sculpture and utilitarian objects. As such, the transit paintings are something of a coda to one chapter of the artist's career, a fact that may help us interpret them more fully.

Initial drawings show us the visual development of Balla's transit paintings. From the most preliminary sketches, three elements are prominent: the disk of the sun and additional circular shapes, several intersecting straight lines, and numerous arcing or curved lines (FIG. 25.3). The tiny black dot of the planet Mercury is visible above centre, represented much as it would have appeared through a telescope against the sun during the transit. The most immediate analogy is to astronomical diagrams and illustrations, where these same elements of circles, straight and curved lines can be found in the depiction of orbits and the movement of bodies through the solar system. Access to both historical and contemporary scientific texts with such illustrations was

---

[7] Gustave Le Bon, *L'Évolution de la Matière* (Paris: Ernest Flammarion, 1905), pp. 1–2.

FIGURE 25.3. Giacomo Balla, *Mercury Passing in Front of the Sun*, 1914, pencil and gouache on paper. Museum of Modern Art, New York. Bequests of Helen Acheson, and Eve Clendenin, gift of A.E. Gallatin (by exchange) and gift of Constance B. Cartwright. © Artists Rights Society (ARS).

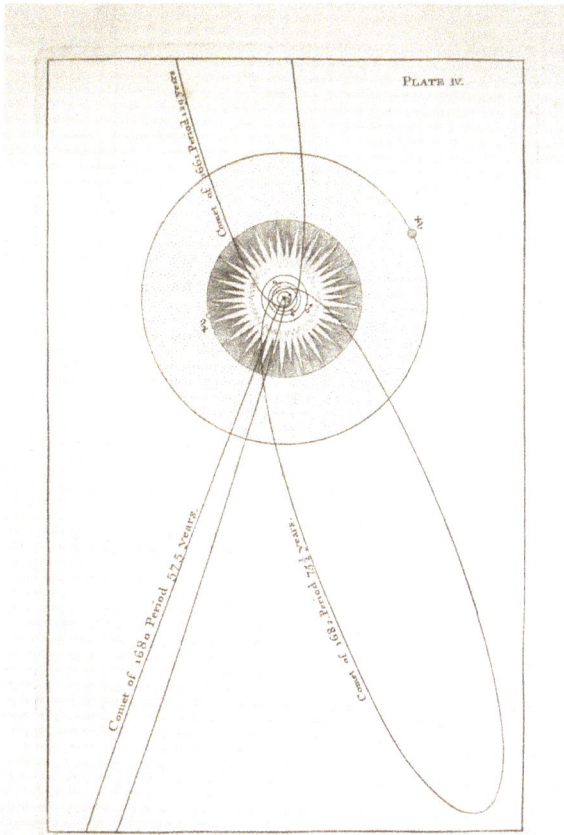

FIGURE 25.4. Thomas Wright, '*The orbits of comets*', *from An original theory or new hypothesis of the universe*, 1750.

fairly easy, and given his interests in science and astronomy, it is likely that Balla was familiar with the tropes of science illustration from a variety of sources (FIG. 25.4).

These visual similarities give us a first clue to reading the Mercury transit images. As with his representations of earthly objects like cars, running figures, or street lamps, Balla was less interested in the object itself than in the forces acting upon the object, or those generated from the movement of that object. These lines of force, which Balla had depicted independently, associated matter and energy in a complex system containing visible and invisible elements. Just as the lines depicting an orbiting comet or planet in an astronomical diagram are not meant to represent a physical thing, but rather to denote a path or trajectory through space, Balla also

321

FIGURE 25.5. William Parsons, 3rd Earl of Rosse, Drawing of *Messier 51*, April 1845.

used the abstraction of movement to suggest the presence of gravity acting upon a body in motion.

The drawings also indicate another feature that translates, in colour, to the paintings – gradations of value or tone that give some of the planes of the composition a sense of depth or three-dimensionality. These careful, even subtle tonal passages are again unexpected if we make the assumption that the drawings and paintings are direct depictions of the event as observed through Balla's telescope. But they do suggest a connection, however tentative, to something about the nature of observing the sky at the margins of visibility in the late nineteenth and early twentieth century.

The debate about the nature of nebulae, which had accelerated through the nineteenth century as visual telescopic observations gave way to photographic recordings through larger and larger instruments, had made its way into the popular imagination through reproductions of subtle and complex drawings by many astronomers, particularly after mid-century. (FIG. 25.5) The filaments and lines rendered by observers hinted at an unexpected dynamism in the cosmos, perhaps evidence of the supposed ether that pervaded otherwise empty space.[8] It was this suggestion of hidden dynamic forces that inspired Vincent van Gogh's 1889 *Starry Night* (Museum of Modern Art, New York), and which might have appealed to the sensibilities of Giacomo Balla, who already had sought to depict the manifestations of universal motion in earthly subjects.

---

[8] Albert Boime, 'Van Gogh's *Starry Night*: A History of Matter and a Matter of History', *Arts Magazine* 59, no. 4 (1984): pp. 92–94.

FIGURE 25.6. Giacomo Balla, *Mercury Passing Before the Sun*, 1914. Tempera and oil on textured paper, mounted on canvas. Yale University Art Gallery, New Haven. Katharine Ordway Fund. Photo credit: Yale University Art Gallery. © Artists Rights Society (ARS).

The carryover of this debate could be found, at the end of the nineteenth century, in the various depictions of markings on the surfaces of the rocky inner planets of the solar system. Mars was the most often depicted world crossed by lines and arcs, supplemented by subtle shading to suggest possible vegetation. But the other inner planets, Venus and Mercury, also had their share of observers claiming to discern marks that characterized their surfaces. These maps also found their way to the public as part of the larger debate about the nature of other worlds, and the possibility of extraterrestrial life.

While these planetary drawings are at best a tentative source for Balla's representation of the transit of Mercury, we must acknowledge the fact that the pervasive appearance of this imagery in the scientific literature, professional and popular, and to which artists of the early twentieth century had access, was a readily available resource with striking parallels to the abstractions of modernist art. While simple visual comparisons can only go so far in establishing a connection, this connection bears more scrutiny, especially among the works of artists known to have a strong interest in science like Balla.

Before we leave this brief discussion of the formal properties of Balla's transit works, we should point out one more element. The arc of green that sweeps up from the bottom of many of the paintings seems, in some ways, out of place in these compositions (FIG. 25.6). The prevailing oranges, reds, blues and black of the works might be explained by the actual appearance of the sun through the smoked glass filter of Balla's telescope. But what is the green arc? It has been suggested that this flare of complementary colour (complementary to the orange, if one uses standard colour theory) is an afterimage or temporary glare that overloaded the eye of the observer.[9] The familiar optical effect of staring at an object of a particular colour, and then quickly shifting one's gaze to a white field in order to see the negative afterimage, was a well-known phenomenon in the nineteenth century, and would have been part of Balla's understanding of Impressionist and Neo-Impressionist colour theory. If the green arc is indeed a careful notation of a transient optical effect experienced by the artist at the eyepiece of his telescope, then we should think carefully about the boundaries of representation and abstraction in this body of paintings.

Transit images are relatively rare in the history of art. In the nineteenth century, the transit of Venus was depicted both as allegorical mythology: (E. Penning-Duprin, *Transit of Venus*, Council Room, Paris Observatory, 1886) and as scientific event (Ford Madox Brown, *Crabtree Observing the Transit of Venus in 1639*, Manchester City Hall, 1881–1888). Since they are phenomena of the telescope, they do not lend themselves readily to visual representation except in special circumstances. Even through the telescope, the lack of compelling visual interest makes Mercury transits poor subjects. The overwhelming size and brightness of the sun reduces the planet

[9] Peggy Guggenheim Collection, image annotations, available at http://www.guggenheim-venice.it/inglese/collections/artisti/dettagli/opere_dett.php?id_art=1&id_opera=58, [accessed 16 June 2015].

itself to a tiny speck (as Balla accurately depicts). Without exaggerating the scale of the transiting planet, or using some other means of enlivening the scene, it is not surprising that artists have not found themselves drawn to the actual spectacle of planetary transit as a subject. More often, we see transits in the form of allegory, with mythological characters playing the role of the sun and planets, which is certainly more interesting if one is not interested in interpreting the actual visual experience into tangible form. Particularly in the eighteenth century, such personifications of heavenly bodies found their way into atlases, frontispieces and even decorative ceiling paintings.

When transits are represented, at least from the nineteenth century onwards, the image is usually focused upon the observer rather than the phenomenon itself. The act of viewing is the interest, instead of the actual sight of the transit. Such images are about the human dimension of science and discovery. Sometimes, as with Ford Madox Brown's depiction of William Crabtree's observation of the 1639 Venus transit (1879–1893, Manchester City Hall, England), a certain national or civic pride guides the choice of subject. These belong properly in the genre of history painting, rather than as illustrations of science. Their values are connected to notions of progress and accomplishment, foundations and origins. They are rarely interesting as visual works on their own merits, but serve as ideological pieces in support of a particular view of civilization and material or intellectual achievement. It is not surprising then that the Horrocks window at St. Michael's Church, Much Hoole, Lancashire (1872), commemorating and representing Jeremiah Horrocks' 1639 Venus transit observation, reads 'oh, most grateful spectacle, the realization of so many ardent desires'.[10] Concepts of celestial movement, reason and science, knowledge and mathematics, are overshadowed by the personal fulfilment of the individual.

To further complicate the picture of the environment in which Balla's transit paintings were produced, we should also observe that traditional allegorical and figurative art representing scientific concepts and phenomena persisted into the early twentieth century, particularly in pubic and semi-private contexts. These works drew inspiration from past forms, revived or made new by historicist or symbolic views of art. Astronomical imagery, in the form of classical gods, mythological characters, or personifications, could still be found in painting and sculpture, as well as commercial art, architectural decoration and popular imagery. It might be easy to dismiss these as carryovers and manifestations of conservative taste if not for the fact that they still expressed particular values and meanings that the public, and artists

---

[10] G. Napier Clark, 'Sketch of the Life and Works of Rev. Jeremiah Horrox', *The Journal of the Royal Astronomical Society of Canada* 10 (1916), p. 530.

as well, collectively understood as 'high art'. Even the most ardent modernist could not simply dismiss these images as irrelevant. The Futurists were notorious for their pronouncements against traditional art, museums and the forms and subjects of the past. [11] Nevertheless, a somewhat subversive view would be that these traditionalist works still exerted an influence upon the thinking of modern artists even as they rejected them.

The mythological Mercury, after all, was associated directly with communication, commerce and ingenuity, among many other fields.[12] He is sometimes depicted in Renaissance and Baroque art as presiding over the arts, or of facilitating the eloquence of artists. Examples such as Andrea Casali's *Mercury and Minerva Presiding Over the Arts,* early 1760s (National Trust, Dryham Park) or Dosso Dossi's striking *Jupiter, Mercury and Virtue,* 1524 (Wawel Castle, Kraków) place the mythological Mercury at the centre of the creative process as facilitator and even guardian of inspiration.

He is also associated with the sciences, especially astronomy and alchemy. As an intermediary or mediator, Mercury bridged realms and ideas, and as the swift messenger of the gods he signified speed and facility.

The special significance that Giacomo Balla gave to the 1914 transit of Mercury might not be explicable in terms of mythology and historical artwork. But it is tempting to think that the traditional associations of Mercury with creativity, movement, energy, and ingenuity, and those of the Sun with enlightenment, life, art and intellect, might not have had some appeal, however unconscious. In his essay on 'The Futurist Reconstruction of the Universe', written a few months after the transit paintings, Balla would proclaim the qualities of the Futurist object: abstract, dynamic, luminous, transformable, dramatic, volatile. Qualities, also, of the mercurial temperament.

Since the planet Mercury is the minor visual player in the spectacle of the transit, the dot against the brilliant sun, the sun is, literally and figuratively, the star. Balla's representation is of the modernist sun, and it represents a set of ideas and observations about nature and humanity infused by the enthusiasms of the early twentieth century, tinged by traditional views of sunlight. Long before Balla, the sun

---

[11] F.T. Marinetti, 'The Foundation and Manifesto of Futurism', *Critical Writings,* ed. Günter Thompson, trans. Doug Thompson (New York: Farrar, Straus and Giroux, 2006), pp. 11–16.

[12] Ovid, *Fasti,* ed. and trans. Anthony J. Boyle and Roger D. Woodard (London and New York: Penguin, 2000), p. 343. A very good summary of the ancient sources on the attributes and character of Hermes/Mercury can be found at 'Hermes', *Theoi Greek Mythology,* available at http://www.theoi.com/Olympios/Hermes.html [accessed 28 October 2016].

was central to the observational shift in nineteenth-century art that moved from allegorical and traditional representations of nature to the artist's experience of the visible world. The context of the sun, in sunsets and sunrise landscape, was framed by sentiment, mood and symbolic allusion. As the origin of colour, the sun's light came to also be associated with the expressive range of the artist's palette. Turner's often quoted exclamation that the 'sun is God' captures the broad character of the early nineteenth-century Romantic artistic experience.[13]

But this view was upset when Impressionism, erasing some of the Romantic residue surrounding the image of the sun, chose the effects of sunlight and the phenomena of light in nature as a means toward an end rather than the end in itself. Even as symbolic iconography clings to the edges of certain Impressionist works, like Monet's famous *Impression: Sunrise* (1872, Musée Marmottan, Paris), the representation and meaning of the sun in modern art shifted dramatically in the closing decades of the nineteenth century. The attempt to reassert some of the often conflicting ideas about moral value and truth associated with the traditional depiction of the sun resulted in the conflicted image of the sun as cipher rather than as object in the early decades of the twentieth century. The leap to abstraction and non-objectivity in the early 1900s thus seems all the more abrupt and shocking. Artists like Robert Delaunay and František Kupka sought to reframe the traditional image of the sun through new terms and new contexts. The prismatic, or, as Guillaume Apollinaire called it with mythic irony, 'Orphic', colorize of Delaunay's *Sun Disks* or Kupka's *Disks of Newton* takes Impressionism's underlying basis for observation and turns it into the subject of painting (FIG. 25.7).

Instead of colour as the product of sunlight and the means to depict the world, colour is light and in itself the subject of art. The sun is subsumed by its own radiance and the broken constituent parts of its spectrum denote the dematerializing of the star into the energy that is its light. As Delaunay said in his 'Notes on the Construction of the Reality of Pure Painting' of 1912 (as quoted by Guillaume Apollinaire): 'We are no longer dealing here either with effects… or with objects… or with images…'.[14] This idea of the transformation of matter to energy was also important to Balla, working at exactly the same moment, and provided a conceptual basis for his Futurist experiments in painting motion and speed.

Colour was tied closely to the exploration of new spiritual ideas, as articulated

---

[13] Attributed to Turner by John Ruskin, *The Works of John Ruskin*, ed. Sir E. T. Cook and A. Wedderburn (London: George Allen, 1903–1912), Vol. xxviii, p. 147.

[14] Leroy C. Breunig, ed., *Apollinaire on Art: Essays and Reviews, 1902-1918*, trans. Susan Suleiman (New York: Viking, 1972), p. 264.

FIGURE 25.7. František Kupka, *Disks of Newton (Study for 'Fugue in Two Colors')*, 1912, oil on canvas. Philadelphia Museum of Art. The Louise and Walter Arensberg Collection, 1950. © Artists Rights Society (ARS).

by artists like Wassily Kandinsky (whose work appeared in Rome in early 1914, and whose writings were appearing in Italian journals at the same time), and implemented by others like Paul Klee.[15] Science could be seen as opening new paths towards understanding not only the material and immaterial worlds, but also the nature of energy and vital forces, the analogies of spirit and the invisible world of the atomic universe. Basic qualities like balance and movement, or states like chaos and order, come to revolve around the knowable and familiar symbols like horizon line, right angle, and solar disk. Kandinsky himself used the analogy of sunspots and the clear disk of the sun in a passage in 'Concerning the Spiritual in Art' to highlight changing attitudes toward knowledge gained though scientific inquiry as opposed to intuition and irrational belief.[16]

The modernist sun, as Delaunay hinted, was neither object nor symbol, but something else. Artists working both immediately before and shortly after the First World War seem to have wanted to redefine the meaning of the sun, even as they stripped away the cultural and historical associations that defined its place in Western art. This struggle to define and redefine led to both Balla's conception of the sun as the focus and embodiment of visible and invisible forces, and of other artists' representations of the sun as enigmatic cipher or the dispassionate relic of a lost past. Balla's sun in his transit paintings is still an astronomical sun, even if the modernist sun of his colleagues is not. Marianne Martin has written thus, 'Balla was asserting the inextinguishable strength and purity of the sun, unaffected by the antics of lesser cosmic bodies, and thus he created a totally unpretentious but profoundly Futurist act of faith'.[17]

Early twentieth century artists tended to cast the sun into broad symbolic terms that were neither part of the Romantic sublime nor anchored in the individual perception of light and colour. This broader vision was not motivated by the same concern for mechanical forces and dynamic energy that prompted Balla to gaze at the tiny dot of a planet crossing the face of the sun. These other artists sought a distilled and essential vision from nature. They tended toward a more primal contrast of dark and light, of colour as raw material for probing the deep recesses of psychological space. The starkness of Arthur G. Dove's *Sunrise III*, for example (1936–37, Yale University Art Gallery, New Haven), is not only in its visual simplicity (contrast this with

---

[15] Giovanni Lista, Paolo Baldacci, and Livia Velani, *Balla, La modernità futurista* (Milano: Skira, 2008), p. 67, n. 95.

[16] Wassily Kandinsky, *Concerning the Spiritual in Art*, trans. M. T. H. Sadler (New York: Dover, 1977) p. 12.

[17] Marianne W. Martin, *Futurist Art and Theory, 1909–1915* (Oxford: Clarendon Press, 1968), p. 200.

Balla's riotous complexity) but also in the brooding colour that chills rather than warms, and which reminds us of the darkness of space and the abyss.

And isn't this what certain science images promise as well? Sir Arthur Eddington's photo of the 1919 solar eclipse, from the expedition that sought to test Einstein's theory of the gravitational curvature of space predicted in his General Theory of Relativity, is also stark and forbidding, although that was not the intent (FIG. 25.8).

The fortuitous flaw that creates an arching line from the bottom of the frame reminds us that the sources for Balla's expressive abstractions need not lie far from the documents of observation. But it the contrast of light and dark that are most striking, and which connect this image more closely to the esoteric tradition emerging in modernism than to Balla's animated dynamism.

How else, then, to talk about the modernist sun, except in extremes? Kazimir Malevich would invoke these fundamental and mysterious dualities in his Suprematist art, first exhibited in 1915 shortly after Balla's transit works.[18] But Malevich had hinted at this radical vision in his 1913 set designs for *Victory Over the Sun*, with text by Alexei Kruchenykh and music by Mikhail Matiushin, an opera inspired directly by the Italian Futurists but pushing further toward the boundaries of the visible and the rational. Later manifestations of *Victory Over the Sun*, such as the 1923 revival depicted by El Lissitzky, draw back from the edge of the artistic abyss, and return us closer to the original Futurist vision of a universe of forces and machines. Whether El Lissitzky had seen Balla's transit paintings by this point is unknown, but the similarity in terms of formal elements is striking.

Once again, we could suggest that Balla was responding to astronomical diagrams, charts and illustrations from the past that distilled and refined, in precise and geometric terms, the essential characteristics of the objects of the sky. These are formal similarities, but they are also conceptual partners to Balla's wonder and faith in the workings of the universe. They promise exact knowledge and clarity. The goal of scientific visualization is to make visible that which is invisible, much as Balla does in his transit paintings. Le Bon, for example, attempted to convey some sense of these invisible forces through photographs of electrical fields, magnetic fields, and so forth. James Nasmyth's observation of so-called 'willow leaf' structures on the sun's surface had hinted at hidden structures and unseen organizing behind the glare of sunlight.[19] Helena Blavatsky, founder of the Theosophical Society, would cite John

[18] Linda S. Boersma, *0,10: The Last Futurist Exhibition of Painting* (Rotterdam: 010 Publishing, 1994), p. 83, n. 61.
[19] James Nasmyth, *James Nasmyth, Engineer, An Autobiograpy,* ed. Samuel Smiles (London: John Murray, 1883), pp. 282–287.

FIGURE 25.8. Arthur Eddington, Photograph of the 1919 Solar Eclipse.

Herschel's speculations about these structures, since Herschel alluded to the possibility that we are witnessing the origin of solar light, floating atop a (possibly opaque and dark) inner body.[20] The striking similarity of Nasmyth's 'willow-leaf' illustration to Balla's *Iridescent Interpenetration* paintings, done just before the Mercury transit series, is very suggestive of a direct or indirect visual link. The Futurists, and many other artists in the early twentieth century, were fascinated with Theosophy, so Balla's contact with and use of these ideas would have been possible.[21]

Balla's transit paintings stand apart from the works of his contemporaries in their specificity, their observational foundation, and their apparent referencing of scientific and astronomical visual culture. Where Balla's paintings do connect to early twentieth–century modernist art, primarily in style and the conventions of Futurist abstraction, they still occupy a unique place as a series and in the richness of their associations. Perhaps only Delaunay's *Sun Disk* series comes close to the ideas that Bal-

---

[20] H.P. Blavatsky, *The Secret Doctrine*, 3rd ed., 2 vols (London: Theosophical Publishing Society, 1893),Vol. 1, pp. 577–79. John Herschel, *Outlines of Astronomy*, 10th ed. (London: Longmans, Green and Co., 1893), pp. 695–96.

[21] Fabio Benzi, *Giacomo Balla: genio futurista* (Milano: Electa, 2007), pp. 132–40. Also Fabio Benzi, 'Giacomo Balla: The Conquest of Speed', in *Italian Futurism, 1909-1944: Reconstructing the Universe*, ed. Vivian Greene (New York: Guggenheim Museum, 2014), pp. 103–6.

la captures. Traces of Balla's style and themes show up in late Italian Futurist painting in the 1920s, in the works of his associate Fortunato Depero and Gerardo Dottori (*Dinamismo di mondi*, 1932, private collection, Rome). A few Russian Futurists also reflected the influence of Balla's ideas, even picking up some of his cosmic themes, as in the work of Ivan Kudriashev (*Trajectory of the Earth's Orbit Around the Sun*, 1926, State Tret'iakov Gallery, Moscow).

Balla described his vision of nature in 'The Futurist Reconstruction of the Universe' in 1915: 'We will give skeleton and flesh to the invisible, the impalpable, the imponderable and the imperceptible. We will find abstract equivalents for every form and element in the universe, and then we will combine them according to the caprice of our inspiration, creating plastic complexes which we will set in motion'.[22] The parallels and similarities to earlier science illustrations, noted earlier, demonstrate a shared, common vocabulary of forms and symbols that Balla acknowledges, if not imitates. As such, Balla's paintings are as much about the visual culture of science as they are about the rise of abstraction in modernist art. But these aspects of appropriating scientific visual culture, whether used as direct sources or not, also suggest a struggle to come to terms with the visualizing ideas about near infinite space, cosmic forces, unseen energies, and the relative smallness of the modern human in the face of an overwhelming universe. Robert Delaunay captured this point of view in a line from his 1912 article 'Light': 'We see as far as the stars'.[23] Dynamic astronomical forms and titles persisted in Balla's art into the 1920s, although he does not depict anything so specific as the transit of Mercury in his later work. By the end of the 1920s, Balla's style shifted radically toward conservative realism, and he abandoned the modernist mode completely for both personal and political reasons.

Contemporary astronomical visualizations of the sky, the planets, and the sun are data-driven rather than photographic, and their arbitrary colors and formal structuring often resemble the abstractions of early modernism more than their black and white scientific predecessors.[24] Seen in this light, Balla's paintings are perhaps closer to narrow band images of the sun than we might at first imagine, with energetic arcing lines and a graphic representation of otherwise invisible forces at work.

---

[22] Apollonio, *Futurist Manifestos*, p. 197.

[23] Herschel B. Chipp, ed., *Theories of Modern Art: A Source Book by Artists and Critics* (Berkeley, CA: University of California Press, 1968), p. 319.

[24] For example, see Nigel Sharp, *High Resolution Solar Spectrum*, 1984 (NOAO/NSO/Kitt Peak FTS/AURA/NSF), available at https://www.noao.edu/image_gallery/html/im0600.html [accessed 11 November 2016]. On data-intensive imaging in astronomy, see Robert J. Brunner et al., 'Massive Datasets in Astronomy', *Handbook of Massive Data Sets*, Vol 4., ed. James Abello et al. (Dordrecht: Springer, 2002), pp. 931–79.

Of course, Balla could only intuit the dynamism of the sun's surface in order to make visible what was invisible to him. Highly detailed spectroscopic imagery, on the other hand, spreads out the sun's visible light and hints at the expressiveness of light as colour. But even here, the scientists who construct marvellous visualizations from raw data are driven by aesthetic considerations that derive, consciously or not, from modernist ideas.

False colour views of Mercury's surface, captured by remote spacecraft in orbit around the planet, speak to an expressive, as well as scientific, desire (FIG. 25.9). And given the mythological Mercury's association with the arts, it is fitting that artists should find their place in the solar system on its surface. The International Astronomical Union has established that the craters on the surface of Mercury be designated after prominent creative persons from history, including visual artists.[25] Among the 200 or so named features, modern artists such as Picasso, Matisse, Derain, and Kandinsky are noted. It would seem appropriate to add Giacomo Balla's name to a Mercurial feature, a century after he observed the tiny planet crossing the face of the sun.

FIGURE 25.9. Enhanced Color Mercury Map, based on MESSENGER data, 2013. NASA/ Johns Hopkins University Applied Physics Laboratory/Carnegie Institution of Washington.

---

[25] See the IAU, 'Categories for Naming Features on Planets and Satellites', available at http://planetarynames.wr.usgs.gov/Page/Categories [accessed 16 June 2015].

ABOUT THE CONTRIBUTORS

JOHN BARSA (1963–2018) was a Meriam man from Mer (Murray Island) in the eastern Torres Strait. He possessed a wealth of knowledge about traditional Meriam astronomy and was a well-known artist and craftsman, particularly with regard to traditional islander masks. His works feature in museums across Australia and the world.

DAN BROWN is a professional astronomer who graduated from Ruhr-Universitaet Bochum, Germany, and completed his PhD at Liverpool John Moores University, UK. He works at Nottingham Trent University and its on-site observatory, where he supports astronomy teaching and outreach work with schools and the general public. This includes working with creative practitioners and theatre groups. The focus of his outreach work is based on archaeoastronomy and the use of the outdoor classroom. He is also a founding member of the 'Horizontastronomie im Ruhrgebiet e.V.', a German private initiative promoting astronomy outreach based on an EU funded Science Park located within the Ruhr area.

ALASTAIR G. BRUCE recently completed his PhD in astronomy at the Institute for Astronomy, University of Edinburgh. He is currently a postdoctoral researcher and project officer for the James Webb Space Telescope public engagement campaign in the UK. His research involves observations of active galactic nuclei transient events.

NICHOLAS CAMPION is a historian and anthropologist, specialising in the cultural contexts and consequences of cosmology, astronomy and astrology. He is Associate Professor in Cosmology and Culture at the University of Wales Trinity Saint David, where he is also programme director for the MAs in Cultural Astronomy and Astrology, and Ecology and Spirituality His books include the two-volume *History of Western Astrology* (London: Continuum 2008/9) and *Astrology and Cosmology in the World's Religions* (New York: New York University Press, 2012) and he is the editor of *Culture and Cosmos*, the journal on the history of cultural astronomy and astrology. He submitted various articles to the *Biographical Dictionary of Astronomers*. His recent papers include 'Cosmos and Cosmology', in Robert Segal and Kocku von Stukrad (eds), *Vocabulary for the Study of Religion* (Leiden: Brill, 2015) and 'The Importance of Cosmology in Culture: Contexts and Consequences', in Jesse Abraao Capistrano de Souza (ed.), *Cosmology* (Rijeka: InTech Open, 2017). His current projects include the six volume *Cultural History of the Universe* (Bloomsbury, forthcoming), for which he is General Editor (with Richard Dunn).

HOWARD CARLTON has recently moved into the rather different world of academia, having spent a number of years as an IT Consultant. In 2013 he was awarded an MA in the History of Christianity by the University of Birmingham and he is currently studying for a PhD in Modern History at the same institution. His thesis will explore a number of nineteenth-century astronomical controversies in order to examine the relationship between the respective epistemological statuses of religion, metaphysics and science. It will also consider the development of the discipline of astronomy during the nineteenth century and reflect on the extent to which it influenced, and was influenced by, the changing cultural, educational, social and political contexts of Victorian society.

LIANA DE GIROLAMI CHENEY is presently a Visiting Scholar in Art History at the Università di Aldo Moro in Bari, Italy, and Investigadora de Historia de Arte, SIELAE, Universidad de Coruña, Spain. Dr. Cheney received her BS/BA in Psychology and Philosophy from the University of Miami, Florida, her MA in History of Art and Aesthetics from the University of Miami, Florida, and her Ph.D. in Italian Renaissance and Baroque from Boston University, MA. Dr. Cheney is coauthor of numerous articles and books, including: *Neoplatonism and the Arts; Neoplatonic Aesthetics in Literature, Music and the Visual Arts; The Homes of Giorgio Vasari* (English and Italian); *Giorgio Vasari's Teachers: Sacred and Profane Love; Giorgio Vasari's Prefaces: Art and Theory; Giorgio Vasari's Artistic and Emblematic Manifestations, Agnolo Bronzino: The Florentine Muse;* and *Edward Burne-Jones' Mythical Paintings.*

CAROLYN CRAWFORD is one of Britain's foremost science communicators. She is Gresham Professor of Astronomy and Outreach Officer at the Institute of Astronomy and a Fellow of Emmanuel College, Cambridge. After receiving her PhD from Newnham College, Cambridge, Professor Crawford went on to a series of fellowships from Balliol College, Oxford, Trinity Hall, Cambridge and the Royal Society. In 2004 she was appointed as a Fellow and College Lecturer at Emmanuel College, Cambridge, where she is now also the undergraduate Admissions Tutor for the Physical Sciences. Since 2005 she has combined her college role with that of Outreach Officer at the Institute of Astronomy at the University of Cambridge.

Professor Crawford's primary research interests are in combining X-ray, optical and near-infrared observations to study the physical processes occurring around massive galaxies at the core of clusters of galaxies. In particular, she observes the complex interplay between the hot intra-cluster medium, filaments of warm ionized gas, cold molecular clouds, star formation and the radio plasma flowing out from the central supermassive black hole. In 2009 Professor Crawford's outstanding abilities at science communication were recognized by a Women of Outstanding Achievement Award by the UK Resource Centre for Women in Science, Engineering and Technology, presented for 'communication of science with a contribution to society'.

CLIVE DAVENHALL has a long-standing interest in the history of astronomy. Since 2004 he has been the Editor of the Society for the History of Astronomy's Bulletin (previously Newsletter) and is currently a member of that Society's Council. He contributed to a previous INSAP meeting (VII) and has written entries for the *Biographical Encyclopaedia of Astronomers*. He is a project manager and software developer in the Wide Field Astronomy Unit, Institute for Astronomy, University of Edinburgh, based at the Royal Observatory Edinburgh. He has degrees from the universities of London and St Andrews.

ANDREW CHRISTOPHER FABIAN, OBE, FRS was educated at King's College London (BSc, Physics) and University College London (PhD). He is a Royal Society Research Professor at the Institute of Astronomy, Cambridge, and Vice-Master of Darwin College, Cambridge. He was the President of the Royal Astronomical Society from May 2008 through to 2010. He was Professor of Astronomy at Gresham College 1982-84, a position in which he delivered free public lectures within the City of London. He was also editor-in-chief of the astronomy journal *Monthly Notices of the Royal Astronomical Society*. His current areas of research include galaxy clusters, active galactic nuclei, strong gravity, black holes and the X-ray background. He has also worked on X-ray binaries, neutron stars and supernova remnants in the past. Much of his research involves X-ray astronomy and high energy astrophysics. His notable achievements include his involvement in the discovery of broad iron lines emitted from active galactic nuclei, for which he was jointly awarded the Bruno Rossi Prize. He is author of over 800 refereed articles and head of the X-ray astronomy group at the Institute of Astronomy. Fabian was awarded the Dannie Heineman Prize for Astrophysics by the American Astronomical Society in 2008 and the Gold Medal of the Royal Astronomical Society in 2012.

JOSÉ FUNES is a Jesuit priest, Director of the Vatican Observatory, and member of the Pontifical Academy of Sciences. His field of research includes kinematics and dynamics of disk galaxies and star formation in the local universe. He has published about 80 papers in referee journals and conference proceedings. He is co-editor of three volumes of the proceedings of two conferences on the formation and evolution of disk galaxies and a study-week on astrobiology.

MICHAEL GEFFERT is astronomer at the Argelander-Institut of Bonn University. He has published over years in astronomical journals like Astronomy and astrophysics and is now responsible for the public outreach activities of the institute. In 2009 Geffert was the coordinator of the German activities of the IYA 2009. He has been working also as an artist (woodcut, painting, installation) with exhibitions in the region of Bonn and is a member of the German organization of artists (BBK).

DUANE HAMACHER is a Lecturer and ARC Discovery Early Career Research Fellow in the Nura Gili Indigenous Programs Unit at the University of New South Wales in Sydney, Australia. His teaching and research focus on cultural astronomy. His Australian Research Council grant involves studying the astronomical knowledge of Torres Strait Islanders. He leads the Indigenous Astronomy Group at Nura Gili and works as a consultant curator at Sydney Observatory. He earned graduate degrees in astrophysics and Indigenous studies, with a doctoral thesis on Australian Aboriginal astronomy.

JOHN G. HATCH is Associate Professor of Art History at Western University (London, Canada) where he also serves as an Associate Dean for the Faculty of Arts and Humanities. He received his PhD from the University of Essex and his area of research is 20th-century European and American art and theory, with a special focus on the influence of the physical sciences on modern art. His recent publications include the articles *Wrestling Proteus: The Role of Science in Modern Art and Architecture's New Images of Nature* (2014*), Seeing and Seen: Acts of the Voyeur in the Works of Francis Bacon* (2012), *Nature, Entropy, and Robert Smithson's Utopian Vision of a Culture of Decay* (2012), and *Some Adaptations of Relativity in the 1920s and the birth of Abstract Architecture* (2010).

CHRIS IMPEY is a University Distinguished Professor and Deputy Head of the Department of Astronomy at the University of Arizona. He has published over 170 refereed publications on observational cosmology, galaxies, and quasars, and his research has been supported by $20 million in grants from NASA and the NSF. He has won eleven teaching awards, and he is currently teaching an online class with over 13,000 people enrolled. Impey is a past Vice President of the American Astronomical Society and he has been an NSF Distinguished Teaching Scholar and the Carnegie Council's Arizona Professor of the Year. He has written over 40 popular articles on cosmology and astrobiology, two introductory textbooks, a novel, and five popular science books: *The Living Cosmos* (2007, Random House), *How It Ends* (2010, Norton), *How It Began* (2012, Norton), *Dreams of Other Worlds* (2013, Princeton), and *Humble Before the Void* (2014, Templeton).

KEVIN J. KILBURN is Vice President, Manchester Astronomical Society, and is currently Acting Chairman of the Society for the History of Astronomy. His discovery of the Manchester Astronomical Society's copy of the Bevis atlas has led to the decades-long investigation of all the Bevis atlases and unpublished versions and plates around the world.

MAREK KUKULA obtained his PhD in Radio Astronomy from Jodrell Bank Observatory then carried out research into galaxy evolution at a number of institutions including the Space Telescope Science Institute, home of the Hubble Space Telescope, and the University of Edinburgh. As the Public Astronomer at the Royal Observatory Greenwich he works alongside a team of astronomy education and outreach specialists to engage with the general public, schools and the media. In the Observatory's exhibitions and public programmes he works extensively with scientists, historians and artists to explore the cultural as well as the scientific impact of astronomical research.

GEORGE LATURA is the author of 'Plato's X & Hekate's Crossroads: Astronomical Links to the Mysteries of Eleusis' (SEAC XXI Proceedings), 'Eternal Rome: Guardian of the Heavenly Gates' (INSAP VIII Proceedings), and 'Plato's Cosmic X: Heavenly Gates at the Celestial Crossroads' (SEAC XX Proceedings).

ANNETTE S. LEE is an astrophysicist and artist who leads the Native Skywatchers research project, which seeks out and celebrates indigenous peoples' connection to the stars. She works closely with Ojibwe and D(L)akota elders and community members to document Native star knowledge. Currently Lee is an Associate Professor of Astronomy & Physics at St. Cloud State University (SCSU) in St. Cloud, Minnesota and Director of the SCSU Planetarium. Her credentials include: B.S. Applied Mathematics 1992 University of California-Berkeley; B.A. Painting 1998 University Illinois-Urbana; M.F.A. Painting 2000 Yale University; M.S. Astrophysics 2008 Washington University-St. Louis. Her academic specialties include extrasolar planets, Native American star knowledge, science/math education.

MICHAEL MENDILLO is Professor of Astronomy at Boston University with research programs in Space Physics, Planetary Astronomy and the History of Astronomy.

DR. PAOLO MOLARO was born in Artegna (Ud) in 1955 and is married and the father of four children. He is Professor astronomer at the Astronomical Observatory of Trieste INAF, of which he was director. He graduated from the University of Trieste with Margherita Hack and did his doctorate at SISSA with Dennis Sciama. He is the author of over 300 publications with interests ranging from observational cosmology to the variability of the fundamental physical constants, to the astronomical instrumentation.

DAVID MORGAN has a PhD in particle physics from The College of William and Mary and for a decade served as a professor of physics and astronomy at The New School in New York City. He is currently the academic director of the Ross School Innovation Lab, an academy for advanced high school students in STEM fields. In 2004, he received a Sloan Foundation/EST commission to co-author a play titled 'The Osiander Preface' which explored the publication of Copernicus's On the Revolutions of the Heavenly Spheres. He is also an amateur composer and pianist with a fondness of the solo piano repertoire of Philip Glass. Dr. Morgan was an attendee at the Second INSAP Conference in Malta.

IAN MORISON is a former Gresham Professor of Astronomy. He began his love of astronomy when, at the age of 12, he made a telescope out of lenses given to him by his optician. He went on to study Physics, Mathematics and Astronomy at Hertford College, Oxford. In September 1965, he became a research student at the University of Manchester's Jodrell Bank Observatory. In 1970 he was appointed to the staff of the Observatory and teaches astronomy at the University of Manchester. In 1990 he helped found the Macclesfield Astronomy Society which meets at the Observatory and later became president of the Society for Popular Astronomy, the UK's largest astronomical society. He remains on the Society's Council and holds the post of instrument advisor helping members with their choice and use of Telescopes. He lectures widely on astronomy, has co-authored books for amateur astronomers and writes regularly for the UK astronomy magazines Astronomy Now and Sky at Night. He also writes a monthly sky guide for the Observatory's web site and produces an audio version as part of the Jodrell Bank Podcast. In 2003 the Minor Planets Committee of the International Astronomical Union named asteroid 15,727 in his honour citing his work with MERLIN, the world's largest linked array of radio telescopes, and that in searching for intelligent life beyond our Solar System in Project Phoenix.

Jay M. Pasachoff is Field Memorial Professor of Astronomy, Williams College, and is currently Past Chair, Historical Astronomy Division of the American Astronomical Society. He received the Education Prize from the American Astronomical Society and the Janssen Prize from the Société Astronomique Deborah France. He is also Chair of the Working Group on Solar Eclipses of the International Astronomical Union.

Segar Passi is an award-winning artist, Meriam elder, and Dauareb man from Mer (Murray Island) in the eastern Torres Strait. He holds a wealth of traditional knowledge about weather, astronomy, and animals, all of which feature in his paintings and drawings. His knowledge of clouds is particularly relevant and his art is featured in UNESCO's Australian Memory of the World Register and in many exhibition books. He is a highly regarded and respected elder and is the only surviving contributor to Margaret Lawrie's landmark cultural project to document culture and traditions in the Torres Strait. His knowledge is featured in the Sir Thomas Brisbane Planetarium exhibit entitled *SKYLORE* and is working with the Royal Australian Mint on a coin commemorating Meriam astronomy.

Ethan Pollock is Associate Professor of History and Associate Professor of Slavic Studies at Brown University. His book *Stalin and the Soviet Science Wars* is in both hardback (2006) and paper (2008) editions.

Martin John Rees, Baron Rees of Ludlow, Kt, OM, FRS, Hon FREng, F MedSci is a British cosmologist and astrophysicist. He has been Astronomer Royal of the Royal Observatory at Greenwich since 1995, was Master of Trinity College, Cambridge from 2004 to 2012 and President of the Royal Society between 2005 and 2010.

Michael Rowan-Robinson was Head of the Astrophysics Group at Imperial College in London from 1993 to 2007. From 2007-12 he continued to teach part-time in the Blackett Lab, Imperial. He was President of the Royal Astronomical Society 2006-8. In addition, he chaired the UK Ground-Based Facilities Review in 2009, and was Chair of the European Southern Observatory's Observation Planning Committee in 2011. From 1981 to 1982, he gave public lectures as professor of astronomy at Gresham College.

Valerie Shrimplin specialises in Renaissance history and art history. Her major publication is on the influence of Copernican heliocentricity on Michelangelo's Last Judgment in the Sistine Chapel. She is a member of the International Executive Committee of INSAP and has presented papers on Byzantine, Medieval and Renaissance art and cosmology at previous conferences in the INSAP series. She is Academic Registrar and Company Secretary of Gresham College.

Alop Tapim is a Meriam elder and Dauareb man from Mer (Murray Island) in the eastern Torres Strait. He is a fluent speaker of the traditional Meriam Mir language and serves on the Cultural Committee in the Torres Shire Council, based on Thursday Island.

He was a key witnesses for the Meriam legal battle for sea rights and is a custodian of traditional astronomical knowledge, which has been featured in the Sir Thomas Brisbane Planetarium exhibit entitled *SKYLORE* and is working with the Royal Australian Mint on a coin commemorating Meriam astronomy.

ROBERTO TROTTA is an astrophysicist at Imperial College London, where he studies dark matter, dark energy and the Big Bang. His award-winning book for the general public, *The Edge of the Sky* explains Universe using only the most common 1,000 words in English. He was named as one of the 100 Global Thinkers 2014 by Foreign Policy, for 'junking astronomy jargon'. Here he describes the challenge of describing the universe and its mysteries in 707 words and asks if we need a new language for science. Roberto is a passionate science communicator and the recipient of numerous awards for his research, outreach and art and science collaborations, including the Lord Kelvin Award of the British Association for the Advancement of Science and the Michelson Prize of Case Western Reserve University.

TARJA TRYGG is a Finnish artist, licentiate of Arts, and a doctoral student in the Department of Art at Aalto University, School of Arts, Design and Architecture, Helsinki Finland, where she has worked as a senior lecturer and taught photography many years. She uses solargraphy to make the invisible movements of the Sun's paths visible in landscapes at various latitudes. Trygg has taken part in international projects such as the Spanish project Time in a Can, which exhibited in Madrid in 2013 and in Mühlheim, Germany in 2014. She also participated in the Australian project Sky Lab, curated by Felicity Spear, in 2011, 2013 and 2015.

MELANIE VANDENBROUCK is curator of art post-1800 at Royal Museums Greenwich, which involves developing, researching, interpreting, displaying and communicating the Museum's superlative art collection. She is the founder and chair of the museum-wide Contemporary Art Forum, and a judge of the Astronomy Photographer of the Year competition. In discussion with the Public Astronomer and the Curator of the Royal Observatory, Melanie is currently drafting a contemporary arts collecting policy for Royal Museums Greenwich which will include artistic responses and contributions to the field of astronomy. Prior to joining royal Museums Greenwich in 2012, Melanie earned a PhD at the Courtauld Institute of Art, and worked in the Sculpture department of the Victoria and Albert Museum.

For thirty years, ELIZABETH FORBES WALLACE owned and operated a visa agency that served international corporate travelers. While she used to help people expand their horizons by getting them across borders, she now would like to help them get 'beyond borders'. Her projects include 'The StarryTelling Festival' in the Metro DC area; Service on the IYA 2009 Cultural Astronomy Committee; the 'I'm a StarryTeller' planetarium show for the National Air and Space Museum, which included stories by over forty students; coaching students to retell NASA stories from Jay O'Callahan's *Forged in the Stars*

(2008); Hands on astronomy workshops in an elder care facility; and presentations at the Next Gen Suborbital Researchers Conference including 'Space Tourism is the New Higher Education'.

GARY WELLS is an associate professor and chair of the Department of Art History at Ithaca College, Ithaca. New York. His areas of research and writing include the intersections of art and science in the nineteenth century, the visual culture of science, and the origins of Modernism in the early twentieth century. He has presented and published on artists such as Marcel Duchamp, Jules Breton and Paul Cézanne. He also teaches and writes about the material history of art and issues of technology in art history, and has developed technology projects on modern art in Paris and late–nineteenth century painting.

THE SOPHIA CENTRE

The Sophia Centre for the Study of Cosmology in Culture is a teaching and research centre within the Faculty of Humanities and the Performing Arts at the University of Wales Trinity Saint David. The Centre works from a humanities and social science perspective and encompasses research styles and methodologies from anthropology, history, philosophy, religious studies and sociology. Its academic goals are:

- 'to pursue research, scholarship and teaching in the relationship between astrological, astronomical and cosmological beliefs and theories, and society, politics, religion and the arts, past and present' and

- 'to undertake the academic and critical examination of astrology and its practice'.

The Centre's wider goal is stated in its title – to 'study cosmology in culture'. This enables us to tackle a wide range of topics, from Egyptian sky religion and Babylonian astrology, to astronomy in surrealist painting, astrology in contemporary culture, UFO abduction and the politics of the space race. The Centre promotes research in the subject area, holds seminars and conferences, is associated with Sophia Centre Press and the publication *Culture and Cosmos*, and supervises PhD students.

To find more about our work please visit http://www.uwtsd.ac.uk/sophia/

INDEX